Ignazio Licata and Sauro Succi (Eds.)

Non-Linear Lattice

MDPI

This book is a reprint of the Special Issue that appeared in the online, open access journal, *Entropy* (ISSN 1099-4300) from 2015–2016, available at:

http://www.mdpi.com/journal/entropy/special_issues/nonlinear_lattice

Guest Editors
Ignazio Licata
ISEM Institute for Scientific Methodology,
School of Advanced International Studies on Applied Theoretical and
Non Linear Methodologies of Physics
Italy

Sauro Succi
Instituto Applicazioni Calcolo "Mauro Picone"
Italy
Institute for Applied Computational Science and Physics
Harvard University
USA

Editorial Office
MDPI AG
St. Alban-Anlage 66
Basel, Switzerland

Publisher
Shu-Kun Lin

Assistant Editor
Yuejiao Hu

1. Edition 2016

MDPI • Basel • Beijing • Wuhan • Barcelona • Belgrade

ISBN 978-3-03842-306-5 (Hbk)
ISBN 978-3-03842-307-2 (electronic)

Table of Contents

List of Contributors VII

About the Guest Editors X

Preface to "Non-Linear Lattice": Living in a Complex World XIII

Sheng Zhang, Jiahong Li and Yingying Zhou
Exact Solutions of Non-Linear Lattice Equations by an Improved Exp-Function Method
Reprinted from: *Entropy* **2015**, *17*(5), 3182–3193
http://www.mdpi.com/1099-4300/17/5/3182 1

Tao Yang, Fa-Kai Wen, Kun Hao, Li-Ke Cao and Rui-Hong Yue
The Effect of a Long-Range Correlated-Hopping Interaction on Bariev Spin Chains
Reprinted from: *Entropy* **2015**, *17*(9), 6044–6055
http://www.mdpi.com/1099-4300/17/9/6044 14

Miller Mendoza Jimenez and Sauro Succi
Short-Lived Lattice Quasiparticles for Strongly Interacting Fluids
Reprinted from: *Entropy* **2015**, *17*(9), 6169–6178
http://www.mdpi.com/1099-4300/17/9/6169 28

Suemi Rodríguez-Romo and Oscar Ibañez-Orozco
Two-Dimensional Lattice Boltzmann for Reactive Rayleigh–Bénard and Bénard–Poiseuille Regimes
Reprinted from: *Entropy* **2015**, *17*(10), 6698–6711
http://www.mdpi.com/1099-4300/17/10/6698 38

Qing Chen, Hongping Zhou, Xuesong Jiang, Linyun Xu, Qing Li and Yu Ru
Extension of the Improved Bounce-Back Scheme for Electrokinetic Flow in the Lattice Boltzmann Method
Reprinted from: *Entropy* **2015**, *17*(11), 7406–7419
http://www.mdpi.com/1099-4300/17/11/7406 54

Gregor Chliamovitch, Orestis Malaspinas and Bastien Chopard
A Truncation Scheme for the BBGKY2 Equation
Reprinted from: *Entropy* **2015**, *17*(11), 7522–7529
http://www.mdpi.com/1099-4300/17/11/7522 70

Yan Wang, Liming Yang and Chang Shu
From Lattice Boltzmann Method to Lattice Boltzmann Flux Solver
Reprinted from: *Entropy* **2015**, *17*(11), 7713–7735
http://www.mdpi.com/1099-4300/17/11/7713 79

Ilya V. Karlin, Fabian Bösch, Shyam S. Chikatamarla and Sauro Succi
Entropy-Assisted Computing of Low-Dissipative Systems
Reprinted from: *Entropy* **2015**, *17*(12), 8099–8110
http://www.mdpi.com/1099-4300/17/12/7867 104

Binghai Wen, Chaoying Zhang and Haiping Fang
Hydrodynamic Force Evaluation by Momentum Exchange Method in Lattice Boltzmann Simulations
Reprinted from: *Entropy* **2015**, *17*(12), 8240–8266
http://www.mdpi.com/1099-4300/17/12/7876 120

Hongsheng Chen, Zhong Zheng, Zhiwei Chen and Xiaotao T. Bi
A Lattice Gas Automata Model for the Coupled Heat Transfer and Chemical Reaction of Gas Flow Around and Through a Porous Circular Cylinder
Reprinted from: *Entropy* **2016**, *18*(1), 2
http://www.mdpi.com/1099-4300/18/1/2 157

Bo Han, Meng Ni and Hua Meng
Three-Dimensional Lattice Boltzmann Simulation of Liquid Water Transport in Porous Layer of PEMFC
Reprinted from: *Entropy* **2016**, *18*(1), 17
http://www.mdpi.com/1099-4300/18/1/17 178

Alexander P. Chetverikov, Werner Ebeling and Manuel G. Velarde
Long-Range Electron Transport Donor-Acceptor in Nonlinear Lattices
Reprinted from: *Entropy* **2016**, *18*(3), 92
http://www.mdpi.com/1099-4300/18/3/92 198

Gentaro Watanabe, B. Prasanna Venkatesh and Raka Dasgupta
Nonlinear Phenomena of Ultracold Atomic Gases in Optical Lattices: Emergence of Novel Features in Extended States
Reprinted from: *Entropy* **2016**, *18*(4), 118
http://www.mdpi.com/1099-4300/18/4/118 220

List of Contributors

Xiaotao T. Bi Fluidization Research Center, Department of Chemical and Biological Engineering, University of British Columbia, Vancouver V6T 1Z3, Canada.

Fabian Bösch Department of Mechanical and Process Engineering, ETH Zurich, Zurich 8092, Switzerland.

Li-Ke Cao School of Physics, Northwest University, Xi'an 710069, China.

Hongsheng Chen School of Materials Science and Engineering, Chongqing University, Chongqing 400044, China.

Qing Chen College of Mechanical and Electronic Engineering, Nanjing Forestry University, Nanjing 210037, China.

Zhiwei Chen Fluidization Research Center, Department of Chemical and Biological Engineering, University of British Columbia, Vancouver V6T 1Z3, Canada.

Alexander P. Chetverikov Department of Physics, Saratov State University, Astrakhanskaya 83, 410012 Saratov, Russia.

Shyam S. Chikatamarla Department of Mechanical and Process Engineering, ETH Zurich, Zurich 8092, Switzerland.

Gregor Chliamovitch Department of Theoretical Physics, and Department of Computer Science, University of Geneva, Route de Drize 7, 1227 Geneva, Switzerland.

Bastien Chopard Department of Computer Science, University of Geneva, Route de Drize 7, 1227 Geneva, Switzerland.

Raka Dasgupta Asia Pacific Center for Theoretical Physics (APCTP), Pohang, Gyeongbuk 37673, Korea; Department of Physics, University of Calcutta, Kolkata 700009, India.

Werner Ebeling Institut für Physik, Humboldt-Universität Berlin, Newtonstraße 15, 12489 Berlin, Germany.

Haiping Fang Division of Interfacial Water and Key Laboratory of Interfacial Physics and Technology, Shanghai Institute of Applied Physics, Chinese Academy of Sciences, Shanghai 201800, China; Shanghai Science Research Center, Chinese Academy of Sciences, Shanghai 201204, China.

Bo Han School of Aeronautics and Astronautics, Zhejiang University, Hangzhou 310027, China.

Kun Hao Institute of Modern Physics, Northwest University, Xi'an 710069, China.

Oscar Ibañez-Orozco Facultad de Estudios Superiores Cuautitlán, Universidad Nacional Autónoma de México, Av. 1 de mayo s/n, 54750 Cuautitlán Izcalli, Edo de México, Mexico.

Xuesong Jiang College of Mechanical and Electronic Engineering, Nanjing Forestry University, Nanjing 210037, China.

Miller Mendoza Jimenez ETH Zürich, Computational Physics for Engineering Materials, Institute for Building Materials,Wolfgang-Pauli-Strasse 27, HIT, CH-8093 Zürich, Switzerland.

Ilya V. Karlin Department of Mechanical and Process Engineering, ETH Zurich, Zurich 8092, Switzerland.

Jiahong Li School of Mathematics and Physics, Bohai University, 19 Keji Road, New Songshan District, Jinzhou 121013, China.

Qing Li School of Energy Science and Engineering, Central South University, Changsha 410083, China.

Orestis Malaspinas Department of Computer Science, University of Geneva, Route de Drize 7, 1227 Geneva, Switzerland.

Hua Meng School of Aeronautics and Astronautics, Zhejiang University, Hangzhou 310027, China.

Meng Ni Department of Building and Real Estate, Hong Kong Polytechnic University, Hung Hom 999077, Hong Kong, China.

Suemi Rodríguez-Romo Facultad de Estudios Superiores Cuautitlán, Universidad Nacional Autónoma de México, Av. 1 de mayo s/n, 54750 Cuautitlán Izcalli, Edo de México, Mexico.

Yu Ru College of Mechanical and Electronic Engineering, Nanjing Forestry University, Nanjing 210037, China.

Chang Shu Department of Mechanical Engineering, National University of Singapore, 10 Kent Ridge Crescent, Singapore 119260, Singapore.

Sauro Succi Istituto per le Applicazioni del Calcolo C.N.R., Via dei Taurini 19, 00185 Rome, Italy; Institute for Applied Computational Science, Harvard University, Oxford Street, 52, Cambridge, 02138 MA, USA.

Manuel G. Velarde Instituto Pluridisciplinar, Universidad Complutense, Paseo Juan XXIII, 1, 28040 Madrid, Spain.

B. Prasanna Venkatesh Institute for Theoretical Physics, University of Innsbruck, Innsbruck A-6020, Austria;Asia Pacific Center for Theoretical Physics (APCTP), Pohang, Gyeongbuk 37673, Korea; Institute for Quantum Optics and Quantum Information, Austrian Academy of Sciences, Innsbruck A-6020, Austria.

Yan Wang Department of Mechanical Engineering, National University of Singapore, 10 Kent Ridge Crescent, Singapore 119260, Singapore.

Gentaro Watanabe University of Science and Technology (UST), Daejeon 34113, Korea; Asia Pacific Center for Theoretical Physics (APCTP), Pohang, Gyeongbuk 37673, Korea; Department of Physics, Pohang University of Science and Technology (POSTECH), Pohang, Gyeongbuk 37673, Korea; Center for Theoretical Physics of Complex Systems, Institute for Basic Science (IBS), Daejeon 34051, Korea; Department of Physics, Zhejiang University, Hangzhou 310027, China.

Binghai Wen Guangxi Key Lab of Multi-source Information Mining & Security, Guangxi Normal University, Guilin 541004, China; Division of Interfacial Water and Key Laboratory of Interfacial Physics and Technology, Shanghai Institute of Applied Physics, Chinese Academy of Sciences, Shanghai 201800, China.

Fa-Kai Wen Institute of Modern Physics, Northwest University, Xi'an 710069, China.

Linyun Xu College of Mechanical and Electronic Engineering, Nanjing Forestry University, Nanjing 210037, China.

Liming Yang Department of Aerodynamics, College of Aerospace Engineering, Nanjing University of Aeronautics and Astronautics, Yudao Street, Nanjing 210016, China.

Tao Yang Institute of Modern Physics, Northwest University, Xi'an 710069, China.

Rui-Hong Yue School of Physical Science and Technology, Yangzhou University, Yangzhou 225002, China.

Chaoying Zhang Guangxi Key Lab of Multi-source Information Mining & Security, Guangxi Normal University, Guilin 541004, China.

Sheng Zhang School of Mathematics and Physics, Bohai University, 19 Keji Road, New Songshan District, Jinzhou 121013, China.

Zhong Zheng School of Materials Science and Engineering, Chongqing University, Chongqing 400044, China.

Hongping Zhou College of Mechanical and Electronic Engineering, Nanjing Forestry University, Nanjing 210037, China.

Yingying Zhou School of Mathematics and Physics, Bohai University, 19 Keji Road, New Songshan District, Jinzhou 121013, China.

About the Guest Editors

Ignazio Licata, born in 1958, is a theoretical physicist, the Scientific Director of the Institute for Scientific Methodology, Palermo; Professor at the School of Advanced International Studies on Theoretical and Non-Linear Methodologies of Physics, Bari, Italy; and Visiting Professor at the International Institute for Applicable Mathematics and Information Sciences (IIAMIS), B.M. Birla Science Centre, Adarsh Nagar, Hyderabad 500, India. His topics of research include the foundation of quantum mechanics, dissipative QFT, space-time at Planck scale, the group approach in quantum cosmology, systems theory, non-linear dynamics, as well as computation in physical systems. He is on the Editorial Board of the *Entropy- Elect. Journal of Theor. Phys.- Quantum Biosystems-Collana "Il Nucleare" (Aracne, Rome); Frontiers in Interdisciplinary Physics Referee for: Found Phys. Int. Jour. of Theor. Phys., European Phys. Journal; Physica Scripta; Advances in Applied Clifford Algebras Asian Journal of Mathematics and Computer Research; Jour. Mod. Phys., Found. of Science; Physical Science International Journal; British Journal of Mathematics and Computer Science.*

Sauro Succi is an Italian scientist, internationally accredited for being one of the founders of the successful Lattice Boltzmann method for fluid dynamics. Since 1995, Succi has been the Research Director at the Istituto Applicazioni Calcolo of the National Research Council (CNR) in Rome. He is also a Research Affiliate to the Physics Department at Harvard University (from 2000), a Fellow of the Freiburg Institute for Advanced Studies (FRIAS) and Senior Fellow of the Erwin Schrödinger International Institute for Mathematical Physics. He is an Alumnus of the University of Bologna, from which he earned a degree in nuclear engineering, and the École Polytechnique Fédérale de Lausanne, from which he obtained a PhD in plasma physics in 1987. He has published extensively in the fields of plasma physics, fluid dynamics, kinetic theory and quantum fluids. He has also authored the well-known monograph "*The Lattice Boltzmann Equation for Fluid Dynamics and Beyond*", (Oxford University Press). Dr Succi has held visiting/teaching appointments at many academic institutions, such as the University of Harvard, Paris VI, University of Chicago, Yale, Tufts, Queen Mary London and Scuola Normale Superiore di Pisa. Dr Succi is an elected Fellow of the American Physical Society (1998). He has received the Humboldt Prize in physics (2002), the Killam Award bestowed by the University of Calgary (2005) and the Raman Chair of the Indian Academy of Sciences (2011). He has also served as an External Senior Fellow at the Freiburg Institute for Advanced Studies (2009–2013) and Senior Fellow of the Erwin Schrödinger International Institute for Mathematical Physics in Vienna (2013). Dr Succi is an elected member of the Academia Europaea (2015). He has also been awarded the 2017 American Physical Society (APS) Aneesur Rahman Prize for Computational Physics.

Preface to "Non-Linear Lattice": Living in a Complex World

Since the epoch-making work of Pasta–Fermi–Ulam–Tsingou on non-linear relaxation of discrete chains, the lattice has proven an invaluable tool to explore the complexities of non-linear phenomena through computer simulation. In this field the complexity is twofold: The nonlinearity allows cooperative phenomena which are simply impossible in the linear regime; the lattice aspect sets a scale of interest, and often changing such scale also changes the type of observed behavior. The development of mathematical techniques, combined with new possibilities of computational simulations, have greatly broadened the study of non-linear lattices, one of the most advanced and interdisciplinary themes of mathematical physics.

Over the years, the role of lattices has extended to virtually all walks of physics, from classical non-linear field theory, to quantum chromodynamics, all the way down to quantum gravity.

From a practical standpoint, the lattice serves as a natural regulator of UV infinities by providing a finite cutoff to otherwise divergent interactions. In this respect, the lattice is a generous friend, which helps in providing finite answers, then leaving the stage in the continuum limit, the place where "true" physics is supposed to take place.

This is the ground of **discretized** systems, those that result from placing a continuum theory on a discrete spacetime for the "mere" matter of computational convenience and viability.

For all the importance of discretized systems, the role of the lattice in modern physics runs far deeper than mere discretization. We refer here to genuinely **discrete** dynamical systems, whose dynamics are formulated ab initio on a lattice, because this is the most natural way of encoding the physics at hand. Among others, discrete chains, Hubbard models, lattice gas and lattice Boltzmann belong to this class. The relevant physics, though, is still believed to live in the continuum limit, where the lattice spacing is sent to zero.

Finally, there is a third class that we call "inherently discrete" (ID), in which the "true" physics is believed to take place at finite mesh spacing; the continuum limit, if existing at all, being a mere idealization. Quantum gravity is quintessential ID, and so is a broad class of cellular automata, sandpile models and similar rule-driven lattice systems. The "transferability" of class two to class three models is one of the most interesting fields of modern mathematical physics research.

Just think of the quantum Bose-Hubbard model as a toy model for emergent spacetime in quantum gravity. In 1611, Johannes Kepler wrote *Strena seu de nive sexangula* (the six cornered snowflake), which suggested that the macroscopic symmetry of a snow flake depended on the structure of its constituents. In this way, he linked the system symmetries with a "nuts and bolts" explanation: a splendid work of theoretical physics and an ideal procedural model! It seems that Nature often likes to play the same game at different scales. In a way, we could say that the renormalization group is a sort of "mathematical zoom device" that allows us to watch this game and distinguish it from other types of scale-dependent behaviors.

This volume deals mostly, but not exclusively, with discrete systems in the second class, with various instances of non-linear lattice systems ranging from non-linear spin chains, to optical lattices, lattice gas and lattice Boltzmann models for fluids. It is hoped that this Special Issue will foster further work in the direction of bringing all the three aforementioned families under the unifying umbrella of "lattice physics", fostering cross-fertilization of new ideas and techniques to further our understanding of the beautiful complexity of non-linear dynamical systems, through a synergistic combination of analytics, experiments and computer simulations.

Ignazio Licata and Sauro Succi
Guest Editors

Exact Solutions of Non-Linear Lattice Equations by an Improved Exp-Function Method

Sheng Zhang, Jiahong Li and Yingying Zhou

Abstract: In this paper, the exp-function method is improved to construct exact solutions of non-linear lattice equations by modifying its exponential function ansätz. The improved method has two advantages. One is that it can solve non-linear lattice equations with variable coefficients, and the other is that it is not necessary to balance the highest order derivative with the highest order nonlinear term in the procedure of determining the exponential function ansätz. To show the advantages of this improved method, a variable-coefficient mKdV lattice equation is considered. As a result, new exact solutions, which include kink-type solutions and bell-kink-type solutions, are obtained.

Reprinted from *Entropy*. Cite as: Zhang, S.; Li, J.; Zhou, Y. Exact Solutions of Non-Linear Lattice Equations by an Improved Exp-Function Method. *Entropy* **2015**, *17*, 3182–3193.

1. Introduction

The work of Fermi, Pasta and Ulam in the 1950s [1] has attached much attention on exact solutions of non-linear lattice equations arising different fields which include condensed matter physics, biophysics, and mechanical engineering. In the numerical simulation of soliton dynamics in high energy physics, some non-linear lattice equations are often used as approximations of continuum models. In fact, the celebrated Korteweg–de Vries (KdV) equation can be considered as a limit of the Toda lattice equation [2]. Non-linear lattice equations can provide models for non-linear phenomena such as wave propagation in nerve systems, chemical reactions, and certain ecological systems (for example, the famous Volterra equation). Unlike difference equations which are fully discretized, lattice equations are semi-discretized with some of their spatial variables discretized while time is usually kept continuous. In the past several decades, many effective methods for constructing exact solutions of non-linear partial differential equations (PDEs) have been presented, such as the inverse scattering method [3], Bäcklund transformation [4], Hirota's bilinear method [5], homogeneous balance method [6], tanh-function method [7], Jacobi elliptic function expansion method [8], Lucas Riccati method [9], differential transform method [10], and others [11–17]. Generally speaking, it is hard to generalize one method for non-linear PDEs to solve non-linear lattice equations because of the difficulty in finding iterative relations from indices n to $n \pm 1$(here n denotes an integer). When the inhomogeneities of media and non-uniformities of boundaries are taken into account, the variable-coefficient equations could describe more realistic physical phenomena than their constant-coefficient counterparts [18], such as seen, e.g., in the super-conductors, coastal waters of oceans, blood vessels, space and laboratory plasmas and optical fiber communications [19]. Therefore, how to solve non-linear lattice equations with variable coefficients is worth studying.

Recently, He and Wu proposed exp-function method [20] to solve non-linear PDEs. It is shown in [20–31] that the exp-function method or its improvement is available for many kinds of nonlinear PDEs, such as Dodd–Bullough–Mikhailov equation [20], sine-Gorden equation [21], combined KdV-mKdV equation [23], Maccari's system [24], variable-coefficient equation [25], non-linear lattice equation [26], stochastic equation [27], and generalized Klein–Gordon equation [31]. For some recent applications of the method itself, we can refer to Fitzhugh–Nagumo equation [32], extended shallow water wave equations [33] and generalized mKdV equation [34]. In [35–37], there are two remarkable developments of the exp-function method. One is that the exp-function method with a fractional complex transform was generalized to deal with fractional differential equations [35,36], and the other is that the method was hybridized with heuristic computation to obtain numerical solution of generalized Burger–Fisher equation [37]. On the other hand, it is necessary to check the solutions obtained by the exp-function method carefully [38] because some authors have been criticized for incorrect results [39,40]. Besides, for a given non-linear PDEs with independent variables $t, x_1, x_2, \cdots, x_s$ and dependent variable u:

$$F(u, u_t, u_{x_1}, u_{x_2}, \cdots, u_{x_s}, u_{x_1 t}, u_{x_2 t} \cdots, u_{x_s t}, u_{tt}, u_{x_1 x_1}, u_{x_2 x_2}, \cdots, u_{x_s x_s}, \cdots) = 0, \quad (1)$$

the exp-function method can also be used to construct different types of exact solutions. This is due to its exponential function ansätz:

$$u(\xi) = \frac{\sum_{n=-f}^{g} a_n \exp(n\xi)}{\sum_{m=-p}^{q} b_m \exp(m\xi)}, \quad \xi = \sum_{i=1}^{s} k_i x_i + wt, \quad (2)$$

where a_n, b_m, k_i and w are undetermined constants, f, p, g and q can be determined by using Equation (2) to balance the highest order non-linear term with the highest order derivative of u in Equation (1). It is He and Wu [20] who first concluded that the final solution does not strongly depend on the choices of values of f, p, g and q. Usually, $f = p = g = q = 1$ is the simplest choice. More recently, Ebaid [41] proved that $f = p$ and $g = q$ are the only relations for four types of nonlinear ordinary differential equations (ODEs) and hence concluded that the additional calculations of balancing the highest order derivative with the highest order non-linear term are not longer required. Ebaid's work is significant, which makes the exp-function method more straightforward. The present paper is motivated by the desire to prove that $f = p$ and $g = q$ are also the only relations when we generalize the exp-function method [20] to solve non-linear lattice equations. Thus, the exp-function method can be further improved because it is not necessary to balance the highest order derivative with the highest order non-linear term in the process of solving non-linear lattice equations.

The rest of this paper is organized as follows. In Section 2, we generalize exp-function method to solve non-linear lattice equations with variable coefficients. In Section 3, a theorem is proved and then used to improve the generalized exp-function method in determining its exponential function ansätz of non-linear lattice equations. In Section 4, we take a variable-coefficient mKdV lattice equation as an example to show the advantages of the improved exp-function method. In Section 5, some conclusions are given.

2. Generalized Exp-Function Method for Non-Linear Lattice Equations

In this section, we outline the basic idea of generalizing the exp-function method [20] to solve a given non-linear lattice equation with variable coefficients, say, in three variables n, x and t:

$$P(u_{nt}, u_{nx}, u_{ntt}, u_{nxt}, \cdots, u_{n-1}, u_n, u_{n+1}, \cdots) = 0, \tag{3}$$

which contains both the highest order nonlinear terms and the highest order derivatives of dependent variables. Here P is a polynomial of u_n, $u_{n-\theta}(\theta = \pm 1, \pm 2, \cdots)$ and the various derivatives of u_n. Otherwise, a suitable transformation can transform Equation (3) into such an equation.

Firstly, we take the following transformation:

$$u_n = U_n(\xi_n), \quad \xi_n = dn + c(x,t) + \omega, \tag{4}$$

where d is a constant to be determined, $c(x,t)$ is the undetermined function of x and t, and ω is the phase. Then, Equation (3) can be reduced to a non-linear ODE with variable coefficients:

$$Q(U_n', U_n'', \cdots, U_{n-1}, U_n, U_{n+1}, \cdots) = 0. \tag{5}$$

Secondly, we suppose that the ansätz of Equation (5) can be expressed as:

$$U_n = \frac{\sum_{N=-f}^{g} a_N(x,t)\exp(N\xi_n)}{\sum_{M=-p}^{q} b_M \exp(M\xi_n)} = \frac{a_{-f}(x,t)\exp(-f\xi_n) + \cdots + a_g(x,t)\exp(g\xi_n)}{b_{-p}\exp(-p\xi_n) + \cdots + b_q\exp(q\xi_n)}. \tag{6}$$

Thirdly, we substitute U_n and $U_{n-\theta}(\theta = \pm 1, \pm 2, \cdots)$ determined by Equation (6) into Equation (5) and then balance the highest order derivative with the highest order nonlinear term in Equation (5) to obtain the integers f, p, g and q. Finally, we determine the coefficients $a_{-f}(x,t), \cdots, a_g(x,t), b_{-p}, \cdots, b_q, d$ and $c(x,t)$ by solving the resulting equations from the substitution of U_n and $U_{n-\theta}(\theta = \pm 1, \pm 2, \cdots)$ along with the obtained values of f, p, g, q into Equation (5).

In order to identify the highest order nonlinear term, we define in this paper the negative order $N(\cdot)$ and the positive order $P(\cdot)$ of ansätz (6) as follows:

$$N(U_n) = -f - (-p) = p - f, \quad P(U_n) = g - q \tag{7}$$

under the condition that the functions $a_{-f}(x,t)$ and $a_g(x,t)$, and the constants b_{-p} and b_q are all nonzero coefficients. Therefore, we can easily obtain $N(U_{n-\theta}) = p - f$ and $P(U_{n-\theta}) = g - q$. For the derivatives of U_n, we have a general formula:

$$U_n^{(r)} = \frac{\tau_r(x,t)\exp[-(f - p + 2^r p)\xi_n] + \cdots + \sigma_r(x,t)\exp[(g - q + 2^r q)\xi_n]}{\delta_r \exp[(-2^r p)\xi_n] + \cdots + \varsigma_r \exp[(2^r q)\xi_n]}, \tag{8}$$

where $\tau_r(x,t)$ and $\sigma_r(x,t)$ are functions of x and t, δ_r and ς_r are constants, and $r \geq 1$ is an integer. If $\tau_r(x,t), \sigma_r(x,t), \delta_r$ and ς_r are nonzero coefficients, then $N(U_n^{(r)}) = p - f$ and $P(U_n^{(r)}) = g - q$.

Since

$$N(U_n) = N(U_{n-\theta}), \quad P(U_n) = P(U_{n-\theta}), \tag{9}$$

we define the product

$$U_n^h U_{n-1}^{i_1} U_{n+1}^{j_1} U_{n-2}^{i_2} U_{n+2}^{j_2} \cdots U_{n-z}^{i_z} U_{n+z}^{j_z} (U_n')^{l_1} (U_n'')^{l_2} \cdots (U_n^{(s)})^{l_s} \tag{10}$$

as the highest order nonlinear term of Equation (5). Here $h, i_1, j_1, i_2, j_2, \cdots, i_z, j_z, l_1, l_2, \cdots, l_s$ are nonnegative integers which satisfy

$$h + i_1 + j_1 + i_2 + j_2 + \cdots + i_z + j_z + l_1 + l_2 + \cdots + l_s \geq 2. \tag{11}$$

With above preparations, we can see that Equations (8) and (10) include all possibilities of the highest order derivative and the highest order nonlinear term of Equation (5). In what follows, we shall proof that $f = p$ and $g = q$ are the only relations when using the exponential function ansätz (6) to balance the highest order derivative (8) with the highest order nonlinear term (10).

Remark 1. *If we let $a_{-f}(x,t), \cdots, a_g(x,t)$ be nonzero constants and take $c(x,t)$ as a linear function $kx + lt$, k and l are undetermined constants, then the generalized exp-function method described in this section is also effective for non-linear lattice equations with constant coefficients. So the starting point of this paper is to generalize the exp-function method [20]* *to solve Equation (3) with variable coefficients. In the next section, we shall further improve this generalized exp-function method.*

3. Theorem and Improvement

Theorem 1. *Suppose that Equations (8) and (10) are respectively the highest order derivative and the highest order nonlinear term of Equation (5), then the balancing procedure using the exponential function ansätz (6) leads to $f = p$ and $g = q$.*

Proof. By contradiction, we suppose that $f \neq p$ and $g \neq q$. Then a computation shows that $\tau_r(x,t)$, $\sigma_r(x,t)$, δ_r, and ς_r in Equation (8) are all nonzero coefficients. Using Equations (6) and (8), we have

$$U_n^h = \frac{a_{-f}^h(x,t)\exp(-hf\xi_n) + \cdots + a_g^h(x,t)\exp(hg\xi_n)}{b_{-p}^h\exp(-hp\xi_n) + \cdots + b_q^h\exp(hq\xi_n)}, \tag{12}$$

$$U_{n-\theta}^i = \frac{a_{-f}^i(x,t)\exp(-if\xi_n)\exp(ifd\theta)+\cdots+a_g^i(x,t)\exp(ig\xi_n)\exp(-igd\theta)}{b_{-p}^i\exp(-ip\xi_n)\exp(ipd\theta)+\cdots+b_q^i\exp(iq\xi_n)\exp(-iqd\theta)}, \tag{13}$$

$$(U_n^{(r)})^l = \frac{\tau_r^l(x,t)\exp[-l(f-p+2^r p)\xi_n]+\cdots+\sigma_r^l(x,t)\exp[l(g-q+2^r q)\xi_n]}{\delta_r^l\exp[(-2^r lp)\xi_n]+\cdots+\varsigma_r^l\exp[(2^r lq)\xi_n]}. \tag{14}$$

With the help of Equations (12)–(14), the left hand side and the right hand side of Equation (8) can be respectively written as:

$$\frac{\vartheta(x,t)\exp\{-[f(h+i_1+j_1+\cdots+i_z+j_z)+l_1(f+p)+\cdots+l_s(f-p+2^s p)]\xi_n\}+\cdots}{\kappa\exp[-p(h+i_1+j_1+\cdots+i_z+j_z+2l_1+\cdots+2^s l_s)\xi_n]+\cdots}, \tag{15}$$

$$\frac{\cdots+\mu(x,t)\exp\{[g(h+i_1+j_1+\cdots+i_z+j_z)+l_1(g+q)+\cdots+l_s(g-q+2^s q)]\xi_n\}}{\cdots+\lambda\exp[q(h+i_1+j_1+\cdots+i_z+j_z+2l_1+\cdots+2^s l_s)\xi_n]}, \tag{16}$$

with nonzero coefficients

$$\vartheta(x,t) = a_{-f}^{h+i_1+j_1+\cdots+i_z+j_z}(x,t)\tau_1^{l_1}(x,t)\cdots\tau_s^{l_s}(x,t)\exp[(i_1-j_1+2i_1-2j_1+\cdots+zi_z-zj_z)fd], \tag{17}$$

$$\mu(x,t) = a_g^{h+i_1+j_1+\cdots+i_z+j_z}(x,t)\sigma_1^{l_1}(x,t)\cdots\sigma_s^{l_s}(x,t)\exp[-(i_1-j_1+2i_1-2j_1+\cdots+zi_z-zj_z)gd], \tag{18}$$

$$\kappa = b_{-p}^{h+i_1+j_1+\cdots+i_z+j_z}\delta_1^{l_1}\cdots\delta_s^{l_s}\exp[(i_1-j_1+2i_1-2j_1+\cdots+zi_z-zj_z)fd], \tag{19}$$

$$\lambda = b_q^{h+i_1+j_1+\cdots+i_z+j_z}\varsigma_1^{l_1}\cdots\varsigma_s^{l_s}\exp[-(i_1-j_1+2i_1-2j_1+\cdots+zi_z-zj_z)gd]. \tag{20}$$

Multiplying Equations (15) and (16) by

$$\frac{\delta_r\exp[(-2^r p)\xi_n]+\cdots+\varsigma_r\exp[(2^r q)\xi_n]}{\delta_r\exp[(-2^r p)\xi_n]+\cdots+\varsigma_r\exp[(2^r q)\xi_n]},$$

we have

$$\frac{\vartheta(x,t)\delta_r \exp\{-[f(h+i_1+j_1+\cdots+i_z+j_z)+l_1(f+p)+\cdots+l_s(f-p+2^s p)+2^r p]\xi_n\}+\cdots}{\kappa\delta_r \exp[-p(h+i_1+j_1+\cdots+i_z+j_z+2l_1+\cdots+2^s l_s+2^r)\xi_n]+\cdots}, \tag{21}$$

$$\frac{\cdots+\mu(x,t)\varsigma_r \exp\{[g(h+i_1+j_1+\cdots+i_z+j_z)+l_1(g+q)+\cdots+l_s(g-q+2^s q)+2^r q]\xi_n\}}{\cdots+\lambda\varsigma_r \exp[q(h+i_1+j_1+\cdots+i_z+j_z+2l_1+\cdots+2^s l_s+2^r)\xi_n]}. \tag{22}$$

We further use

$$\kappa \exp[-p(h+i_1+j_1+\cdots+i_z+j_z+2l_1+\cdots+2^s l_s)\xi_n]+\cdots$$

$$+\lambda \exp[q(h+i_1+j_1+\cdots+i_z+j_z+2l_1+\cdots+2^s l_s)\xi_n]$$

to multiply the numerator and denominator of Equation (8), then the left hand side and the right hand side of Equation (8) can be respectively written as:

$$\frac{\kappa\tau_r(x,t)\exp\{-[p(h+i_1+j_1+\cdots+i_z+j_z+2l_1+\cdots+2^s l_s)+(f-p+2^r p)]\xi_n\}+\cdots}{\kappa\delta_r \exp[-p(h+i_1+j_1+\cdots+i_z+j_z+2l_1+\cdots+2^s l_s+2^r)\xi_n]+\cdots}, \tag{23}$$

$$\frac{\cdots+\lambda\sigma_r(x,t)\exp\{[q(h+i_1+j_1+\cdots+i_z+j_z+2l_1+\cdots+2^s l_s)+(g-q+2^r q)]\xi_n\}}{\cdots+\lambda\varsigma_r \exp[q(h+i_1+j_1+\cdots+i_z+j_z+2l_1+\cdots+2^s l_s+2^r)\xi_n]}. \tag{24}$$

Balancing the lowest order of the exponential function in Equations (21) and (23) and the highest order of the exponential function in Equations (22) and (24) yields

$$(p-f)(h+i_1+j_1+\cdots+i_z+j_z+l_1+\cdots+l_s-1)=0, \tag{25}$$

$$(q-g)(h+i_1+j_1+\cdots+i_z+j_z+l_1+\cdots+l_s-1)=0. \tag{26}$$

It is easy to see from Equation (11) that

$$h+i_1+j_1+\cdots+i_z+j_z+l_1+\cdots+l_s-1\neq 0, \tag{27}$$

then Equations (25) and (26) give $f=p$ and $g=q$. This contradicts with our assumption that $f\neq p$ and $g\neq q$. Thus we complete the proof of Theorem 1. □

Theorem 1 shows that $f=p$ and $g=q$ are the only relations when using the exponential function ansätz (6) to balance the highest order derivative (8) with the highest order nonlinear term (10). Therefore, the simplest choice $f=p=g=q=1$ is often selected so that some additional calculations in determining the exponential function ansätz (6) are not longer required. Thus, Theorem 1 improves the generalized exp-function method described in Section 2.

4. Application

To give a concrete application of our improved exp-function method in Sections 2 and 3, we consider in this section the mKdV lattice equation with variable coefficient [42]:

$$\frac{du_n}{dt} = [\alpha(t) - u_n^2](u_{n+1} - u_{n-1}), \quad n \in Z, \tag{28}$$

where $u_n = u(n,t)$, $\alpha(t)$ is an arbitrary differentiable function of t. When $\alpha(t) = 0, 1, \alpha(\text{const.})$, Equation (28) can give three known constant-coefficient versions of the mKdV lattice equation.

Using the transformation

$$u_n = U_n(\eta_n), \quad \eta_n = dn + c(t) + \eta_0, \tag{29}$$

where d is a constant to be determined, $c(t)$ is the undermined function of t, and η_0 is the phase, we transform Equation (28) into

$$\frac{\mathrm{d}c(t)}{\mathrm{d}t}U_n' = [\alpha(t) - U_n^2](U_{n+1} - U_{n-1}). \tag{30}$$

According to the exp-function method improved in Sections 1 and 2, we directly suppose that:

$$U_n = \frac{a_{-1}(t)\exp(-\eta_n) + a_0(t) + a_1(t)\exp(\eta_n)}{b_{-1}\exp(-\eta_n) + b_0 + b_1\exp(\eta_n)}, \tag{31}$$

$$U_{n-1} = \frac{a_{-1}(t)\exp(d)\exp(-\eta_n) + a_0(t) + a_1(t)\exp(-d)\exp(\eta_n)}{b_{-1}\exp(d)\exp(-\eta_n) + b_0 + b_1\exp(-d)\exp(\eta_n)}, \tag{32}$$

$$U_{n+1} = \frac{a_{-1}(t)\exp(-d)\exp(-\eta_n) + a_0(t) + a_1(t)\exp(d)\exp(\eta_n)}{b_{-1}\exp(-d)\exp(-\eta_n) + b_0 + b_1\exp(d)\exp(\eta_n)}, \tag{33}$$

Substituting Equations (31)–(33) into Equation (30), and using Mathematica, equating the coefficients of all powers of $\exp(j\eta_n)(j = 0, \pm1, \pm2, \pm3)$ to zero yields a set of equations for $a_1(t)$, $a_0(t)$, $a_{-1}(t)$, b_1, b_0, b_{-1} and $c(t)$. Solving the system of equations by the use of Mathematica, we have:

$$a_0(t) = 0, \quad a_1(t) = \pm b_1\sqrt{\alpha(t)}\tanh(d), \quad a_{-1}(t) = \mp b_{-1}\sqrt{\alpha(t)}\tanh(d), \tag{34}$$

$$b_0 = 0, \quad c(t) = 2\tanh(d)\int \alpha(t)\mathrm{d}t, \tag{35}$$

and

$$a_0(t) = \pm 2\sqrt{-b_1 b_{-1}\alpha(t)}\tanh(\frac{d}{2}), \quad a_1(t) = \pm b_1\sqrt{\alpha(t)}\tanh(\frac{d}{2}), \tag{36}$$

$$a_{-1}(t) = \mp b_{-1}\sqrt{\alpha(t)}\tanh(\frac{d}{2}), \quad b_0 = 0, \quad c(t) = 4\tanh(\frac{d}{2})\int \alpha(t)\mathrm{d}t, \tag{37}$$

where b_1 and b_{-1} are arbitrary constants.

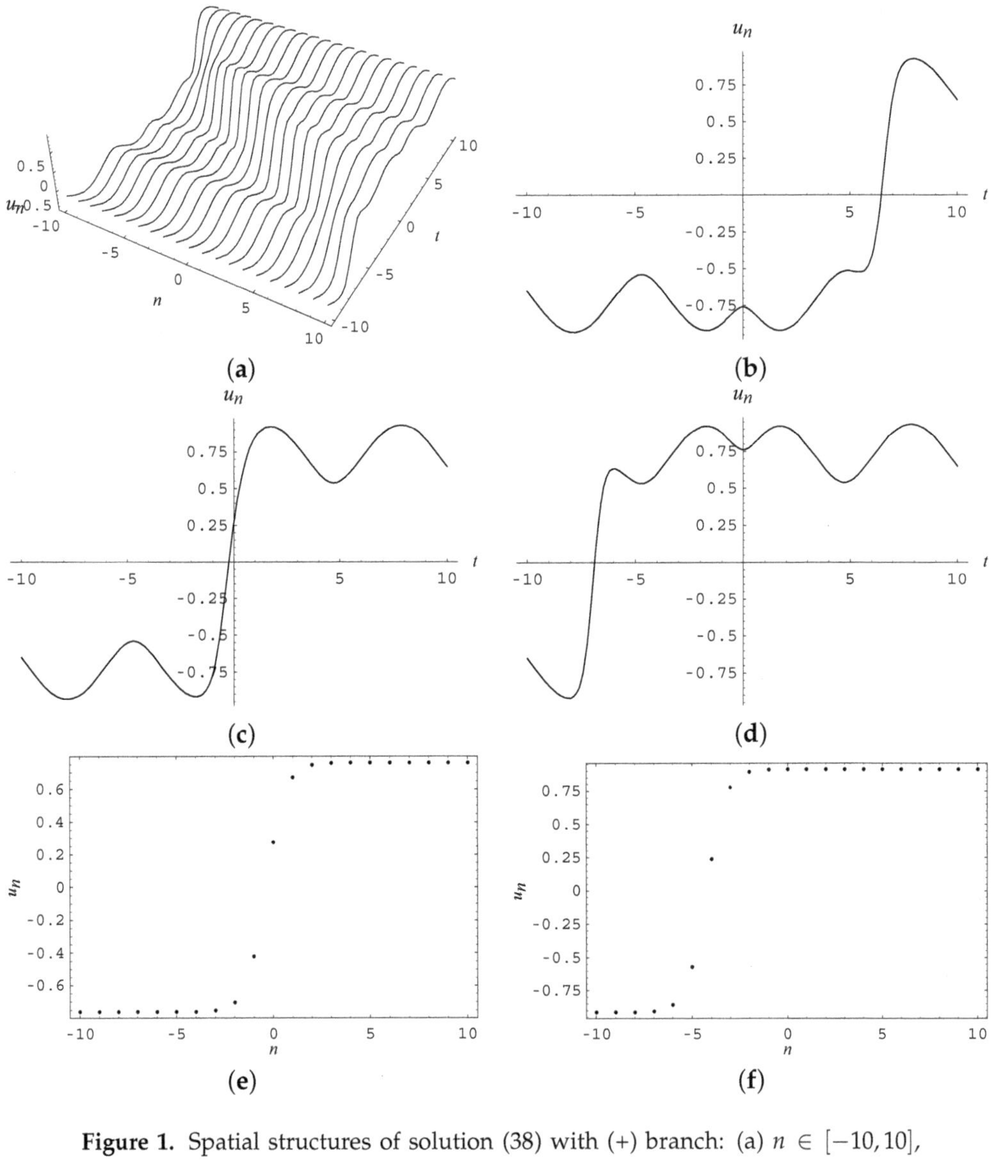

Figure 1. Spatial structures of solution (38) with (+) branch: (a) $n \in [-10, 10]$, $t \in [-10, 10]$; (b) $n = -10$, $t \in [-10, 10]$; (c) $n = 0$, $t \in [-10, 10]$; (d) $n = 10$, $t \in [-10, 10]$; (e) $n \in [-10, 10]$, $t = 0$; (f) $n \in [-10, 10]$, $t = 2$.

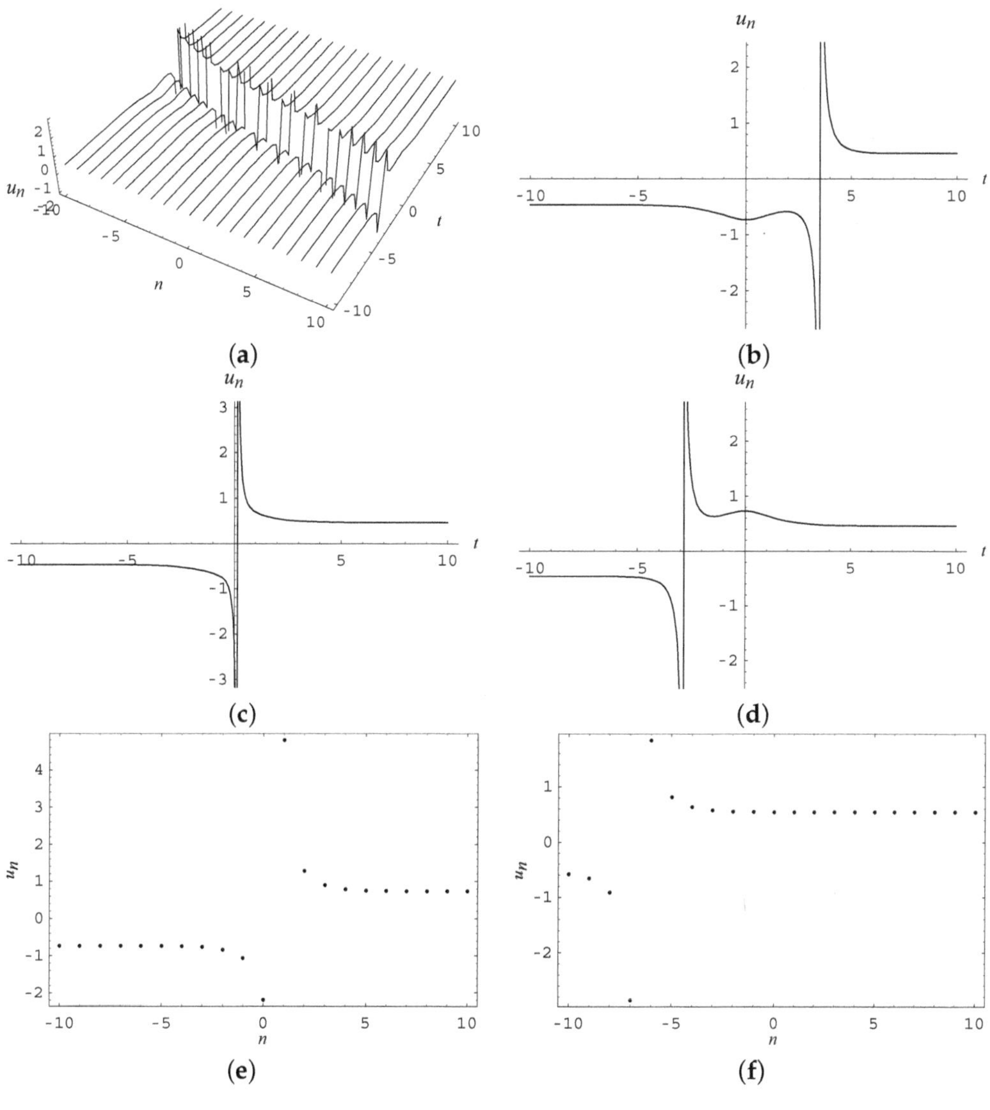

Figure 2. Spatial structures of solutions (39) with (+,+) branch: (a) $n \in [-10, 10]$, $t \in [-10, 10]$; (b) $n = -10$, $t \in [-10, 10]$; (c) $n = 0$, $t \in [-10, 10]$; (d) $n = 10$, $t = [-10, 10]$; (e) $n \in [-10, 10]$, $t = 0$; (f) $n \in [-10, 10]$, $t = 2$.

We, therefore, obtain from Equations (29), (31), (34) and (35) a pair of new kink-type solutions of Equation (28):

$$u_n = \pm\sqrt{\alpha(t)}\tanh(d)\frac{b_1\exp(\eta_n) - b_{-1}\exp(-\eta_n)}{b_1\exp(\eta_n) + b_{-1}\exp(-\eta_n)}, \tag{38}$$

where $\eta_n = dn + 2\tanh(d)\int \alpha(t)\mathrm{d}t + \eta_0$. If set $b_1 = 1$, then solutions (38) become the known solutions [42].

With the help of Equations (29), (31), (36) and (37), we obtain two pairs of new bell-kink-type solutions of Equation (28):

$$u_n = \pm\sqrt{\alpha(t)}\tanh(\frac{d}{2})\frac{b_1\exp(\eta_n) \pm 2\sqrt{-b_1 b_{-1}} - b_{-1}\exp(-\eta_n)}{b_1\exp(\eta_n) + b_{-1}\exp(-\eta_n)}, \quad (39)$$

where $\eta_n = dn + 4\tanh(\frac{d}{2})\int \alpha(t)\mathrm{d}t + \eta_0$.

In Figure 1, the spatial structures of solutions (38) with (+) branch are shown, where the parameters are selected as $\alpha(t) = 1 + 0.5\sin t\,\mathrm{sech} t$, $b_1 = -1.5$, $b_{-1} = -2$, $d = 1$, $\eta_0 = 0$. Figs. 1(a)–(d) show that the amplitude of wave changes periodically in the process of propagation. It is shown in Figure 1c that the "breather"-like phenomena has occurred at the location $n = 0$. In Figure 2, we show the structures of solutions (39) with (+,+) branch, where $\alpha(t) = 1 + \mathrm{sech} t$, $b_1 = 1.5$ and the other parameters are same as those in Figure 1. From Figure 2c, we can see that u_0 has a singularity in the interval $t \in (0, 1)$. It is easy to see that when $b_1 = 1.5$ and $b_{-1} = -2$, solutions (39) are unbounded. Such unbounded solutions develop singularity at a finite time, i.e. for any fixed $n = n_0$, there always exists $t = t_0$ at which these solutions "blow-up". In view of the physical significance, they do not exist after "blow-up". In the actual experimental physical system, there is no "blow-up", but a sharp spike [43]. Thus, the finite time "blow-up" can provide an approximation to the corresponding physical phenomenon.

5. Conclusions

In summary, we have improved the exp-function method [20] for solving non-linear lattice equations by modifying its exponential function ansätz. In order to show the advantages of the improved method, the variable-coefficient mKdV lattice equation (28) is considered. As a result, kink-type solutions (38) and bell-kink-type solutions (39) are obtained. To the best of our knowledge, they have not been reported in the literature. Solutions (38) and (39) contain arbitrary function $\alpha(t)$ and arbitrary constants b_1 and b_{-1}, which provide enough freedom for us to describe rich spatial structures of these obtained solutions. Applying the improved exp-function method to some other non-linear lattice equations with variable coefficients are worthy of study. This is our task in the future.

Acknowledgments: We would like to express our sincerest thanks to the editor and the referees for their valuable suggestions and comments, and to English teacher Yifei Li at Bohai University for her help in English writing. This work was supported by the PhD Start-up Funds of Liaoning Province of China (20141137) and Bohai University (bsqd2013025), the Natural Science Foundation of Educational Committee of Liaoning Province of China (L2012404), the Liaoning BaiQianWan Talents Program (2013921055) and the Natural Science Foundation of China (11371071).

Author Contributions: Sheng Zhang, Jiahong Li and Yingying Zhou conceived and designed the study. Sheng Zhang and Yingying Zhou wrote the paper. Sheng Zhang and Jiahong Li reviewed and edited the manuscript. All authors have read and approved the final manuscript.

Conflicts of Interest: The authors declare no conflict of interest.

Bibliography

1. Fermi, E.; Pasta, J.; Ulam, S. *Collected Papers of Enrico Fermi II*. University of Chicago Press: Chicago, IL, USA, 1965.
2. Toda, M. *Theory of Nonlinear Lattics*, 2nd ed.; Springer: Berlin, Germany, 1989.
3. Garder, C.S.; Greene, J.M.; Kruskal, M.D.; Miura, R.M. Method for solving the Korteweg–de Vries equation. *Phys. Rev. Lett.* **1965**, *19*, 1095–1097.
4. Miurs, M.R. *Bäcklund Transformation*. Springer: Berlin, Germany, 1978.
5. Hirota, R. Exact solution of the Korteweg–de Vries equation for multiple collisions of solitons. *Phys. Rev. Lett.* **1971**, *27*, 1192–1194.
6. Wang, M.L. Exact solutions for a compound KdV-Burgers equation. *Phys. Lett. A.* **1996**, *213*, 279–287.
7. Malfliet, M. Solitary wave solutions ofnonlinear wave equations. *Am. J. Phys.* **1992**, *60*, 650–654.
8. Liu, S.K.; Fu, Z.T.; Liu, S.D.; Zhao, Q. Jacobi elliptic function expansion method and periodic wave solutions of nonlinear wave equations. *Phys. Lett. A* **2001**, *289*, 69–74.
9. Abdel-Salam, E. A-B.; Al-Muhiameed, Z.I.A. Exotic localized structures based on the symmetrical lucas function of the (2+1)-dimensional generalized Nizhnik–Novikov–Veselov system. *Turk. J. Phys.* **2011**, *35*, 241–256.
10. Tabatabaei, K.; Celik, E.; Tabatabaei, R. The differential transform method for solving heat-like and wave-like equations with variable coeffcients. *Turk. J. Phys.* **2012**, *36*, 87–98.
11. Fan, E.G. Travelling wave solutions in terms of special functions for nonlinear coupled evolution systems. *Phys. Lett. A* **2002**, *300*, 243–249.
12. Fan, E.G. An algebraic method for finding a series of exact solutions to integrable and nonintegrable nonlinear evolution equations. *J. Phys. A: Math. Gen.* **2003**, *36*, 7009–7026.
13. Dai, C.Q.; Wang, Y.Y.; Tian, Q.; Zhang, J.F. The management and containment of self-similar rogue waves in the inhomogeneous nonlinear Schrödinger equation. *Ann. Phys.* **2012**, *327*, 512–521.
14. Dolapci, I.T.; Yildirim, A. Some exact solutions to the generalized Korteweg–de Vries equation and the system of shallow water wave equations. *Nonlinear Anal. Model. Control* **2013**, *18*, 27–36.
15. He, J.H. Asymptotic methods for solitary solutions and compactons. *Abstr. Appl. Anal.* **2012**, *2012*, 916793(130pages).
16. Dai, C.Q.; Wang, X.G.; Zhou, G.Q. Stable light-bullet solutions in the harmonic and parity-time-symmetric potentials. *Phys. Rev. A* **2014**, *89*, 013834.
17. Dai, C.Q.; Zhu, H.P. Superposed Kuznetsov–Ma solitons in a two-dimensional graded-index grating waveguide. *J. Opt. Soc. Am. B* **2013**, *30*, 3291–3297.

18. Liu, Y.; Gao, Y.T.; Sun, Z.Y.; Yu, X. Multi-soliton solutions of the forced variable-coefficient extended Korteweg–de Vries equation arisen in fluid dynamics of internal solitary vaves. *Nonlinear Dyn.* **2011**, *66*, 575–587.
19. Zhang, S.; Cai, B. Multi-soliton solutions of a variable-coefficient KdV hierarchy. *Nonlinear Dyn.* **2014**, *78*, 1593–1600.
20. He, J.H.; Wu, X. H. Exp-function method for nonlinear wave equations. *Chaos Solitons Fractals* **2006**, *30*, 700–708.
21. He, J.H.; Abdou, M.A. New periodic solutions for nonlinear evolution equations using Exp-function method. *Chaos Solitons Fractals* **2006**, *34*, 1421–1429.
22. He, J.H.; Zhang, L.N. Generalized solitary solution and compacton-like solution of the Jaulent–Miodek equations using the Exp-function method. *Phys. Lett. A* **2008**, *372*, 1044–1047.
23. Ebaid, A. Exact solitary wave solutions for some nonlinear evolution equations via Exp-function method. *Phys. Lett. A* **2007**, *365*, 213–219.
24. Zhang, S. Exp-function method for solving Maccari's system. *Phys. Lett. A* **2007**, *371*, 65–71.
25. Zhang, S. Application of Exp-function method to a KdV equation with variable coefficients. *Phys. Lett. A* **2007**, *365*, 448–453.
26. Zhang, S. Exp-function method for constructing explicit and exact solutions of a lattice equation. *Appl. Math. Comput.* **2008**, *199*, 242–249.
27. Dai, C.Q.; Zhang, J.F. Application of He's exp-function method to the stochastic mKdV equation. *Int. J. Nonlinear Sci. Num. Simul.* **2009**, *10*, 675–680.
28. Marinakis, V. The exp-function method and *n*-soliton solutions, *Z. Naturforsch. A* **2008**, *63*, 653–656.
29. Zhang, S. Application of Exp-function method to high-dimensional nonlinear evolution equation. *Chaos Solitons Fractals* **2008**, *38*, 270–276.
30. Ebaid, A. Generalization of He's exp-function method and new exact solutions for Burgers equation. *Z. Naturforsch. A* **2009**, *64*, 604–608.
31. Ebaid, A. Exact solutions for the generalized Klein–Gordon equation via a transformation and Exp-function method and comparison with Adomian's method. *J. Comput. Appl. Math.* **2009**, *223*, 278–290.
32. Mohyud-Din, S.T.; Khan, Y.; Faraz, N.; Yildirim, A. Exp-function method for solitary and periodic solutions of Fitzhugh–Nagumo equation. *Int. J. Numer. Methods H* **2012**, *22*, 335–341.
33. Bekir, A.; Aksoy, E. Exact solutions of extended shallow water wave equations by exp-function method. *Int. J. Numer. Method. H* **2013**, *23*, 305–319.
34. Chai, Y.Z.; Jia, T.T.; Hao, H.Q.; Zhang, J.W. Exp-function method for a generalized mKdV equation. *Discrete Dyn. Nat. Soc.* **2014**, *2014*, 153974.
35. He, J.H. Exp-function method for fractional differential equations, *Int. J. Nonlinear Sci. Num. Simul.* **2013**, *14*, 363–366.
36. Yan, L.M. Generalized exp-function method for non-linear space-time fractional differential equations. *Therm. Sci.* **2014**, *18*, 1573–1576.

37. Malik, S.A; Qureshi, I.M.; Amir, M.; Malik, AN; Haq, I. Numerical solution to generalized Burgers'–Fisher equation using exp-function method hybridized with heuristic computation. *PLoS ONE* **2015**, *10*, 1–15.
38. Kudryashov, N.A.; Loguinova, N.B. Be careful with the Exp-function method. *Commun. Nonlinear Sci. Numer. Simul.* **2009**, *14*, 1881–1890.
39. Aslan, I; Marinakis, V. Some remarks on exp-function method and its applications. *Commun. Theor. Phys.* **2011**, *56*, 397–403.
40. Aslan, I. Some remarks on exp-function method and its applications-a supplement. *Commun. Theor. Phys.* **2013**, *60*, 521–525.
41. Edaid, A. An improvement on the Exp-function method when balancing the highest order linear and nonlinear terms. *J. Math. Anal. Appl.* **2012**, *392*, 1–5.
42. Zhang, S.; Zhou, Y.Y. Kink-type solutions of the mKdV lattice equation with an arbitrary function. *Adv. Mater. Res.* **2014**, *989–994*, 1716–1719.
43. Zhang, S.; Zhang, H.Q. Discrete Jacobi elliptic function expansion method for nonlinear differential-difference equations. *Phys. Scr.* **2009**, *80*, 045002.

The Effect of a Long-Range Correlated-Hopping Interaction on Bariev Spin Chains

Tao Yang, Fa-Kai Wen, Kun Hao, Li-Ke Cao and Rui-Hong Yue

Abstract: We introduce a long-range particle and spin interaction into the standard Bariev model and show that this interaction is equivalent to a phase shift in the kinetic term of the Hamiltonian. When the particles circle around the chain and across the boundary, the accumulated phase shift acts as a twist boundary condition with respect to the normal periodic boundary condition. This boundary phase term depends on the total number of particles in the system and also the number of particles in different spin states, which relates to the spin fluctuations in the system. The model is solved exactly via a unitary transformation by the coordinate Bethe ansatz. We calculate the Bethe equations and work out the energy spectrum with varying number of particles and spins.

Reprinted from *Entropy*. Cite as: Yang, T.; Wen, F.-K.; Hao, K.; Cao, L.-K.; Yue, R.-H. The Effect of a Long-Range Correlated-Hopping Interaction on Bariev Spin Chains. *Entropy* **2015**, *17*, 6044–6055.

1. Introduction

One dimensional (1D) (quasi-1D) systems exhibit some of the most diverse and intriguing physical phenomena seen in all of condensed matter physics, such as charge (spin) density waves, quantum wires, quantum Hall bars, Josephson junction arrays, polymers and 1D Bose-Einstein condensates. The complete description of a solid is a complex many body problem. The particles are strongly correlated and cannot be understood by removing the interactions between them or by considering the effects of interactions as a perturbation. However, for some realistic low-dimensional strongly correlated systems a proper understanding has yet to be established through the examination of simplified exactly solvable models, in which the integrability has been considered to be one of the striking properties from the points of view of physics and mathematics. The 1D Hubbard model, in which the electron hopping is strongly disturbed by the on-site Coulomb interaction, has been mainly investigated with regard of Mott-transition through its exact solution [1]. The supersymmetric $t-J$ model [2], which includes the spin fluctuations via antiferromagnetic coupling, is relevant to the description of electronic mechanisms in high-T_c superconductivity. The 1D Bariev (interacting XY) chain [3,4] is also a Hubbard-like integrable model of special interest, as it supports Cooper type hole pairs. Motivated by the inclusion of additional interactions, whether through internal impurities or external boundary fields, many works have been carried out to generalize these models for different boundary fields [5–16]. This provides a non-perturbative method to study the boundary impurity effects in one-dimensional quantum systems in condensed matter physics. Bariev model has been generalized in many ways. The Hamiltonian studied in [17] included the onsite interaction and pair hopping processes. The Bariev chains with correlated single-particle and uncorrelated pair hopping were studied in [18], but there is only one type of particle. Bariev *et al.* [19,20] have considered the situations with multi-particle hopping and interchain tunneling, respectively. However, most of the investigated systems include only the nearest neighbor interactions; the question of how to find an integrable system with long range interaction is an interesting topic.

Schulz and Shastry [21] presented a class of lattice and continuum fermion models which are exactly solvable by a pseudo-unitary transformation, leading to nontrivial and non-Fermi-liquid behavior, with an exponential dependence upon the interaction. The idea behind this approach is the finding of a basis (through a unitary transformation of the original Fock basis [22,23]) in which the model takes the form of the original Hubbard or XXZ model up to boundary twists which do not affect their solvability. Furthermore, the Schultz-Sharstry model was generalized by introducing an exponential interaction involving two spins with same orientation [24].

In this paper, we generalize the Bariev model by introducing an Schultz-Sharstry-like exponential interaction which is dependent on the spin orientations of particles in the system. We note that the applied long-range spin-dependent interaction in the hopping term can be treated as a boundary phase twist. The phase change is in turn a function of number of particles and spins. When the Aharonov-Bohm effect is added to a 1D Hubbard chain with periodic boundary conditions it contributes to an extra phase shift related to the external magnetic flux [25,26], however our model can be applied both in the situation with external magnetic field and with internal field induced by impurities or spin fluctuations. We find the charge and spin excitations in our generalized model is a function of band filling, which is similar to the model proposed by Hirsch [27] for studying the high-T_c superconductivity. The latter, however, is not integrable in 1D. By applying an unitary transformation we prove the integrability of our model. The model is solved in the framework of the coordinate Bethe ansatz [28,29]. All charge and spin momenta are determined by a set of Bethe equations. The energy spectrum is listed based on the classification of varying number of particles. These may be useful in the systems where the long-range interactions cannot be ignored by only taking account of the nearest neighbour interactions.

2. From Long-Range Interactions to a Twist Boundary Condition

To include long-range spin interactions, we introduce some coordinate dependent parameters, α, β and κ into the Hamiltonian of the standard Bariev model. The Hamiltonian of the generalized Bariev model to be studied is in the form

$$H = -t\sum_{j=1}^{L}\sum_{\sigma=\uparrow,\downarrow}\left\{c_{j+1,\sigma}^{\dagger}c_{j,\sigma}e^{i\kappa_j(\sigma)}e^{i\sum_{l=1}^{L}[\alpha_{j,l}(\sigma)n_{l,-\sigma}+\beta_{j,l}(\sigma)n_{l,\sigma}]}+h.c.\right\}\times e^{-\eta\sum_{\sigma'\neq\sigma}n_{j+\theta(\sigma-\sigma'),\sigma'}}, \quad (1)$$

with $c_{j,\sigma}^{\dagger}$ ($c_{j,\sigma}$) being the creation (annihilation) operator of a particle with spin σ (σ being either $\uparrow$ or $\downarrow$) located at the j_{th} site, $n_{j,\sigma} \doteq c_{j,\sigma}^{\dagger}c_{j,\sigma}$ being the number operator, and $\theta(x)$ being a step function, *i.e.*, $\theta(x) = 1$ if $x > 0$ and $\theta(x) = 0$ if $x < 0$. The anti-commutation relation is satisfied by

$$\{c_{j,\sigma}, c_{l,\sigma'}^{\dagger}\} = \delta_{j,l}\delta_{\sigma,\sigma'}, \quad \{c_{j,\sigma}^{\dagger}, c_{l,\sigma'}^{\dagger}\} = \{c_{j,\sigma}, c_{l,\sigma'}\} = 0. \quad (2)$$

η is a coupling constant that influences the hopping amplitude of particles. Positive and negative values of η correspond to attractive and repulsive inter-particle interactions, respectively. It is clear that the system is reduced to standard Bariev model and is integrable if α, β and κ all vanish. The exponential term of $\alpha_{j,l}$ and $\beta_{j,l}$ is a generalized Jordan-Wigner transformation which includes interactions between the particle on the j_{th} site and the occupation state of all sites on the spin chain. This can be seen clearly if we make an expansion around small α and β. If we set $\alpha_{j,l} = \beta_{j,l} = \pi$ and take the summation of l from 1 to $j-1$, the generalized Jordan-Wigner transformation will degenerate into Jordan-Wigner transformation.

For arbitrary values of α, β and κ the system described by Equation (1) is not integrable by direct coordinate Bethe ansatz because all particles in the system are coupled through the long-range interaction. So the question now turns into how to determine these free parameters but keep the integrability. For this purpose, we introduce a special unitary transformation

$$U \doteq \exp\sum_{l,m=1}^{L+1}\sum_{\mu,\nu=\uparrow,\downarrow}\left[i(\xi_{l,m}^{\mu,\nu}n_{l,\mu}n_{m,\nu}+\zeta_{l,\mu}n_{l,\mu})\right], \quad (3)$$

where $\xi_{l,m}$ is the spin interaction strength between two sites and $\zeta_l \in R$ is a parameter related to the local chemical potential and magnetic field. They are all free parameters to be confirmed by specific physical models. The subscripts l and m are coordinate indices, and the superscripts μ, ν are spin indices. This is similar to choosing a different basis for the coordinate Bethe ansatz calculations. We will show later that

α, β and κ can be expressed in the form of ξ and ζ. Under the transformation $c_{j,\sigma} \xrightarrow{U} Uc_{j,\sigma}U^{-1}$, the hopping term in Hamiltonian Equation (1) turns into

$$c^{\dagger}_{j+1,\sigma}c_{j,\sigma} \xrightarrow{U} c^{\dagger}_{j+1,\sigma}c_{j,\sigma}\exp\left[2i(\xi^{\sigma,\mu}_{j+1,m} - \xi^{\sigma,\mu}_{j,m})n_{m,\mu} + i(\zeta_{j+1,\sigma} - \zeta_{j,\sigma} - 2\xi^{\sigma,\sigma}_{j,j+1})\right], \quad (4)$$

while the particle number operator keeps unchanged.

For normal periodic boundary conditions, there will be a phase change of pkL when one particle hops from site L to $L+1=1$ ($j=L$ in Hamiltonian Equation (1)), where $k=2\pi/L$ and p is an integer. We will give up the original boundary condition of the standard Bariev model but apply new boundary conditions which can keep the integrability of model Equation (1). For the transformation Equation (3), the phase shift across the boundary is determined by the relations $\xi_{1,m} \leftrightarrow \xi_{L,m}$, $\xi_{L,1} \leftrightarrow \xi_{L,L+1}$ and $\zeta_{1,m} \leftrightarrow \zeta_{L,m}$. Without loss of generality, we can set the phase change across the boundary to be

$$\xi^{\sigma,-\sigma}_{L+1,m} \doteq \xi^{\sigma,-\sigma}_{1,m} - \Phi_{\perp}(\sigma), \quad (5)$$
$$\xi^{\sigma,\sigma}_{L+1,m} \doteq \xi^{\sigma,\sigma}_{1,m} - \Phi_{\parallel}(\sigma), \quad (6)$$
$$\zeta_{L+1,\sigma} - 2\xi^{\sigma,\sigma}_{L,L+1} \doteq \zeta_{1,\sigma} - 2\xi^{\sigma,\sigma}_{L,1} - \Phi(\sigma). \quad (7)$$

If we set

$$\alpha_{j,m}(\sigma) \doteq 2(\xi^{\sigma,-\sigma}_{j,m} - \xi^{\sigma,-\sigma}_{j+1,m}), \quad (8)$$
$$\beta_{j,m}(\sigma) \doteq 2(\xi^{\sigma,\sigma}_{j,m} - \xi^{\sigma,\sigma}_{j+1,m}), \quad m \in \{1,\cdots,j-1,j+2,\cdots,L\}, \quad (9)$$
$$\kappa_j(\sigma) \doteq (\zeta_{j,\sigma} - \zeta_{j+1,\sigma} + 2\xi^{\sigma,\sigma}_{j,j+1}), \quad (10)$$

we can see easily that the Hamiltonian Equation (1) can reduce to the original Bariev model by the unitary transformation UHU^{-1}, up to a set of boundary conditions, and is integrable if α, β and κ all vanish. Through Equations (5)–(10), we obtain

$$\alpha_{L,m}(\sigma) = 2(\xi^{\sigma,-\sigma}_{L,m} - \xi^{\sigma,-\sigma}_{1,m}) + \Phi_{\perp}(\sigma), \quad (11)$$
$$\beta_{L,m}(\sigma) = 2(\xi^{\sigma,\sigma}_{L,m} - \xi^{\sigma,\sigma}_{1,m}) + \Phi_{\parallel}(\sigma), \quad m \in \{2,\cdots,L-1\}, \quad (12)$$
$$\kappa_L(\sigma) = (\zeta_{L,\sigma} - \zeta_{1,\sigma} + 2\xi^{\sigma,\sigma}_{L,1}) + \Phi(\sigma). \quad (13)$$

Then for the given boundary phase shift $\Phi_\perp$, $\Phi_\parallel$ and Φ the specific expressions for α, β and κ are obtained by Equations (8)–(10) with the constraints

$$\sum_{j=1}^{L} \alpha_{j,m}(\sigma) = \Phi_\perp(\sigma), \tag{14}$$

$$\sum_{\substack{j=1 \\ j\neq m,m-1}}^{L} \beta_{j,m}(\sigma) + \beta_{m,m-1}(\sigma) + \beta_{m-1,m+1}(\sigma) = \Phi_\parallel(\sigma), \tag{15}$$

$$\sum_{j=1}^{L} \left[\kappa_j(\sigma) + \beta_{j,j}(\sigma)\right] = \Phi(\sigma). \tag{16}$$

The total boundary twist is given by

$$\gamma_\sigma = \Phi(\sigma) + \Phi_\perp(\sigma)N_{-\sigma} + \Phi_\parallel(\sigma)(N_\sigma - 1), \tag{17}$$

where N_σ is number of particles with spin σ. We note that the coefficient of the last term in Equation (17) is $N_\sigma - 1$ because the terms for $m = j$ and $m = j+1$ in constraint Equation (9) do not exist. This boundary condition we will call a twist boundary condition. When $\Phi_\perp$, $\Phi_\parallel$ and Φ all take a value pkL, the twist boundary condition reduces to the trivial periodic boundary condition. For any chosen twist boundary condition other than the normal periodic boundary condition, if one can find parameters α, β and κ satisfying the constraints Equations (8)–(10) and (14)–(16), the Hamiltonian Equation (1) is then solvable. The transform of Hamiltonian Equation (1) under U is

$$\begin{aligned} H \xrightarrow{U} UHU^{-1} &= -t\sum_{j=1}^{L-1}\sum_{\sigma=\uparrow,\downarrow}\left\{c^\dagger_{j+1,\sigma}c_{j,\sigma} + h.c.\right\}\exp\left[-\eta\sum_{\sigma'\neq\sigma} n_{j+\theta(\sigma-\sigma'),\sigma'}\right] \\ &\quad -t\sum_{\sigma=\uparrow,\downarrow}\left\{c^\dagger_{1,\sigma}c_{L,\sigma}\exp[i\gamma_\sigma] + h.c.\right\}\exp\left[-\eta\sum_{\sigma'\neq\sigma} n_{j+\theta(\sigma-\sigma'),\sigma'}\right], \end{aligned} \tag{18}$$

with the boundary term

$$c^\dagger_{L+1,\sigma}c_{L,\sigma} = \exp[i\gamma_\sigma]c^\dagger_{1,\sigma}c_{L,\sigma}. \tag{19}$$

Generally, the sites of the chain are chosen with a homogeneous distribution. So it is natural to think the effect of this boundary phase term γ_σ as an average phase shift $\delta_\sigma = \gamma_\sigma/L$ when a particle hops from one site to its neighbour sites. The corresponding Hamiltonian is then

$$H' = -t\sum_{j=1}^{L}\sum_{\sigma=\uparrow,\downarrow}\left\{e^{i\delta_\sigma}\tilde{c}^\dagger_{j+1,\sigma}\tilde{c}_{j,\sigma} + h.c.\right\}\exp\left[-\eta\sum_{\sigma'\neq\sigma}\tilde{n}_{j+\theta(\sigma-\sigma'),\sigma'}\right], \tag{20}$$

where

$$\widetilde{c}_{j,\sigma}^{\dagger} = e^{-ij\delta_\sigma} c_{j,\sigma}^{\dagger}, \quad \widetilde{c}_{j,\sigma} = c_{j,\sigma} e^{ij\delta_\sigma}, \quad \widetilde{n}_{j,\sigma} = \widetilde{c}_{j,\sigma}^{\dagger} \widetilde{c}_{j,\sigma}, \tag{21}$$

and the basic commutation relations are kept unchanged. δ is a function of σ, N and $N_\downarrow$, which is different from the case of a periodic chain. By comparing this with the standard Bariev model ($\alpha = \beta = \kappa = 0$ in Equation (1)), one can see clearly that the introduced long-range spin interactions are equivalent to applying a twist boundary condition. However, we note that the phase shift between the neighbour sites may as well be distributed in any other way such that the sum equals γ_σ without changing any results.

3. Bethe Equations and Energy Spectrum

In the standard Bethe ansatz approach, modified for the twist boundary condition, any eigenfunction of the Hamiltonian takes a form similar to tensor products of plane waves [30]. We consider the eigenstate corresponding to N particles

$$|\Psi\rangle = \sum_{x_q=1}^{L} f(x_{q_1}, \cdots, x_{q_N}) \prod_{j=1}^{N} c_{x_j,\sigma_j}^{\dagger} |\Omega\rangle \tag{22}$$

in which the number of spin-down particles is $N_\downarrow$. In the region $x_{q_1} \leq \cdots \leq x_{q_N}$, the function f can be written as [31]

$$f(x_{q_1}, \cdots, x_{q_N}) = \epsilon_P A_{\sigma_{q_1}, \cdots, \sigma_{q_N}}(k_{p_1}, \cdots, k_{p_N}) \times \exp\left[i \sum_{j=1}^{N} k_{p_j} x_{q_j} \right] \theta(x_{q_1} \leq \cdots \leq x_{q_N}). \tag{23}$$

By solving the Schrödinger equation $H|\Psi\rangle = E|\Psi\rangle$, the energy eigenvalue of the Hamiltonian Equation (18) is given by

$$E = -2t \sum_{j=1}^{N} \cos k_j . \tag{24}$$

We note that the form of the energy eigenvalue does not change from the standard Bariev chain. However, we will see later that the momentum k_j is now spin dependent. Two-particle scattering matrices are given by

$$\begin{aligned} S_{\alpha_1\, \alpha_2}^{\alpha_1\, \alpha_2} &= \frac{sin[(k_1 - k_2)/2]}{sin[(k_1 - k_2)/2 + i\eta]} \quad (\alpha_1 \neq \alpha_2) \\ S_{\alpha_2\, \alpha_1}^{\alpha_1\, \alpha_2} &= \frac{sin[i\eta]}{sin[(k_1 - k_2)/2 + i\eta]} \exp\left[i \frac{k_1 - k_2}{2} sign(\alpha_1 - \alpha_2) \right] \quad (\alpha_1 \neq \alpha_2) \\ S_{\alpha_0\, \alpha_0}^{\alpha_0\, \alpha_0} &= 1 \quad (\alpha_1 = \alpha_2 = \alpha_0), \end{aligned} \tag{25}$$

which are similar to the $R-$matrices of the standard 6-vertex models. The two are related via a simple gauge transformation as [32]

$$S_{12}(\lambda) = V_1(\lambda) R_{12}(\lambda) V_1^{-1}(\lambda), \quad V(\lambda) = diag\left(e^{i\lambda/4}, e^{-i\lambda/4}\right). \tag{26}$$

for $\lambda = (k_1 - k_2)$.

In general, we have

$$S_{12}(\lambda_1 - \lambda_2) = V_1(\lambda_1) V_2(\lambda_2) R_{12}(\lambda_1 - \lambda_2) V_1^{-1}(\lambda_1) V_2^{-1}(\lambda_2). \tag{27}$$

It is easy to show

$$\begin{aligned} & S_{12}(\lambda_1 - \lambda_2) S_{13}(\lambda_1 - \lambda_3) S_{23}(\lambda_2 - \lambda_3) \\ = \; & V_1(\lambda_1) V_2(\lambda_2) V_3(\lambda_3) R_{12}(\lambda_1 - \lambda_2) R_{13}(\lambda_1 - \lambda_3) R_{23}(\lambda_2 - \lambda_3) V_1^{-1}(\lambda_1) V_2^{-1}(\lambda_2) V_3^{-1}(\lambda_3), \end{aligned} \tag{28}$$

which will satisfy the Yang-Baxter equation. So the integrability of the present model is kept. One can also solve the problem by constructing the R-matrix of this model following the the techniques in [33].

The charge momentum k_j and spin momentum Λ_μ satisfy the Bethe equations

$$e^{ik_j L} = e^{-i\gamma_\uparrow} \prod_{\mu=1}^{N_\downarrow} \frac{\sin\left[\frac{(k_j - \Lambda_\mu)}{2} + \frac{i\eta}{2t}\right]}{\sin\left[\frac{(k_j - \Lambda_\mu)}{2} - \frac{i\eta}{2t}\right]}, \tag{29}$$

$$\prod_{\nu=1, \nu\neq\mu}^{N_\downarrow} \frac{\sin\left[\frac{(\Lambda_\mu - \Lambda_\nu)}{2} + \frac{i\eta}{t}\right]}{\sin\left[\frac{(\Lambda_\mu - \Lambda_\nu)}{2} - \frac{i\eta}{t}\right]} = e^{-i(\gamma_\downarrow - \gamma_\uparrow)} \prod_{j=1}^{N} \frac{\sin\left[\frac{(\Lambda_\mu - k_j)}{2} + \frac{i\eta}{2t}\right]}{\sin\left[\frac{(\Lambda_\mu - k_j)}{2} - \frac{i\eta}{2t}\right]}. \tag{30}$$

The structure of roots for these equations depends strongly on the hopping amplitude η. The interactions between particles are repulsive when $\eta > 0$, all particles with different spins cannot form a pair. In this situation all k_j must be real, which can be proved under the thermodynamical limit. $\eta < 0$ corresponds to attractive interactions in the system. Particles with different spins tend to exist in the form of cooper-pairs. The solution $k_j = \Lambda + i|\eta|$ corresponds to these bound states for charge excitations. If we choose two sets of quantum numbers I_j and J_μ and set $\theta(x;a) = \arctan\Big(\tan(x/2)\coth(a/2)\Big)$, Equations (29) and (30) then take the form

$$\begin{aligned} Lk_j &= -\gamma_\uparrow + 2\pi I_j' + \sum_{\mu=1}^{N_\downarrow}\{\pi - 2\theta(k_j - \Lambda_\mu;\eta)\} \\ &= -\gamma_\uparrow + 2\pi I_j - \sum_{\mu=1}^{N_\downarrow}\{2\theta(k_j - \Lambda_\mu;\eta)\}\,, \end{aligned} \tag{31}$$

$$\begin{aligned} \gamma_\downarrow - \gamma_\uparrow &= 2\pi J_\mu' - \sum_{\nu=1;\nu\neq\mu}^{N_\downarrow}[\pi - 2\theta(\Lambda_\mu - \Lambda_\nu;2\eta)] + \sum_{j=1}^{N}\left[\pi - 2\theta(\Lambda_\mu - k_j;\eta)\right] \\ &= 2\pi J_\mu + \sum_{\nu=1;\nu\neq\mu}^{N_\downarrow}[2\theta(\Lambda_\mu - \Lambda_\nu;2\eta)] - \sum_{j=1}^{N}\left[2\theta(\Lambda_\mu - k_j;\eta)\right]\,, \end{aligned} \tag{32}$$

where I_j' and J_μ' are both common integers. The quantum numbers $I_j = I_j' + N_\downarrow/2$ and $J_\mu = J_\mu' + (N_\uparrow + 1)/2$ depend upon the charge and spin property in the system. From Equations (31) and (32) we can see that I_j and J_μ are either integer or half-integer, according to the number of total particles and the number of spin-up (spin-down) particles. There are four cases,

- I_j and J_μ are both integers if N and $N_\uparrow$ are both odd;
- I_j and J_μ are both half-integers if N is odd and $N_\uparrow$ is even;
- I_j is a half-integer and J_μ is an integer if N is even and $N_\uparrow$ is odd;
- I_j is an integer and J_μ is a half-integer if N and $N_\uparrow$ are both even.

By taking the summation of Equations (31) and (32) over the coordinate and spin indices respectively, the momentum of the system is given as

$$P = \sum_{j=1}^{N}\left(k_j - \frac{1}{L}\gamma'\right) = \frac{2\pi}{L}\left(\sum_{j=1}^{N} I_j + \sum_{\mu=1}^{N_\downarrow} J_\mu\right)\,, \tag{33}$$

where $\gamma' = (\gamma_\downarrow - \gamma_\uparrow) - (N_\uparrow\gamma_\downarrow + N_\downarrow\gamma_\uparrow)/N$. In the above calculations we have used the relation $\sum_{\mu=1}^{N_\downarrow}\sum_{\nu=1}^{N_\downarrow} 2\theta(\Lambda_\mu - \Lambda_\nu, 2\eta) = 0$.

In the limiting case $\eta \longrightarrow \infty$ ($\coth(\eta) \longrightarrow 1$), we get

$$Lk_j = 2\pi\left[I_j + \frac{N_\downarrow}{N(L+N_\downarrow)}\sum_{l=1}^{N}(I_l - I_j) + \frac{1}{N}\sum_{\lambda=1}^{N_\downarrow} J_\lambda\right] + \gamma'\,. \tag{34}$$

Substituting it into Equation (24) we obtain the energy of the system

$$E_0(\gamma) = -2tD\cos\left[\frac{2\pi}{L}\left(\frac{1}{N}\sum_{\lambda=1}^{N_\downarrow} J_\lambda + \bar{I} + \frac{\gamma'}{2\pi}\right)\right], \tag{35}$$

where $D = \sin(N\pi/L)/\sin(\pi/L)$ and $\bar{I} = (I_{min} + I_{max})/2$. We will not consider the situation where $N = L$, which occurs when the chain is half filled and D is 0. Then for arbitrary combinations of N and $N_\downarrow$ we have four different cases

$$\begin{aligned}
E_{even\ N}^{even\ N_\downarrow} &= -2tD\cos\left[\tfrac{2\pi}{L}\left(\tfrac{\gamma'}{2\pi} + g + \tfrac{1}{2} + \tfrac{N_\downarrow}{N}h\right)\right], \\
E_{even\ N}^{odd\ N_\downarrow} &= -2tD\cos\left[\tfrac{2\pi}{L}\left(\tfrac{\gamma'}{2\pi} + g + \tfrac{N_\downarrow}{N}h\right)\right], \\
E_{odd\ N}^{even\ N_\downarrow} &= -2tD\cos\left[\tfrac{2\pi}{L}\left(\tfrac{\gamma'}{2\pi} + g + \tfrac{N_\downarrow}{N}h + \tfrac{N_\downarrow}{2N}\right)\right], \\
E_{odd\ N}^{odd\ N_\downarrow} &= -2tD\cos\left[\tfrac{2\pi}{L}\left(\tfrac{\gamma'}{2\pi} + g + \tfrac{1}{2} + \tfrac{N_\downarrow}{N}h + \tfrac{N_\downarrow}{2N}\right)\right],
\end{aligned} \tag{36}$$

where g and h are any integers and also quantum numbers which describe the charge and spin excitations.

If we treat γ' as the phase shift induced by an external field, for a given N and $N_\downarrow$ the external field can only vary within a small range

$$\frac{2\mathcal{L}-1}{2N} < \frac{\gamma'}{2\pi} + \bar{I} < \frac{2\mathcal{L}+1}{2N}, \quad \mathcal{L} = -\sum_{\lambda=1}^{N_\downarrow} J_\lambda . \tag{37}$$

otherwise the spin inversion will occur.

4. Results for General η

Generally, for a finite η it is difficult to get exact solutions of the Bethe equations. In this section we try to discuss some properties of the system when $L \to \infty$. The root distribution of Bethe equations turns to a continuous density distribution, σ, in this case. We define functions

$$Z_c(k) = Lk + \gamma_\uparrow + \sum_{\mu=1}^{N_\downarrow} 2\theta(k - \Lambda_\mu; \eta), \tag{38}$$

$$Z_s(\Lambda) = \gamma_\uparrow - \gamma_\downarrow - \sum_{\mu=1}^{N_\downarrow} 2\theta(\Lambda - \Lambda_\nu; 2\eta) + \sum_{j=1}^{N} 2\theta(\Lambda - k_j; \eta). \tag{39}$$

In the limiting case of large L Equations (38) and (39) can be expressed in the form of integrals

$$2\pi\sigma_c(k) = \lim_{L\to\infty}\frac{1}{L}\frac{d}{dk}Z_c(k), \quad 2\pi\sigma_s(\Lambda) = \lim_{L\to\infty}\frac{1}{L}\frac{d}{dk}Z_s(\Lambda)\,, \tag{40}$$

where

$$n = \int_{-K}^{K}\sigma_c(k)dk = \frac{N}{L}, \quad n_\downarrow = \int_{-\Lambda_0}^{\Lambda_0}\sigma_s(\Lambda)d\Lambda = \frac{N_\downarrow}{L}\,. \tag{41}$$

Now, we can see from Equations (31) and (32) that the relations

$$Z_c(k_j) = 2\pi I_j, \quad Z_s(\Lambda) = 2\pi J_\lambda\,, \tag{42}$$

$$I_{j+1} - I_j = 1, \quad J_{\mu+1} - j_\mu = 1\,, \tag{43}$$

must be satisfied by roots of the Bethe equations. We are ready to obtain a set of integral equations

$$2\pi\sigma_c(k) = 1 + \int_{-\Lambda_0}^{\Lambda_0} 2\theta(k-\Lambda;\eta)\sigma_s(\Lambda)d\Lambda\,, \tag{44}$$

$$2\pi\sigma_s(\Lambda) = -\int_{-\Lambda_0}^{\Lambda_0} 2\theta'(\Lambda-\Lambda';2\eta)\sigma_s(\Lambda')d\Lambda' + \int_{-K}^{K} 2\theta'(\Lambda-k;\eta)\sigma_c(k)dk\,, \tag{45}$$

where $\theta'(x;y) = d\theta/dx = -sin(y)/4[cos(x)+cosh(y)]$.

Through a Fourier transform we obtain

$$\widetilde{\theta}'(\omega;y) = \int_{-\pi}^{\pi}\theta'(x;y)e^{-i\omega x}dx = \frac{1}{2}e^{-\omega y}\,. \tag{46}$$

For $\Lambda_0 = K = \pi$, we have $\sigma_c(\omega) = 1/\pi$ and $\sigma_s(\omega) = 1/2\pi$. The corresponding particle densities in Equation (41) are $n = 2$ and $n_\downarrow = 1$, which means that the numbers of two spin species are the same if there is no external field.

For a more general situation, the values of K and Λ_0 are determined by Equation (41). When the external field vanishes, the value of Λ_0 must be π. From Equations (44) and (45) we get

$$2\pi\sigma_c(k) = 1 + \sum_{\omega=0}^{\infty}\frac{e^{-ik\omega}}{1+e^{2\eta\omega}}\int_{-K}^{K}e^{i\mu\omega}\sigma_c(\mu)d\mu\,. \tag{47}$$

The numerical solutions of the Bethe Equations (29) and (30) and the corresponding eigenvalues of the Hamiltonian Equation (18) for $L = 2$ and $L = 3$ with different occupation numbers are shown in Tables 1 and 2 respectively. As mentioned before, we will only consider the cases where $N < L$. By analyzing the structure of Bethe equations, we can see that if $N_\downarrow = 0$ or $N = N_\downarrow$ the roots of the Bethe equations do not depend on η. These numerical results coincide with those obtained from the exact diagonalization of the Hamiltonian Equation (18) and the analytical results in the limiting case, $\eta \to \infty$, obtained through Equation (36). For the $L = 2$ chain, we also give the eigenvalues E with varying γ_σ, which do not change with η, as shown in Figure 1.

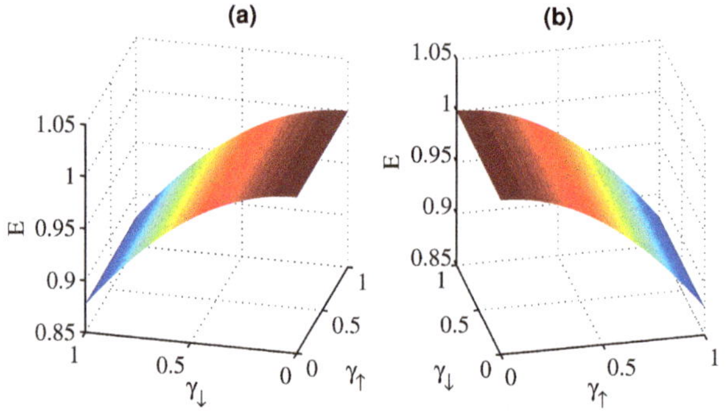

Figure 1. The eigenvalues E calculated from the exact diagonalization of the Hamiltonian Equation (18) with respect to γ_σ for $L = 2$. (**a**) $N = N_\uparrow = 1$. (**b**) $N = N_\downarrow = 1$.

Table 1. The numerical results calculated from Equations (29) and (30) for the parameters, $L = 2$, $t = 0.5$, $\eta = 0.3$, $\gamma_\uparrow = 0.2$, and $\gamma_\downarrow = 0.1$.

Occupation Number	k_1	Λ_1	E
$N = N_\uparrow = 1$	−0.100000	N/A	−0.995004
$N = N_\uparrow = 1$	3.041593	N/A	0.995004
$N = N_\downarrow = 1$	−0.050000	3.754988	−0.998750
$N = N_\downarrow = 1$	3.091593	0.613395	0.998750

Table 2. The numerical results calculated from Equations (29) and (30) for the parameters, $L = 3$, $t = 0.5$, $\eta = 0.3$, $\gamma_\uparrow = 0.4$, and $\gamma_\downarrow = 0.2$.

Occupation Number	k_1	k_2	Λ_1	Λ_2	E
$N = N_\uparrow = 1$	−0.066667	N/A	N/A	N/A	−0.997779
$N = N_\uparrow = 1$	2.027728	N/A	N/A	N/A	0.441197
$N = N_\uparrow = 1$	−2.161062	N/A	N/A	N/A	0.556582
$N = N_\downarrow = 1$	−0.033333	N/A	−2.834687	N/A	−0.999444
$N = N_\downarrow = 1$	2.061062	N/A	−0.740291	N/A	0.470860
$N = N_\downarrow = 1$	−2.127728	N/A	1.354104	N/A	0.528584
$N = N_\uparrow = 2$	−0.133333	1.961062	N/A	N/A	−0.610690
$N = N_\uparrow = 2$	−2.227728	−0.133333	N/A	N/A	−0.380434
$N = N_\uparrow = 2$	1.961062	−2.227728	N/A	N/A	0.991124
$N = N_\downarrow = 2$	−0.066667	2.027728	$-0.405569 + 1.839956i$	$-0.405569 - 1.839956i$	−0.556582
$N = N_\downarrow = 2$	−0.066667	−2.161062	$-2.499964 - 1.839956i$	$-2.499964 + 1.839956i$	−0.441197
$N = N_\downarrow = 2$	2.027728	−2.161062	$1.688826 + 1.839956i$	$1.688826 - 1.839956i$	0.997779

5. Conclusions

In this paper we have constructed a generalized 1D Bariev model which describes spin-1/2 particles with long-range interaction on a lattice. By employing an unitary transformation we find the Hamiltonian is equivalent to a standard Bariev Hamiltonian with twist boundary conditions. This phase twist may be used to explain the effects of an external magnetic potential and the internal fluctuations on the system. For a strong external magnetic field, spin inversions can occur in the system. By solving the Bethe equations in the limiting case where $\eta \to \infty$ we give the specific forms of energy spectrum in different system configurations with respect to total particle number and spin distributions. More general cases have been discussed but the analytical result can only be obtained in the situation where there is no external field. We also solve Bethe equations numerically and conduct the exact diagnalization of the transformed Hamitonian. The numerical results and the analytical results coincide with each other very well. To relate our model with real physical systems we need to determine the specific values of ξ, ζ and η and calculate some other physical properties of the system. This could be an interesting topic for further study.

Acknowledgments: We thank Andrew J. Henning and Dan T. Peng for valuable discussions and suggestions. T. Yang acknowledges support through NSFC11347025 and the Science Foundation of Northwest University (No. 13NW16). K. Hao is supported by NSFC11447239. R.H. Yue acknowledges support through NSFC11275099, NSFC11435006.

Author Contributions: This study was proposed by Rui-Hong Yue. The analytical results were worked out by Tao Yang, Li-Ke Cao and Kun Hao. The numerical simulations were run by Fa-Kai Wen. The manuscript was prepared by Tao Yang and Rui-Hong Yue. All authors have read and approved the final manuscript.

Conflicts of Interest: The authors declare no conflict of interest.

References

1. Essler, F.H.L.; Frahm, H.; Göhmann, F.; Klümper, A.; Korepin, V.E. *The One-Dimensional Hubbard Model*; Cambridge University Press: Cambridge, UK, 2010.
2. Zhang, F.C.; Rice, T.M. Effective Hamiltonian for the Superconducting Cu Oxides. *Phys. Rev. B* **1988**, *37*, 3759–3761.
3. Bariev, R.Z. Integrable Spin Chain with Two- and Three-Particle Interactions. *J. Phys. A* **1991**, *24*, L549, doi:10.1088/0305-4470/24/10/010.
4. Bariev, R.Z.; Klumper, A.; Schadschneider, A.; Zittartz, J. Excitation Spectrum and Critical Exponents of a One-Dimensional Integrable Model of Fermions with Correlated Hopping. *J. Phys. A* **1993**, *26*, 4863, doi:10.1088/0305-4470/26/19/019.
5. Guan, X.W.; Wang, M.S.; Yang, S.D. Lax Pair Formulation for One-Dimensional Hubbard Open Chain with Chemical Potential. *J. Phys. A* **1997**, *30*, 4161, doi:10.1088/0305-4470/30/12/008.

6. Fan, H.; Wadati, M.; Wang, X. Exact Diagonalization of the Generalized Supersymmetric t-J Model with Boundaries. *Phys. Rev. B* **2000**, *61*, 3450–3469.
7. Foerster, A.; Guan, X.-W.; Links, J.; Roditi, I.; Zhou, H.-Q. Exact Solution for the Bariev Model with Boundary Fields. *Nucl. Phys. B* **2001**, *596*, 525–547.
8. Yue, R.H.; Schlottmann, P. Integrable One-Dimensional N-Component Fermion Model with Correlated Hopping and Hard-Core Repulsion. *Nucl. Phys. B* **2002**, *647*, 539–564.
9. Yue, R.; Schlottmann, P. Exact Solution of the Bariev Model for Correlated Hopping with Hard-Core Repulsion. *Phys. Rev. B* **2002**, *66*, 085114.
10. Yang, W.L.; Zhang, Y.Z.; Zhao, S.Y. Drinfeld Twists and Algebraic Bethe Ansatz of the Supersymmetric t-J Model. *J. High Energy Phys.* **2004**, *2004*, 038, doi:10.1088/1126-6708/2004/12/038.
11. Alcaraz, F.C.; Lazo, M.J. Exactly Solvable Interacting Vertex Models. *J. Stat. Mech.* **2007**, *2007*, P08008, doi:10.1088/1742-5468/2007/08/P08008.
12. Frolov, S.; Quinn, E. Hubbard-Shastry Lattice Models. *J. Phys. A* **2011**, *45*, 095004, doi:10.1088/1751-8113/45/9/095004.
13. Galleas, W. Functional Relations from the Yang-Baxter Algebra: Eigenvalues of the XXZ Model with Non-diagonal Twisted and Open Boundary Conditions. *Nucl. Phys. B* **2008**, *790*, 524–542.
14. Cao, J.; Yang, W.L.; Shi, K.; Wang, Y. Off-Diagonal Bethe Ansatz Solution of the XXX Spin Chain with Arbitrary Boundary Conditions. *Nucl. Phys. B* **2013**, *875*, 152–165.
15. Cao, J.; Yang, W.L.; Shi, K.; Wang, Y. Off-Diagonal Bethe Ansatz Solutions of the Anisotropic Spin-Chains with Arbitrary Boundary Fields. *Nucl. Phys. B* **2013**, *877*, 152–175.
16. Li, Y.Y.; Cao, J.; Yang, W.L.; Shi, K.; Wang, Y. Exact Solution of the One-Dimensional Hubbard Model with Arbitrary Boundary Magnetic Fields. *Nucl. Phys. B* **2014**, *879*, 98–109.
17. Bariev, R.Z.; Klumper, A.; Zittartz, J. A New Integrable Two-Parameter Model of Strongly Correlated Electrons in One Dimension. *Europhys. Lett.* **1995**, *32*, 85, doi:10.1209/0295-5075/32/1/015.
18. Alcaraz, F.C.; Bariev, R.Z. Integrable Models of Strongly Correlated Particles with Correlated Hopping. *Phys. Rev. B* **1999**, *59*, 3373–3376.
19. Bariev, R.Z.; Klumper, A.; Schadschneider, A.; Zittartz, J. A One-Dimensional Integrable Model of Fermions with Multi-particle Hopping. *J. Phys. A* **1995**, *28*, 2437, doi:10.1088/0305-4470/28/9/007.
20. Bariev, R.Z.; Klümper, A.; Schadschneider, A.; Zittartz, J. Exact Solution of a One-Dimensional Fermion Model with Interchain Tunneling. *Phys. Rev. B* **1994**, *50*, 9676–9679.
21. Schulz, H.J.; Sriram Shastry, B. A New Class of Exactly Solvable Interacting Fermion Models in One Dimension. *Phys. Rev. Lett.* **1998**, *80*, 1924–1927.
22. Sriram Shastry, B.; Sutherland, B. Twisted Boundary Conditions and Effective Mass in Heisenberg-Ising and Hubbard Rings. *Phys. Rev. Lett.* **1990**, *65*, 243–246.

23. Sutherland, B.; Sriram Shastry, B. Adiabatic Transport Properties of an Exactly Soluble One-Dimensional Quantum Many-Body Problem. *Phys. Rev. Lett.* **1990**, *65*, 1833–1837.
24. Osterloh, A.; Amico, L.; Eckern, U. Exact Solution of Generalized Schulz-Shastry Type Models. *Nucl. Phys. B* **2000**, *588*, 531–551.
25. Stafford, C.A.; Millis, A.J.; Shastry, B.S. Finite-Size Effects on the Optical Conductivity of a Half-Filled Hubbard Ring. *Phys. Rev. B* **1991**, *43*, 13660–13663.
26. Fye, R.M.; Martins, M.J.; Scalapino, D.J.; Wagner, J.; Hanke, W. Drude Weight, Optical Conductivity, and Flux Properties of One-Dimensional Hubbard Rings. *Phys. Rev. B* **1991**, *44*, 6909–6915.
27. Hirsch, J.E. Hole Superconductivity. *Phys. Lett. A* **1989**, *134*, 451–455.
28. Bethe, H. Zur Theorie der Metalle. *Z. Phys.* **1931**, *71*, 205–226.
29. Yang, C.N.; Yang, C.P. One-Dimensional Chain of Anisotropic Spin-Spin Interactions. I. Proof of Bethe's Hypothesis for Ground State in a Finite System. *Phys. Rev.* **1966**, *150*, 321–327.
30. Korepin, V.E.; Wu, A.C.T. Adiabatic Transport Properties and Berry's Phase in Heisenberg-Ising Ring. *Int. J. Mod. Phys. B* **1991**, *5*, 497–507, doi:10.1142/S0217979291000304.
31. Deguchi, T.; Yue, R. Exact Solution of 1-D Hubbard Model with Open Boundary Conditions and the Conformal Dimensions under Boundary Magnetic Fields. **1997**, arXiv:cond-mat/9704138.
32. Doikou, A.; Evangelisti, S.; Feverati, G.; Karaiskos, N. Introduction to Quantum Integrability. *Int. J. Mod. Phys. A* **2010**, *25*, 3307–3351, doi:10.1142/S0217751X10049803.
33. Fonseca, T.; Frappat, L.; Ragoucy, E. R Matrices of Three-State Hamiltonians Solvable by Coordinate Bethe Ansatz. *J. Math. Phys.* **2015**, *56*, 013503, doi:10.1063/1.4905893.

Short-Lived Lattice Quasiparticles for Strongly Interacting Fluids

Miller Mendoza Jimenez and Sauro Succi

Abstract: It is shown that lattice kinetic theory based on short-lived quasiparticles proves very effective in simulating the complex dynamics of strongly interacting fluids (SIF). In particular, it is pointed out that the shear viscosity of lattice fluids is the sum of two contributions, one due to the usual interactions between particles (collision viscosity) and the other due to the interaction with the discrete lattice (propagation viscosity). Since the latter is negative, the sum may turn out to be orders of magnitude smaller than each of the two contributions separately, thus providing a mechanism to access SIF regimes at ordinary values of the collisional viscosity. This concept, as applied to quantum superfluids in one-dimensional optical lattices, is shown to reproduce shear viscosities consistent with the AdS-CFT holographic bound on the viscosity/entropy ratio. This shows that lattice kinetic theory continues to hold for strongly coupled hydrodynamic regimes where continuum kinetic theory may no longer be applicable.

Reprinted from *Entropy*. Cite as: Jimenez, M.M.; Succi, S. Short-Lived Lattice Quasiparticles for Strongly Interacting Fluids. *Entropy* **2015**, *17*, 6169–6178.

1. Introduction

The study of transport properties of strongly interacting fluids (SIF) has gained central stage in modern condensed matter research, with many fascinating connections with quantum-relativistic hydrodynamics, high-energy physics and string theory [1]. By definition, SIF are moving states of matter in which the interactions are strong to the point of preventing the microscopic degrees of freedom from propagating freely over significant distances as compared with the range of their interactions. More precisely, the mean-free path becomes smaller than the interaction range, and, eventually, the De Broglie length. Remarkable examples in point are found mostly in the framework of quantum fluids, namely quark–gluon plasmas [2], electrons in graphene [3] and Bose–Einstein condensates in optical lattices [4]. Note that these extreme and exotic states of matter cover a breathtaking range of densities and temperatures, from one particle/fm^3 and trillions Kelvin degrees in ultra hot-dense quark–gluon plasmas, to about 10^{12} particles/cm^3 and tens of nanokelvins in ultra-cold Bose–Einstein condensates. Thus the SIF regimes show an impressive degree of universality, whose understanding stands as a great challenge at the crossroad of statistical fluid mechanics, condensed matter and high-energy physics. Under SIF conditions, most of the familiar and powerful notions of kinetic theory call for a profound revision. In particular, it is argued that not only Boltzmann's kinetic theory, but also the very notion of quasiparticles as weakly-interacting collective degrees of freedom, would lose meaning due their inconspicuously short lifetime [5]. Hence, new ideas and methods, both analytical and numerical, are in great demand.

Among the analytical ones, a most prominent role has been taken by the AdS-CFT (Anti-de Sitter Conformal Field Theory) duality between gravity in $(d+1)$-dimensions and conformal field theory in d-dimensions [6]. The beauty, and practical import, of the dual-holographic picture is that one can solve an otherwise intractable strongly coupled CFT by dealing with its weakly coupled gravitational analogue, an approach sometimes known as holographic principle. One of the major outcomes of the holographic approach is the so-called minimum-viscosity bound [7]

$$\frac{\eta}{s} \geq \frac{1}{4\pi} \frac{\hbar}{k_B} \quad , \tag{1}$$

where η is the shear viscosity and s the entropy per unit volume. It has been found that many quantum relativistic fluids, such as quark–gluon plasmas and electrons in graphene, come much closer to matching the above bound than any previously known fluid, including superfluid Helium-III. Moreover, the holographic AdS-CFT picture shows that hydrodynamics continues to apply even though kinetic theory does (may) not. Besides the conceptual challenge of developing statistical mechanics without quasiparticles, the ultra-low values of η/s open up the exciting possibility of observing dynamical instabilities and the ensuing (pre) quantum turbulence phenomena in these extreme states of matter [8].

It is hereby maintained that a proper lattice formulation of effective kinetic theory can effectively describe the dynamics of SIF by making use of very short-lived quasiparticles. The two main conceptual ingredients are as follows. (1) Top-down approach: design kinetic equation solely based on symmetry and conservation properties of the macroscopic field-theory (hydrodynamics), including entropic constraints. This just the reverse of the canonical route of deriving kinetic theory from first-principles, *i.e.*, by coarse-graining underlying microscopic dynamics (which is the process generating quasiparticles). (2) The existence of negative propagation viscosity, as an emergent property due to the interaction of the free-streaming quasiparticles with the discrete lattice. Item (2) is crucial to sustain ephemeral quasiparticles, with lifetimes as little as one millionth of the free-flight time, hence to achieve ultra-low viscosities, well below the natural kinematic lattice viscosity $\nu_L = \Delta x^2/\Delta t$, Δx and Δt being the lattice spacing and hopping time, respectively.

2. Results and Discussion

The typical lattice kinetic equation reads as follows [9]:

$$f_i(\vec{x},t) \quad -f_i(\vec{x}-\vec{c}_i\Delta t, t-\Delta t) = -\tfrac{\Delta t}{\tau}[f_i - f_i^{\mathrm{eq}}](\vec{x}-\vec{c}_i\Delta t, t-\Delta t), \tag{2}$$

where $f_i(\vec{x},t)$ is the probability of finding a particle at position $\vec{x}$ in the lattice at time t with discrete velocity $\vec{v} = \vec{c}_i$. The left hand side represents the free-streaming from site to site while the right hand side encodes collisional relaxation to a local equilibrium distribution f_i^{eq} on a time scale τ (see Figure 1). The discrete Boltzmann distributions ("populations") are quintessential quasiparticles, as they encode the dynamics of a collection of molecules in a lattice of size Δx^3 (in three dimensions), moving along the direction $\vec{c}_i$. The local equilibrium is a universal function (e.g., Maxwell–Boltzmann, Bose–Einstein, Fermi–Dirac, *etc.*) of the locally conserved quantities (hydrodynamic modes), such as flow density $\rho = m\sum_i f_i$ and current $J = \rho\vec{u} = m\sum_i \vec{c}_i f_i$, and dictates the structure of the inviscid macroscopic equations (Euler). Here m denotes a characteristic mass. The relaxation time, on the other hand, corresponds to the rate at which this equilibrium is reached, and dictates the fluid transport coefficients.

In the Boltzmann kinetic theory, in the strong-coupling regime $\tau \to 0$, viscosity is directly proportional to this relaxation time, $\nu \sim c_s^2\tau$, $c_s = \sqrt{k_B T/m}$ being the thermal speed [10]. In the lattice, however, this expression receives a crucial contribution from the discrete free-streaming. Indeed, by performing the appropriate Chapman–Enskog asymptotic expansion in the Knudsen number $Kn = c_s\tau/L$, L being a typical hydro-scale, one obtains:

$$\nu = c_s^2(\tau - \Delta t/2) \quad , \tag{3}$$

where the negative term at the right hand side is the contribution of the free-streaming. Given the minus sign, it is immediately seen that ultra-low viscosities of order $\epsilon\nu_L$ can be achieved by near-matching the free-flight and collisional time, *i.e.*, by choosing $\tau = \Delta t(1+\epsilon)/2$. In lattice units, one can write Equation (3) as

$$\nu = C_s^2\nu_L(1/\omega - 1/2) \quad , \tag{4}$$

where $\nu_L = \frac{\Delta x^2}{\Delta t}$ is the natural lattice viscosity and we have set $\omega = \Delta t/\tau$. In the above C_s is the sound speed in lattice units, typically $1/\sqrt{3}$ in most common lattices. Crucial for the correct recovery of the hydrodynamic equations is that the lattice exhibits enough symmetry to ensure the proper conservation laws as well as rotational invariance [11].

To further appreciate the point of ultra-low viscosity, let us now recast the kinetic Equation (2) in the most compact stream-collide form:

$$f_i(\vec{x}, t) = f_i'(\vec{x} - \vec{c}_i\Delta t, t - \Delta t) \quad , \tag{5}$$

where

$$f_i' \equiv (1 - \omega)f_i + \omega f_i^{eq} \quad , \tag{6}$$

is the post-collisional distribution (see Figure 1). The latter expression highlights three distinguished updates:

1. $\omega = 0, f_i' = f_i$. The particles are fully uncoupled (no collision), perform ballistic motion and never equilibrate, corresponding to infinite diffusivity.
2. $\omega = 1, f_i' = f_i^{eq}$. The particles reach equilibrium in a single time-step. According to Equation (4), this gives a viscosity $\nu = \nu_L C_s^2/2$, *i.e.*, comparable to the lattice viscosity ν_L.
3. $\omega = 2, f_i' = f_i - 2f_i^{neq}$. Collisions send the populations exactly "on the other side", to the mirror state $f_i^* \equiv f_i^{eq} - f_i^{neq}$, defined by complete reversal of the non-equilibrium component $f_i^{neq} \equiv f_i - f_i^{eq}$. Based on Equation (4), this yields formally zero viscosity, *i.e.*, infinitely strong coupling.

Hence, by choosing $\tau = \Delta t/2$, the quasiparticles have literally zero lifetime, even though both free-streaming and collisional time-scales are finite. Far from being a mere mathematical nicety, this regime is crucial to the operation of the Lattice Boltzmann (LB) scheme in the very-low viscous regime relevant, say, to fluid turbulence. Indeed, given the current computer resolutions, (iii) is precisely the regime that needs to be approached in order to simulate strongly coupled classical fluids, such as turbulent flows at Reynolds number above a few thousands. To clarify the point, let us recall that the strength of fluid turbulence is measured by the Reynolds number, $Re = UL/\nu$, where U and L are macroscopic velocity and length scales, respectively. In LB units, $\Delta x = \Delta t = 1$, this gives $Re = MaN/\nu_{lb}$, where $Ma = U/c_s$ is the thermal Mach Number, $N = L/\Delta x$ is the number of lattice sites per linear dimension and $\nu_{lb} \equiv \nu/\nu_L = (1/\omega - 1/2)$ is the viscosity in LB units. Current computers allow at best $N \sim 10^4$, so that with Mach numbers of order 1, reaching $Re = 10^7$ (air flowing around a standard car), requires $\nu_{lb} \sim 10^{-3}$, *i.e.*, three orders of magnitude below the natural lattice viscosity ν_L. This simple example highlights the necessity of operating with very short-lived quasiparticles, where short means very small as compared with the physical free-streaming and collisional time scales, Δt and τ, both $O(1)$ in LB units.

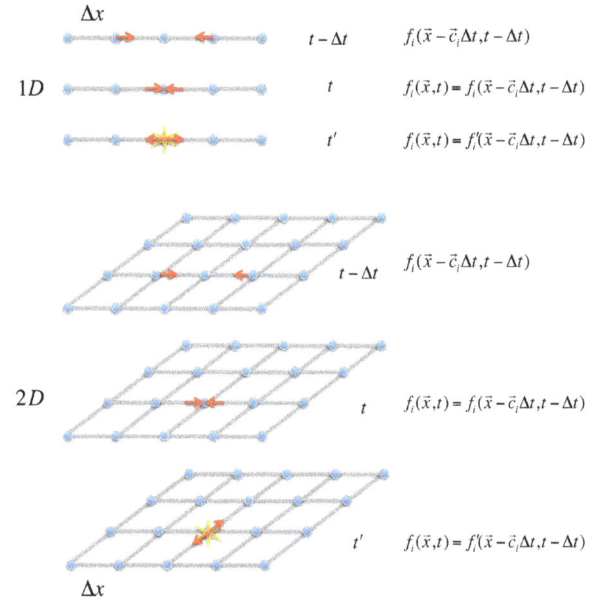

Figure 1. Schematics of the Lattice Boltzmann (LB) scheme: the quasiparticle distributions hop along the links defined by the discrete velocities and, once on the same site, they scatter into the post-collisional state. One- (top) and two- (bottom) dimensional lattices. Here Δx is the size of the lattice.

These quasiparticles function pretty well in the discrete world they live in, if only very shortly. In fact, by promoting the relaxation parameter ω to the status of a dynamic field responding self-consistently to the local constraints of the second principle (Boltzmann's H-theorem), the lifetime of these "ephemeral" quasiparticles can be made as small as one millionth of their natural value, *i.e.*, $\omega \simeq 2(1-10^{-6})$ [12–14]. In order to appreciate why this is remarkable, let us consider the lattice kinetic update in compact form, Equation (5). The realisability constraint imposes that the discrete populations be non-negative, hence, starting with a non-negative pair (f_i, f_i^{eq}), the collisional update should return a non-negative f_i'. A moment thought shows that this is indeed the case in the safe under-relaxation regime $0 < \omega < 1$. However, we have just shown that such regime does not provide access to the large Reynolds numbers typical of most turbulent flows.Hence, the LB update needs to operate deeply into the potentially unprotected over-relaxation regime, $1 < \omega < 2$. The remarkable fact is that this is indeed possible, thanks to the existence of a lattice analogue of the H-theorem [12–14]. Note that the regime $1 < \omega < 2$ is called over-relaxation because the time step is larger than the relaxation time, $\tau < \Delta t$, and consequently, the post-collisional distribution is no longer bounded between the pre-collisional one and the local equilibrium, which is a potential threat to positive-definiteness.

2.1. Propagation Viscosity in Luttinger Liquids

Let us apply this concept to Luttinger liquids. By dealing with a unitary Fermi gas, one should distinguish two different regimes, weak and strong coupling. Since the present work is focused on strongly interacting fluids, we consider the regime of large values of the Lieb–Liniger parameter, $\gamma \equiv mg/(\hbar^2 n) \gg 1$, where m is the atomic mass, n the atomic density, and g the coupling constant [15]. It proves expedient to define the dimensionless quantity $K \equiv \hbar n\pi/(mc_s)$. Note that γ and K relate to each other through an expression that, for $1 < \gamma < 10$, takes the form $K \simeq \pi \left[\gamma - (1/2\pi)\gamma^{3/2}\right]^{-1/2}$ [15].

In the strongly interacting regime and for weak external potentials, a quantum phase transition from Mott insulator to superfluid occurs for $\gamma > \gamma_c$ and $K_c = 2$ [15,16].

Let us next define, from Equation (3), $\nu_c = c_s^2\tau$ and $\nu_p = c_s^2\Delta t/2$, denoting "collision" and "propagation" viscosities, respectively. In order to estimate the propagation viscosity for a strongly interacting fluid at quantum criticality, we recast Equation (3) in the form:

$$\nu_p/\nu_c = \Delta t/2\tau \quad . \tag{7}$$

Assuming that $\Delta t = l/c$, where l is the lattice periodicity and c the light speed, and $\tau = \lambda/c_s$, where λ is the characteristic mean-free path, the above relation reads:

$$\nu_p/\nu_c = (1/2)(c_s/c)(l/\lambda) \quad . \tag{8}$$

In the sequel, we shall take $c \simeq c_s$. Since at criticality, $K_c = 2$, taking into account that the density $n \simeq 1/l$ (commensurate density [15]), it follows that $\lambda_B = 4l$, where λ_B is the De Broglie length. Inserting this into the viscosity ratio, we obtain

$$\left.\frac{\nu_p}{\nu_c}\right|_{crit} = \frac{1}{8}\frac{\lambda_B}{\lambda}. \tag{9}$$

Note that SIF regimes are characterized by the condition $\lambda < \lambda_B$. To determine the ratio λ_B/λ of the lattice fluid, we follow the same procedure adopted in [17] to obtain the AdS-CFT viscosity bound. Based on the Heisenberg principle, $k_B T\tau > \hbar/2$, so that $\tau = \lambda/c = \hbar B/(2k_B T)$, $B > 1$ being a constant.

Consequently,

$$\lambda = \frac{B}{4\pi}\lambda_B \quad . \tag{10}$$

The relations between the lattice size l, the mean-free path λ, the De Broglie length λ_B and the predicted lattice mean-free path λ_{eff}, can be appreciated from Figure 2.

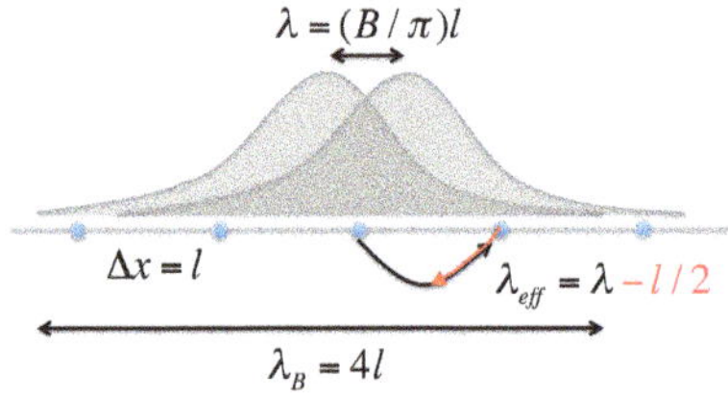

Figure 2. Representation of the Luttinger liquid. Shown are the mean-free path λ, the De Broglie length λ_B, the lattice size $\Delta x = l$, and the effective mean-free path predicted by the lattice λ_{eff}. Note that due to the negative contribution $(-l/2)$, lattice kinetic fluids can attain ultra-low viscosity even though the standard collisional viscosity is of the same order of the lattice viscosity ν_L.

Note that the constant B is related to the AdS-CFT bound by [4],

$$\frac{\eta}{s} = \frac{B}{4\pi}\frac{\hbar}{k_B} \quad . \tag{11}$$

Inserting Equation (10) into Equation (9), we obtain

$$\left.\frac{\nu_p}{\nu_c}\right|_{crit} = \frac{\pi}{2B} \quad . \tag{12}$$

This result calls for a number of comments. First, we note that the condition of non-negative viscosity, $\nu_p/\nu_c \leq 1$, implies $B \geq \pi/2 \simeq 1.6$, which corresponds to a bound $\eta/s \simeq 0.13\hbar/k_B$. In this case, the lattice quasiparticles can model even more "perfect" fluids than the Luttinger liquid. On the other hand, if we assume that $\tau = \Delta t$, which means that the quasiparticles thermalise in a single time step, we obtain $B = \pi$, yielding a bound $\eta/s \simeq 0.25\ \hbar/k_B$, which is very close to the value measured in experiments for two- and three-dimensional Fermi gases at unitarity [18]. Note that in order to get Equation (12), we have not only used lattice kinetic theory but also the Heisenberg principle, which shows that we still need quantum mechanics contributions for this particular example. However, this does not imply that lattice kinetic theory cannot handle quantum systems by itself. In fact, a quantum lattice Boltzmann model exists since long [19] to solve the Dirac equation. It is based on the same stream-collide paradigm of the classical lattice Boltzmann equation, although with a skew-symmetric collision matrix, where τ is related to $\hbar$ via $\tau = \hbar/mc^2$.

3. Conclusions

In summary, one can conclude that lattice kinetic schemes endowed with an H-theorem support ultra-short-lived computational quasiparticles, down to viscosities several orders of magnitude below their natural scale (the lattice viscosity). This is a strict consequence of the fact that, due the discrete nature of the lattice, free-streaming also contributes a viscosity and, most importantly, a negative one. It is as if, in the process of hopping from site to site, the lattice quasiparticles would "surrender" part of their collisional viscosity to the lattice itself, and precisely in proportion of half of the flight time. This is plausible, since negative viscosity corresponds to dynamic instability, namely the system moving away from local equilibrium, a sort of backward move in time (rejuvenation), as opposed to the standard forward move due to collisions. It is also plausible to speculate that this ultra-low viscosity could be regarded as an effect of lattice criticality, in the sense that the two competing processes, collisions and streaming, come to a near exact balance through the interaction of the quasiparticles with the lattice.

This process follows in the footsteps of similar phenomena in condensed matter: electrons in graphene behave like effective near mass-free excitations precisely because, due to the special symmetry of the honeycomb lattice, the electrons near the Dirac point "release" most of their mass to the lattice [20]. Since mass dictates the (Compton) collision frequency, $\omega_c = mc^2/\hbar$, and the collisional viscosity is inversely proportional to ω_c, it is clear that negative viscosity and near mass-free electrons in graphene belong somehow to the same family of lattice-induced phenomena. In the continuum, $\Delta t/\tau \to 0$, the negative viscosity is totally negligible, but in the discrete world it becomes crucial to attain ultra-low kinematic viscosities.

By using lattice kinetic theory, one can indeed achieve zero viscosity, and the same can be done in the continuum theories, such as continuum Boltzmann and Navier–Stokes equations. However, when the physical system possesses intrinsic discrete properties, e.g., quantum liquids in optical lattices, the viscosity can take very small values due to the presence of the negative viscosity induced by the lattice. For these particular systems, the lattice kinetic theory may describe the transport properties better than the continuum kinetic theory, as is the case of the lower bound in the shear viscosity of quantum liquids, which is overestimated by the continuum Boltzmann equation. On the other hand, in the classical context, a perfect fluid has zero viscosity, which in our case corresponds to $\omega = 2$. Thus, in the continuum limit, the lattice kinetic theory becomes just a numerical tool and recovers the Euler equations. However, in strongly coupled fluids, which are the focus of this paper and usually are governed by quantum mechanics, fluids do not have zero but a finite value of viscosity. Here we show that considering $\omega = 1$ one can get instantaneous relaxation and at the same time a lower bound for the viscosity of the physical system.

A few concluding remarks regarding strongly coupled fluids are in order. One may argue that most of these fluids are quantal, hence not captured by classical lattice kinetic models, for instance, the LB schemes discussed in this work. To this regard, it is worth pointing out that the lattice Boltzmann method has been already used to study two-dimensional Fermi gas at unitarity [21], providing a procedure to calculate transport coefficients precisely. Furthermore, the LB versions for quantum wave functions are available since long, and they are based on exactly the same stream-collide paradigm, if only with a suitable scattering matrix for collisions [19]. Incidentally, such quantum LB versions provide exact realisations of the Dirac equation in one spatial dimension and can be extended to $d > 1$ by proper operator splitting [22]. Finally, we wish to point out that matrix extensions of the quantum LB scheme have been recently connected with the Hubbard model for strongly correlated quantum fluids [23], thereby lending further weight to the main idea proposed in this paper, namely that lattice kinetic theory may provide a new angle of investigation for strongly interacting fluid regimes where continuum kinetic theory may no longer hold.

Acknowledgments: The authors are grateful to Ilya Karlin for many discussions over the years. Sauro Succi wishes to acknowledge Juan Maldacena for valuable exchanges on the AdS-CFT duality. We acknowledge financial support from the European Research Council (ERC) Advanced Grant 319968-FlowCCS.

Author Contributions: Both authors conceived and designed the research, analysed the data, worked out the theory, and wrote the manuscript. Both authors have read and approved the final manuscript.

Conflicts of Interest: The authors declare no conflict of interest.

Bibliography

1. Sachdev, S. *Quantum Phase Transitions*, 2nd ed.; Cambridge University Press: Cambridge, UK, 2011.
2. Shuryak, E. Why does the quark–gluon plasma at RHIC behave as a nearly ideal fluid? *Prog. Part. Nucl. Phys.* **2004**, *53*, 273–303.
3. Müller, M.; Schmalian, J.; Fritz, L. Graphene: A nearly perfect fluid. *Phys. Rev. Lett.* **2009**, *103*, 025301.
4. Gelman, B.A.; Shuryak, E.V.; Zahed, I. Ultracold strongly coupled gas: A near-ideal liquid. *Phys. Rev. A* **2005**, *72*, 043601.
5. Sachdev, S.; Müller, M. Quantum criticality and black holes. *J. Phys. Condens. Matter* **2009**, *21*, 164216.
6. Maldacena, J. The large-N limit of superconformal field theories and supergravity. *Int. J. Theor. Phys.* **1999**, *38*, 1113–1133.
7. Policastro, G.; Son, D.T.; Starinets, A.O. Shear viscosity of strongly coupled N = 4 supersymmetric Yang-Mills plasma. *Phys. Rev. Lett.* **2001**, *87*, 081601.
8. Mendoza, M.; Herrmann, H.; Succi, S. Preturbulent regimes in graphene flow. *Phys. Rev. Lett.* **2011**, *106*, 156601.
9. Benzi, R.; Succi, S.; Vergassola, M. The lattice Boltzmann equation: theory and applications. *Phys. Rep.* **1992**, *222*, 145–197.

10. Cercignani, C. *Theory and Application of the Boltzmann Equation*; Scottish Academic Press: Edinburgh, UK, 1975.
11. Pomeau, B.H.Y.; Frisch, U. Lattice-gas automata for the Navier-Stokes equation. *Phys. Rev. Lett.* **1986**, *56*, 1505–1508.
12. Succi, S.; Karlin, I.V.; Chen, H. Colloquium: Role of the H theorem in lattice Boltzmann hydrodynamic simulations. *Rev. Mod. Phys.* **2002**, *74*, 1203–1220.
13. Karlin, I.V.; Ferrante, A.; Öttinger, H.C. Perfect entropy functions of the lattice Boltzmann method. *Europhys. Lett.* **1999**, *47*, 182–188.
14. Chikatamarla, S.S.; Karlin, I.V. Entropy and Galilean invariance of lattice Boltzmann theories. *Phys. Rev. Lett.* **2006**, *97*, 190601.
15. Büchler, H.P.; Blatter, G.; Zwerger, W. Commensurate-Incommensurate Transition of Cold Atoms in an Optical Lattice. *Phys. Rev. Lett.* **2003**, *90*, 130401.
16. Haller, E.; Hart, R.; Mark, M.J.; Danzl, J.G.; Reichsöllner, L.; Gustavsson, M.; Dalmonte, M.; Pupillo, G.; Nägerl, H.C. Pinning quantum phase transition for a Luttinger liquid of strongly interacting bosons. *Nature* **2010**, *466*, 597–600.
17. Gelman, B.A.; Shuryak, E.V.; Zahed, I. Ultracold strongly coupled gas: A near-ideal liquid. *Phys. Rev. A* **2005**, *72*, 043601.
18. Cao, C.; Elliott, E.; Joseph, J.; Wu, H.; Petricka, J.; Schäfer, T.; Thomas, J. Universal quantum viscosity in a unitary Fermi gas. *Science* **2011**, *331*, 58–61.
19. Succi, S.; Benzi, R. Lattice Boltzmann equation for quantum mechanics. *Phys. D Nonlinear Phenom.* **1993**, *69*, 327–332.
20. Novoselov, K.S.; Geim, A.K.; Morozov, S.V.; Jiang, D.; Katsnelson, M.I.; Grigorieva, I.V.; Dubonos, S.V.; Firsov, A.A. Two-dimensional gas of massless Dirac fermions in graphene. *Nature* **2005**, *438*, 197–200.
21. Brewer, J.; Mendoza, M.; Young, R.E.; Romatschke, P. Lattice Boltzmann simulations of a two-dimensional Fermi gas at unitarity. 2015, arXiv:1507.05975.
22. Fillion-Gourdeau, F.; Herrmann, H.J.; Mendoza, M.; Palpacelli, S.; Succi, S. Formal analogy between the Dirac equation in its Majorana form and the discrete-velocity version of the Boltzmann kinetic equation. *Phys. Rev. Lett.* **2013**, *111*, 160602.
23. Fürst, M.L.R.; Mendl, C.B.; Spohn, H. Matrix-valued Boltzmann equation for the nonintegrable Hubbard chain. *Phys. Rev. E* **2013**, *88*, 012108.

Two-Dimensional Lattice Boltzmann for Reactive Rayleigh–Bénard and Bénard–Poiseuille Regimes

Suemi Rodríguez-Romo and Oscar Ibañez-Orozco

Abstract: We perform a computer simulation of the reaction-diffusion and convection that takes place in Rayleigh–Bénard and Bénard–Poiseuille regimes. The lattice Boltzmann equation (LBE) is used along with the Boussinesq approximation to solve the non-linear coupled differential equations that govern the systems' thermo-hydrodynamics. Another LBE, is introduced to calculate the evolution concentration of the chemical species involved in the chemical reactions. The simulations are conducted at low Reynolds numbers and in terms of steady state between the first and second thermo-hydrodynamics instability. The results presented here (with no chemical reactions) are in good agreement with those reported in the scientific literature which gives us high expectations about the reliability of the chemical kinetics simulation. Some examples are provided.

Reprinted from *Entropy*. Cite as: Rodríguez-Romo, S.; Ibañez-Orozco, O. Two-Dimensional Lattice Boltzmann for Reactive Rayleigh–Bénard and Bénard–Poiseuille Regimes. *Entropy* **2015**, *17*, 6698–6711.

1. Introduction

The Rayleigh–Bénard convection phenomenon is common in nature, governing various phenomena such as atmospheric fronts [1], thermal inversion (Bénard–Marangoni), ocean currents [2], circulation of the mantle [3], *etc.* In the chemical industry, convective rolls are present in the chemical and electrochemical reactors [4], fuel cells, distillation columns, selective membranes, ion exchange, among others. To characterize these phenomena in complex systems, a wide variety of numerical methods have been used: such as the finite element [5–8] and finite differences [9,10], among others.

Other methods of statistical physics have also been successfully used, such as Monte Carlo algorithms in the study of microscopic aspects of the Rayleigh–Bérnard dynamics as a phase transition [11,12]. Recently, the Lattice Boltzmann Method (LBM) has emerged [13,14] as one of the most powerful tools to describe complex thermos-hydrodynamic system [15–17]. The introduction of the BGK collision operator approach permits to significantly simplify the LB algorithm, although at some expenses of numerical stability in very low viscous regimes. Most importantly, in LBM the transport (free-streaming) is linear and can be dealt with exactly, while diffusion emerges from the relaxation dynamics of the collision operator, which is fully local. These are major advantages over macroscopic formulations of hydrodynamics.

When a fluid is heated from below and the Reynolds number is small, the heat transfer takes place by conduction. However, if a critical value is exceeded, natural convection is the predominant phenomena. The so-called Rayleigh–Bénard convection occurs due to the competition between the gravity and buoyant forces, creating instabilities in the fluid, which induces convective currents.

There are two thermo-convective instabilities in the Rayleigh–Bénard systems [18]. The first corresponds to the transition from a steady-state thermal conduction towards a state of thermal convection, formed by well-defined bidimensional stationary rollers. This transition occurs, theoretically, at a Rayleigh number of 1707.76, regardless of the value of the Prandtl number. The second transition occurs when the Rayleigh value increases until it can be considered a function of the Prandtl number. It consists of a bifurcation from a single frequency oscillating state, to a quasiperiodic double frequency flow. An additional increment of the Rayleigh number leads to a chaotic regime in a fully developed turbulence. Recently, a large number of articles with regard to the LBM applied on Rayleigh–Bénard convection, mainly between the first and second transition, where the Boussinesq approximation is valid, have been published. The first approach to Rayleigh–Bérnard systems by LBM is introduced by 1993 [19,20].

Many theoretical and experimental results restricted to low Reynolds and Rayleigh numbers have shown that layered Bénard–Poiseuille flow remains also stable, as long as the Prandtl number does not exceed a critical value. There are several articles based on finite differences [3,4,10] and in LBM [21] that have to do with the theory and numeric of this type of flow analysis. However, so far, there are few publications related to reactive systems under Rayleigh–Bénard and Bénard–Poiseuille regimes despite their importance among the applications above mentioned [22].

In order to simulate the dynamics of chemical reactions in two-dimensional systems with thermal and convective phenomena, a D2Q9 LBM is used in this paper at mesoscopic scale by taking into account (virtual) particles which move synchronously in a regular lattice, in accordance with discrete time steps. Then, when these particles reach simultaneously the same node of this lattice, they interact among themselves following collision rules that comply with the principles of mass, momentum and energy conservation, recovering in the macroscopic limit (through multiscale expansion), a number of differential equations, relying on the described phenomena. After the interaction, which is assumed to be instantaneous, the particles are scattered through first or second neighbors in the lattice. These define a distribution function of pseudo-particles f_i; namely, the probability that a particle is in a site-specific lattice node at time t, moving with speed c_i [23,24]. There are several other researches on coupled chemical reactions (electrokinetic) and reactive transport within micropores using mesoscopic modeling [25]. Pioneer work in the study of chemical kinetics using LBM has also been considered in this paper [26–28]. We perform a simulation in the Rayleigh–Bénard and Bénard–Poiseuille regimes for the Nitrous Oxide decomposition where Boussinesq approximation is valid, introducing the chemical kinetics as a function of the temperature via the Arrhenius law.

2. Thermal Decomposition of Nitrous Oxide

In this section, we address two different cases; the first pertains to the thermal decomposition of nitrous oxide (so called laughing gas) in the bulk of a fluid containing air plus the different species involved in this well characterized chemical reaction. This fluid performs a laminar flow through a rectangular open channel. The second case is similar but in a closed cavity; namely, the Rayleigh–Bénard and Bénard–Poiseuille regimes. See Figure 1.

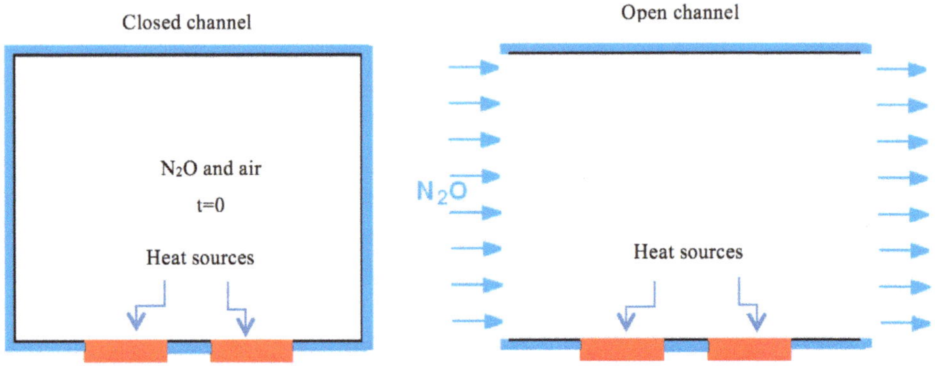

Figure 1. The non-isothermal closed and open cavities used in this paper.

In 1903, Boussinesq noted that when the differences of temperature were small, the thermodynamic properties of the fluid, such as the viscosity, the thermal diffusivity and the specific heat were also small, thus the fluid was approximately incompressible, while buoyant fluid effects were significant. This is due to the fact that the acceleration of the fluid is considerably less than the gravitational acceleration. Namely, the product of the gravity acceleration g and a small density difference is not negligible compared to other terms in the vertical movement component of the fluid in the Navier–Stokes equation. Boussinesq proposed an equation of state in which the fluid density is a linear function of the temperature and does not depend on the pressure. Following this hypothesis, the nonlinear coupled partial differential equations governing the thermos-hydrodynamic instability are as follows [29].

- The continuity equation

$$\partial_t \rho + \Delta \cdot \rho u = 0 \tag{1}$$

- The Navier–Stokes equation with buoyant forces (G).

$$\rho[\partial_t u + u\nabla \cdot u] = -\nabla \cdot \tau - \nabla \cdot P + G \tag{2}$$

- The infinitesimal balance of heat transfer, based on the Fourier's law, including effects of natural convection and viscous dissipation.

$$\partial_t \left(\rho C_p T\right) = \nabla \cdot k_T \nabla T - \nabla \cdot \rho u T - \tau : \nabla u \tag{3}$$

- The linear Bousinesq equation of state.

$$\rho = \rho_0[1 - \beta (T - T_0)] \tag{4}$$

In these equations ρ, C_p and k_T are the density, heat capacity and thermal conductivity of the fluid, respectively. Besides, u, P and T are the speed, pressure and temperature of the fluid; τ is the molecular momentum flux tensor. In this paper, we use the dimensionless version of Equations (1)–(3), namely;

$$\nabla^* \cdot u^* = 0 \tag{5}$$

$$\frac{1}{\Pr}\left[\frac{\delta u^*}{\delta t^*} + u^* \cdot \nabla u^* + \nabla P^*\right] - \nabla^{*^2} u^* - \sqrt{Ra} T^* y^* = 0 \tag{6}$$

$$\frac{\delta T^*}{\delta t} + u^* \cdot \nabla^* T^* - \sqrt{Ra} u^* \cdot y^* - \nabla^{*^2} T^* = 0. \tag{7}$$

where Ra, Pr are the Reynolds and Prandtl dimensionless numbers.

The chemical dynamics are governed by the following equation of reaction-diffusion and convection is used in this paper.

$$\partial_t C_i = \nabla \cdot D_i \nabla C_i - u \cdot \nabla C_i + R_Q \tag{8}$$

The left side of this equation represents the changes of the chemical species in accordance to the concentration of i. Meanwhile, on the right side for each species i, the first term corresponds to the molar diffusion flux, with D_i as the diffusion coefficient; the second term represents the convective transport drew by currents of fluid moving with speed u; finally, the third term corresponds to the speed of the different species production by the chemical reaction. This equation is considered in dimensionless form.

Equations (5)–(8) are a set of nonlinear differential equations, regardless of the nature of the chemical reaction, that pose a serious challenge to its resolution.

The decomposition of the laughing gas is considered as an example in this paper:

$$N_2O \rightarrow N_2 + \frac{1}{2}O_2 \tag{9}$$

Experimental evidence indicates that the corresponding chemical kinetics is given by [28]

$$R_Q = -r_{N_2O} = \frac{k_1\,[N_2O]^2}{1 + k_2\,[N_2O]}. \tag{10}$$

here k_1 and k_2 are constants obtained through experiments. Obviously, the N_2O decomposition is not elemental; *i.e.* if we assume that two molecules of N_2O collide and spontaneously disappear, at certain required conditions, becoming two molecules of N_2 and one of O_2, then the kinetics should be $R_Q = -r_{N_2O} = k\,[N_2O]^2$, which experimentally is not the case. Thus, this is a non-elemental chemical reaction and needs a new proposed mechanism for itself; namely, a set of elemental reactions take place at the decomposition rate. At a fixed temperature, it is proportional to the collision of molecules (or the spontaneous decomposition of one molecule), and also proportional to the concentration of the species involved. In such a case, we must introduce some molecular non-stable structures (they are produced and immediately consumed in another elemental reaction) whose concentration cannot be measured by experiments and are small enough to have a rate of production (or decomposition) equal to zero. These structures are called intermediate complex chemical states.

We propose the following reaction mechanism that suits this kinetics as follows:

$$2N_2O \underset{k_{2'}}{\overset{k_{1'}}{\rightleftarrows}} X^* + N_2O \tag{11}$$

$$X^* \overset{k_{3'}}{\rightarrow} N_2 + \frac{1}{2}O_2. \tag{12}$$

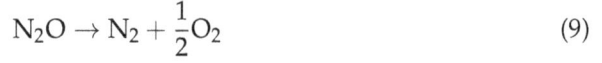

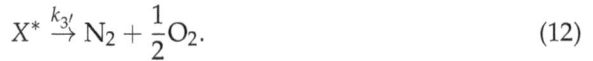

here, $k_{1'}, k_{2'}, k_{3'}$ are theoretical constants (k_1 and k_2 are related with these hypothetical constants). Equations (11) and (12) are added together to build the chemical reaction (9) experimentally obtained and are assumed to be elemental, so we have four differential equations (one for each rate of production or decomposition of the species; N_2O, X^*, N_2 and O_2) and the corresponding algebraic equations obtained by stoichiometry; thus it is possible to write, in this mechanism, the global N_2O decomposition rate only in terms of the N_2O concentration. From the full set of available equations all species' concentrations and rates follow. In the first step of our mechanism (Equation (11)), N_2O disappears following Equation (13), and it may be noted that we deal with this as elemental kinetics.

$$-\frac{d\,[N_2O]}{dt} = k_{1'}\,[N_2O]^2 - k_{2'}\,[N_2O]\,[X^*] \tag{13}$$

By the hypothesis of non-stable intermediate steady-state ($r_{X^*} = 0$), the speed of production of the intermediate complex chemical state X^* is zero. Considering elemental kinetics:

$$\frac{d[X^*]}{dt} = k_{1'}\,[N_2O]^2 - k_{2'}\,[N_2O]\,[X^*] - k_{3'}X^* = 0 \tag{14}$$

From Equations (13) and (14) and the algebraic equations from stoichiometry follow the kinetics given by (10).

In this study, the following simplifying considerations are taken into account.

1. The gaseous fluid (air and different species involved in the thermal decomposition of the laughing gas) is incompressible and Newtonian.
2. The thermal conductivity, viscosity, coefficient of thermal expansion and coefficient of diffusion of chemical species are all constant throughout the studied temperature range.
3. The variation of density is only significant in terms of buoyant forces.
4. Viscous dissipation is negligible.
5. The Reynolds number is small (for Bénard–Poiseuille flow).
6. The laughing gas decomposition is the only chemical reaction that takes place in the system.
7. In the top and bottom plates, temperature is constant. This means that the heat generated by the chemical reaction is assumed to be efficiently removed from the system.
8. In the bottom plates, there are two heat sources (or hot spots) for both the opened and closed channels at a higher temperature than the low constant temperature set in 7. These heat sources trigger the (N_2O decomposition) reaction.
9. The concentrations of chemical species are kept sufficiently small so that they dominate the properties of air during the phenomenon of natural convection.

There are experimental values for the kinetic constants k_1 and k_2 in (10), which we assume follow the Arrhenius law [30].

$$k_1 = 10^{19.39} e^{-81800/RT} \tag{15}$$

$$k_2 = 10^{8.69} e^{-28400/RT} \tag{16}$$

We chose temperatures in order that the chemical reaction rate is relatively slow so the heat generated by it is also small.

3. The Lattice Botlzmann Model

In this section, we introduce the general lattice Boltzmann framework used in this paper.

3.1. Open Channel

As mentioned above, three lattice Boltzmann structures are used in this paper. The first one concerns the thermo-hydrodynamics issue. We proposed for the momentum distribution function the following equation:

$$f_i\left(x+c_i\nabla t,t+\nabla t\right)-f_i\left(x,t\right)=-\frac{\Delta t}{\tau_F}\left[f_i\left(x,t\right)-f_i^{eq}\left(x,t\right)\right]+J_i \tag{17}$$

For J_i, the buoyant term, we use the Boussinesq approximation, namely

$$J_i=3w_i g_y\beta\cdot\left[T\left(x,t\right)-T_\infty\right]\cdot\rho\left(x,t\right)\cdot c_{iy} \tag{18}$$

A second LBE is formulated for the temperature field. The equation for the thermal distribution function is:

$$g_i\left(x+c_i\Delta t,t+\Delta t\right)-g_i\left(x,t\right)=-\frac{\Delta t}{\tau_T}\left[g_i\left(x,t\right)-g_i^{eq}\left(x,t\right)\right] \tag{19}$$

A third LBE is formulated for the chemical reaction kinetics that rules the species' concentration evolution.

$$h_i^\alpha\left(x+c_i\Delta t,t+\Delta t\right)-h_i^\alpha\left(x,t\right)=-\frac{\Delta t}{\tau_D}\left[h_i^\alpha\left(x,t\right)-h_i^{\alpha,eq}\left(x,t\right)\right]+\frac{\Delta t}{Q}R_Q \tag{20}$$

here, τ_D is the diffusion relaxation time $\tau_D=\frac{D_\alpha}{c_s^2}+\frac{1}{2}$, being D_α the α's chemical specie Fick diffusion coefficient. The corresponding (three LBE) distribution equilibrium functions are defined by the following equations:

$$f_i^{eq}\left(x,t\right)=w_t\rho\left[1+\frac{c_{iA}u_A}{c_s^2}+\frac{u_A u_B}{2c_s^2}\left(\frac{c_{iA}c_{iB}}{c_s^2}-\delta_{AB}\right)\right], \tag{21}$$

$$g_i^{eq}\left(x,t\right)=w_t T\left[1+\frac{c_{iA}u_A}{c_s^2}+\frac{u_A u_B}{2c_s^2}\left(\frac{c_{iA}c_{iB}}{c_s^2}-\delta_{AB}\right)\right], \tag{22}$$

$$h_i^{\alpha,eq}\left(x,t\right)=w_\alpha c_i\left[1+\frac{c_i^\alpha\cdot u}{c_s^2}+\frac{\left(c_i^\alpha\cdot u\right)^2}{2c_s^4}-\frac{u^2}{2c_s^2}\right]. \tag{23}$$

In a reactive system, we have as many equations ((20) and (23)) as needed (the same number as the chemical species involved in the reactions). The macroscopic variables are obtained as usual in the LBM. For example, the different species´ concentrations are obtained from the mesoscopic distribution functions as follows:

$$C_\alpha = \sum_i h_i^\alpha. \tag{24}$$

In Figure 2, we schematically introduce the so-called open channel system. The N_2O current (mixed with air) enters by the open rectangular channel on the left side with a fixed and uniform speed and laminar regime (low Reynolds numbers). The internal part the channel interacts with heat sources which triggers the N_2O thermal decomposition, following the mechanism's reaction proposed in this paper by Equations (13) and (14). Simultaneously, the very same heat sources cause the appearance of convection rolls, and the reaction evolves at different rates, depending on the local temperature achieved by the reactive molecules in the fluid at each site of the channel.

As for the boundary conditions we set the following scheme in the open channel case:

1. For the top plate, bounce back boundary conditions for the fluid dynamic LBE and a low constant isothermal boundary conditions for the thermic LBE are used.
2. Inlet and outlet periodic boundary conditions, for the moment and thermal LBE, are set.
3. For the bottom plate, bounce back boundary conditions are imposed in all the nodes for the hydrodynamic LBE, while for thermal LBE we used bounce back conditions as well, except in the heat sources sites where Dirichlet boundary conditions are set.

After obtaining the profiles of velocity and temperature in steady-state, we assess the evolution of the concentration of the chemical species for the reactive diffusional LBE. This procedure results from the fact that kinetics is coupled with the temperature field through the Arrhenius law, namely the concentration rate changes due to kinetic law by both concentration and temperature fields.

These boundary conditions are:

1. Top and bottom bounce back boundary conditions.
2. Invariable concentration in the inlet of the channel.
3. Null gradient concentration in the outlet of the channel (von Neumann condition).

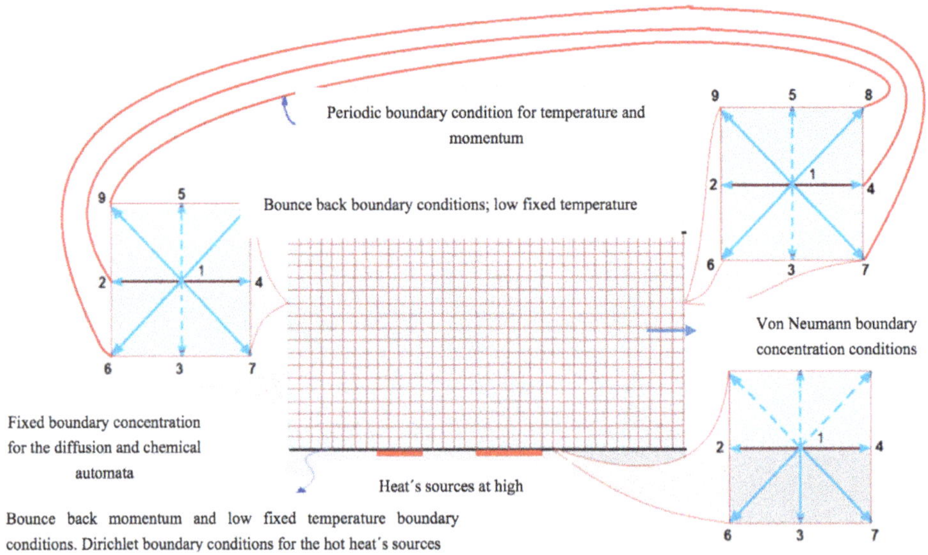

Figure 2. The lattice Boltzmann model for a reactive flow (a mixture of nitrous oxide and air), with heat sources that promote its thermal decomposition in an open channel.

3.2. Closed Channel

Now consider nitrous oxide initially confined to a vessel at 250 K and suddenly the sources located in the bottom plate rise their temperature to 300 K (by the hot spots) producing the thermal decomposition of the laughing gas (Rayleigh–Bénard flow). All the settings given by the open channel case are kept except for the boundary conditions. In the closed channel case, these are the following:

1. Bounce back boundary conditions at all the sites for the fluid and thermal LBE on all borders, except at the heat's sources.
2. Diritchlet boundary conditions at all high temperature sites (heat sources).
3. Initial state with uniformly distributed nitrous oxide in the closed vessel, with initial concentration C_0 and initial temperature T_0, smaller than the temperature of the sources.
4. Bounce back boundary conditions at all the sites for the reaction-diffusion LBE on all borders.

Figure 3 is a schematic representation of this system.

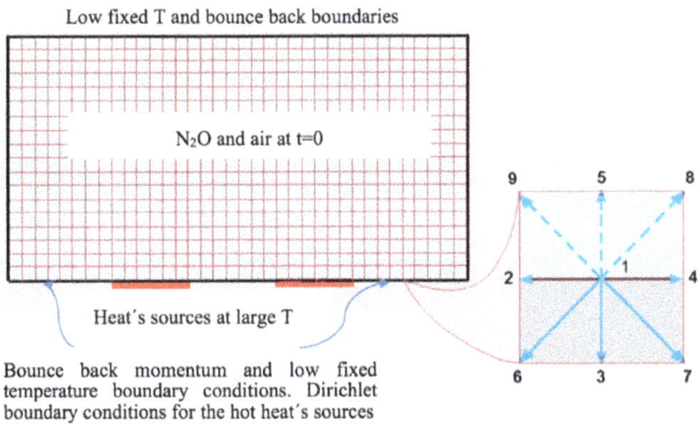

Figure 3. The lattice Boltzmann model for a reactive flow containing nitrous oxide mixed with air, with heat sources that promote its thermal decomposition in a closed vessel.

4. The Lattice Botlzmann Algorithm

In the following we describe the LBM algorithm used in this paper to perform the simulation of our models that allowed us to obtain our results.

4.1. Open Channel

After the discretization of the rectangular 2D domain, a D2Q9 lattice is implemented and then the following standard LBM steps are applied:

1. The distribution functions f and g for the velocity u and temperature T fields, respectively, were evaluated simultaneously in a coupled fashion, taking into account iterative calculations performed by the following steps:

   1.1 Propagation (streaming) of the f and g distribution functions.
   1.2 Calculation of the distribution functions at equilibrium; f_{eq} and g_{eq}.
   1.3 Actualization of the f and g distribution functions.
   1.4 Introduction of thermal and fluid dynamics boundary conditions.
   1.5 Calculation of the u and T fields from the new f and g distribution functions.
   1.6 Assessment of the new and preceding values of u and T, if they are closed enough, finish the iterations; else, return to step 1.1.

2. Once obtaining the steady temperature and velocity profiles, the temporal evolution of the distribution functions h_i for each one of the three chemical species is calculated, solving the LBM equations for the reaction-diffusion advection phenomena. This is a non-iterative procedure, structured practically by the same steps as the ones in the thermo-hydrodynamic analysis, except

for that in the formulation of the diffusional distribution functions for each chemical species. The chemical kinetics term (Arrhenius law also considered) is introduced, keeping the 1:1:1/2 stoichiometric relationships between reactants and products.

After a short period of time, the concentrations of the three chemical species reach a steady-state, due to the constant in-and-out reagent.

4.2. Closed Channel

In this case, we use the open channel algorithm modifying a few instructions in the algorithm concerning the new boundary conditions and establishing a speed equal to zero in the input stream. The remaining instructions are kept unchanged in the algorithm.

5. Results and Conclusions

The results of our calculations are presented in Figures 4 and 5 for the open channel and in Figures 6 and 7 for the closed channel.

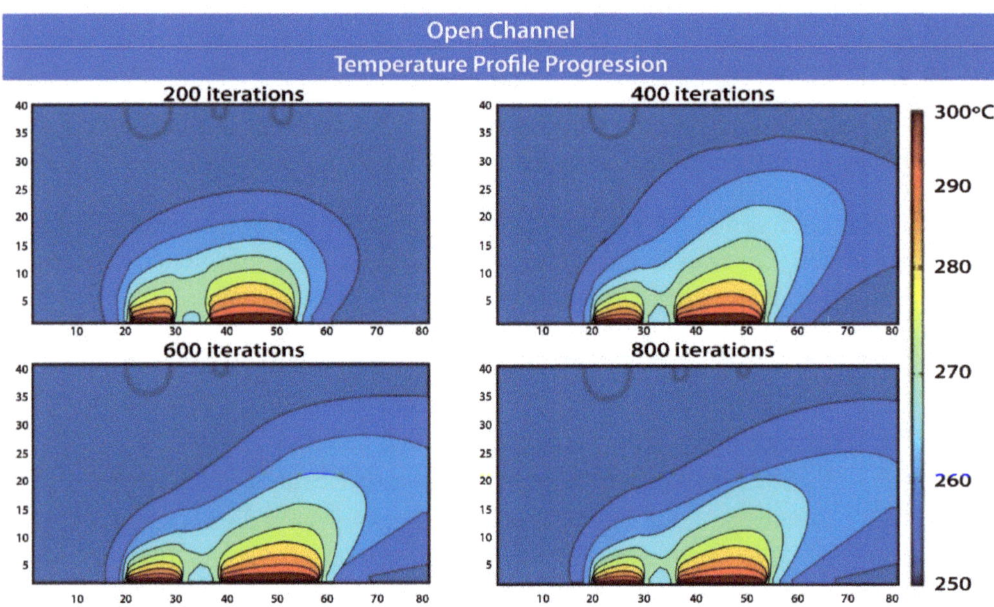

Figure 4. Temperature field evolution for the Bernard–Poiseuille reactive flow as iterations are carried on up to reach the thermo-hydrodynamic steady state.

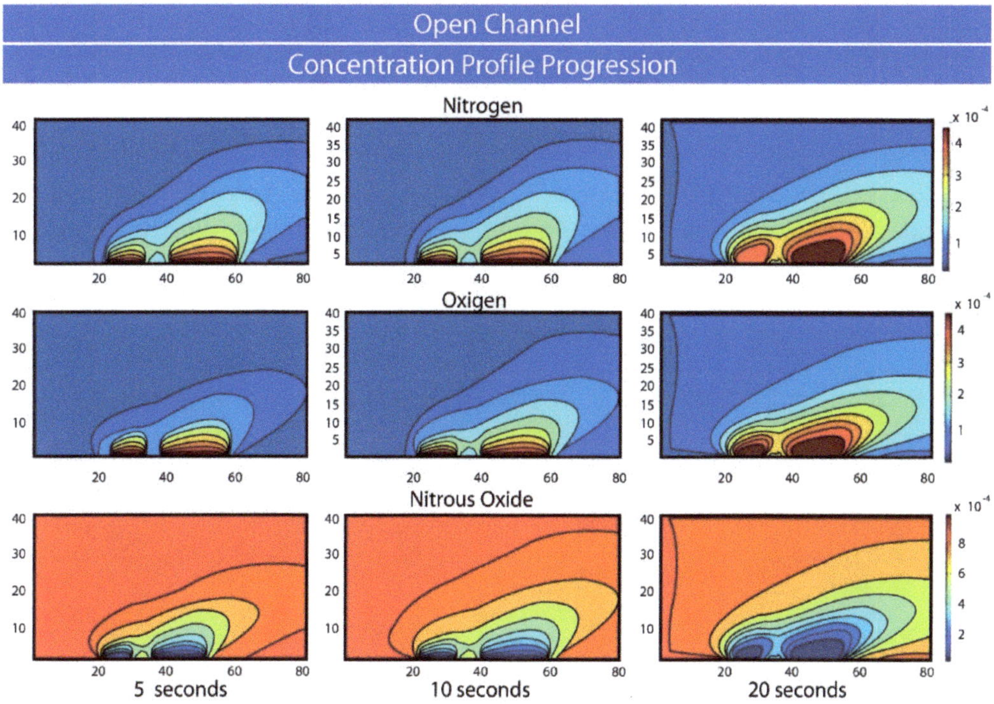

Figure 5. Species concentration field evolution for the Bénard–Poiseuille reactive flow as iterations that are carried out for 5, 10 and 20 s.

One goal of this paper is the analysis of the chemical reactions' dynamics that are carried out on systems presenting thermo-hydrodynamic instabilities, using Boussinesq approximation since this is considered a main stream topic. The study presented here is a new step towards this direction. The model presented in this paper includes a non-stationary thermo-hydrodynamic system where the heat generated by the chemical reactions has an important role in the temperature interval studied. The reaction enthalpy data of several chemical reactions are available in the literature and can be plugged into the algorithm. The heat generated would lead to a different evolution of the Bénard convection rolls. In this case, the solution of the Boussinesq approach and the associated coupled chemical is quite different and shall be addressed elsewhere.

Finally, data numerical analysis can be performed by Fourier series on the bidimensional images (or rough data on the plane) providing precise functions from the contours depicted in Figures 4–7. For example, in Figure 5, for the Nitrogen produced after 5 s, the upper contour of the lowest and second lowest concentration areas can be approached by the following function; $y = -204 + 166.71\cos(.0034x) + 361.2\sin(.0034x)$ with a correlation of 0.994, being x and y space coordinates in the channel. Further work should be done in the near future to take full advantage of our approach.

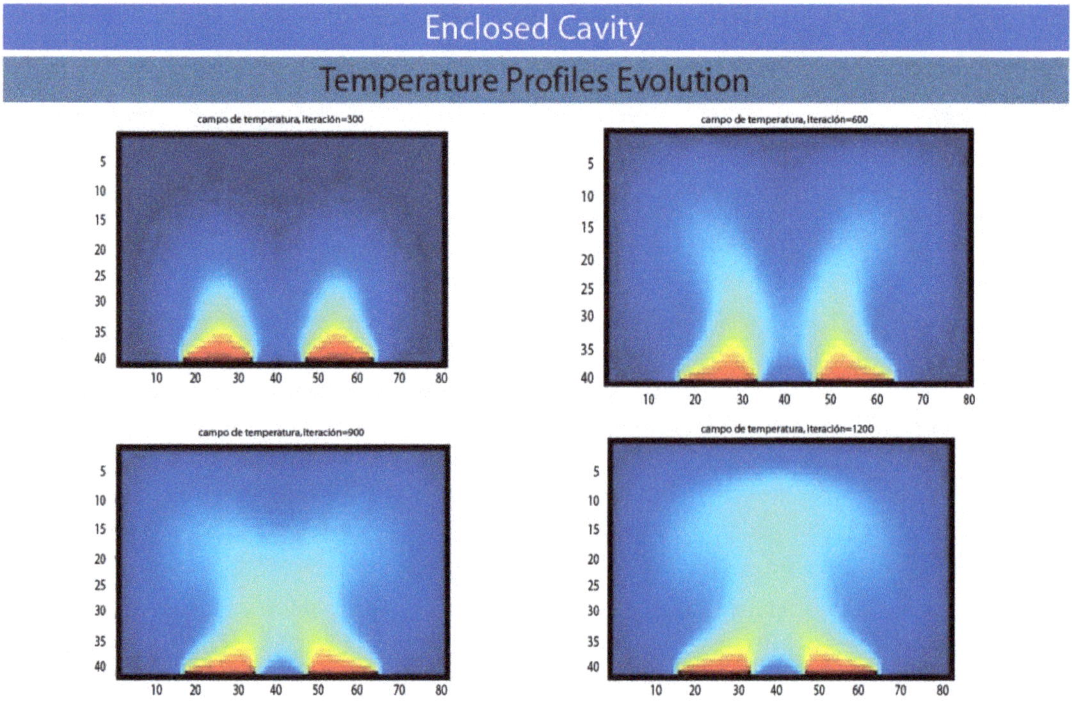

Figure 6. Temperature field evolution for the Rayleigh–Bénard reactive flow as iterations that are carried out to reach the thermo-hydrodynamic steady state, x and y are the space coordinates in the cavity.

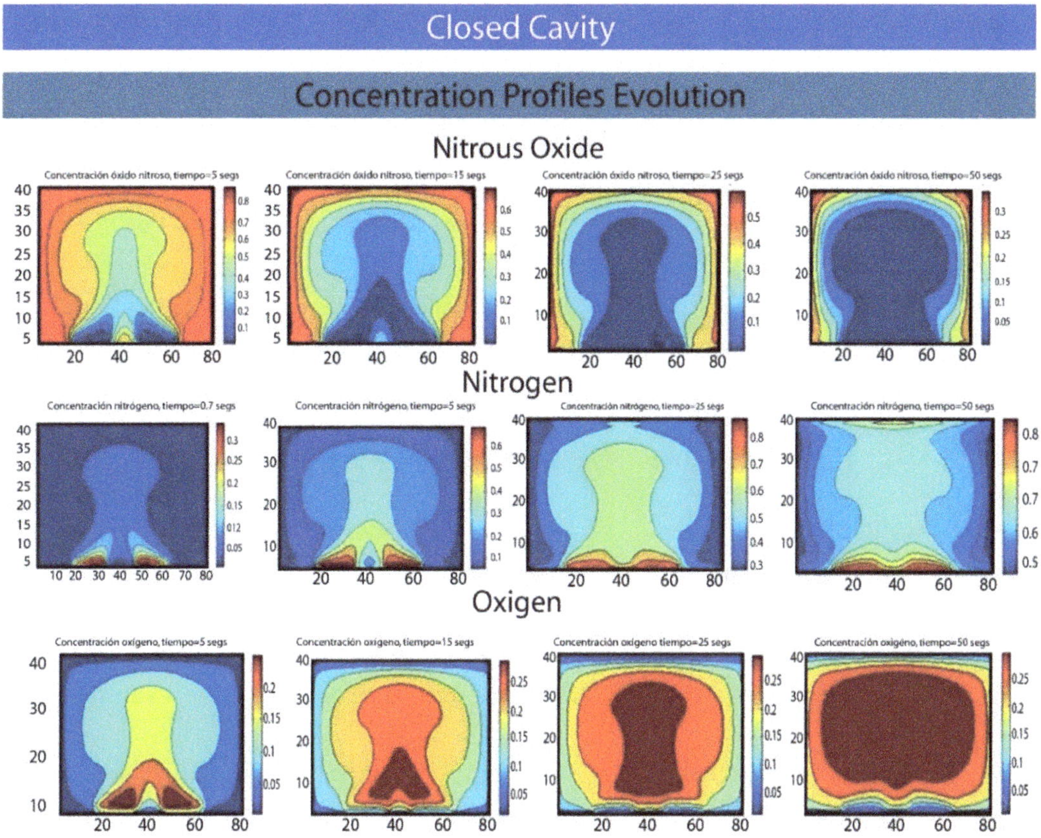

Figure 7. Species concentration field evolution for the Rayleigh–Bénard reactive flow as iterations that are carried out for 5, 15, 25 and 50 seconds, here x and y are the space coordinates in the cavity.

Acknowledgments: S. Rodríguez-Romo thanks the German Academic Exchange Service (DAAD) (Germany) for financial support to visit Hümbolt University Berlin, where part of this paper was written.

Author Contributions: S. Rodríguez-Romo produced the model and the algorithm. O. Ibañez-Orozco obtained the results provided the figures and the quantitative approach. All authors have read and approved the final manuscript.

Conflicts of Interest: The authors declare no conflict of interest.

References

1. Dragani, W.C. A feature model of surface pressure and wind fields associated with passage of atmospheric cold fronts. *Comp. Geosci.* **1999**, *25*, 1149–1157.
2. Agee, E.M. Mesoscale cellular convection over the oceans. *Dyn. Atmos. Oceans* **1987**, *10*, 317–341.

3. Nicolas, X. Revue bibliographique sur les écoulements de Poiseuille–Rayleigh–Bénard: écoulements de convection mixte en conduites rectangulaires horizontales chauffées par le bas. *Int. J. Therm. Sci.* **2002**, *41*, 961–1016.
4. Bograchev, D.A.; Davidov, A.D.; Volgin, V.M. Linear stability of Rayleigh–Benard–Posieuille convection for electrochemical system. *Int. J. Heat Mass Transf.* **2008**, *51*, 4886–4891.
5. Kakuda, K.; Miura, S. Finite element simulation of three- dimensional Rayleigh–Bénard convection. *Int. J. Comput. Fluid Dyn.* **2001**, *15*.
6. Murty, V.D. A numerical investigation of Benard convection using finite elements. *Comp. Fluids* **1986**, *14*, 379–391.
7. Lan, C.H.; Ezekoye, O.A.; Howell, J.R.; Ball, K.S. Stability analysis for three-dimensional Rayleigh–Bénard convection with radioactively participating medium using spectral methods. *Int. J. Heat Mass Transf.* **2003**, *46*, 1371–1383.
8. Howle, L.E. A comparison of the reduced galerkin and pseudo-spectral methods for simulation of steady Rayleigh–Bénard convection. *Int. J. Heat Mass Transf.* **1996**, *39*, 2401–2407.
9. Luo, X.; Chen, W.-K. A discussion on finite-difference schemes for low Prandtl number Rayleigh–Bénard convection. In *Advanced Computational Methods in Heat Transfer*; Sundén, B., Brebbia, C.A., Eds.; WIT Press: Ashurst, UK, 1994.
10. Park, H.M.; Ryu, D.H. Rayleigh–Bènard convection of viscoelastic fluids in finite domains. *J. Non-Newton. Fluid Mech.* **2001**, *98*, 169–184.
11. Tzeng, P.-Y.; Liu, M.-H. Direct-simulation Monte Carlo modeling on two-dimensional Rayleigh–Bénard instabilities of rarefied gas. *Numer. Heat Transf. A Appl.* **2005**, *47*, 805–823.
12. Zhang, J.; Fan, J. Kinetic study of the Rayleigh–Bénard flows. *Chin. Sci. Bull.* **2009**, *54*, 364–368.
13. Higuera, F.J.; Succi, S. Simulating the flow around a circular cylinder with a Lattice Boltzmann equation. *Europhys. Lett.* **1989**, *8*, 517–521.
14. Benzi, R.; Succi, S.; Vergassola, M. The lattice Boltzmann equation: theory and applications. *Phys. Rep.* **1992**, *222*, 145–197.
15. Shan, X.W. Simulation of Rayleigh–Bénard convection using a lattice Boltzmann method. *Phys. Rev. E* **1997**, *55*, 2780–2788.
16. Kao, P.-H.; Yang, R.-J. Simulating oscillatory flow in Rayleigh–Bénard convection using the lattice Boltzmann method. *Int. J. Heat Mass Trans.* **2007**, *50*, 3315–3328.
17. Barrios, G.; Rechtman, R.; Rojas, J.; Tovar, R. The lattice Boltzmann equation for natural convection in a two-dimensional cavity with a partially heated wall. *J. Fluid Mech.* **2005**, *522*, 91–100.
18. Chandrasekar, S. *Hydrodynamic and Hydromagnetic Stabilities*; Clarendon Press: Oxford, UK, 1961.
19. Massaioli, F.; Benzi, R.; Succi, S. Exponential tails in 2-dimensional Rayleight–Bérnard convection. *Europhys. Lett.* **1993**, *21*, 305–310.

20. Benzi, R.; Tripiccione, R.; Massaioli, F.; Succi, S.; Ciliberto, S. On the scaling of the velocity and temperature structure functions in Rayleigh–Bénard convection. *Europhys. Lett.* **1994**, *25*, 341–346.
21. Moussaoui, M.A.; Jami, M.; Mezrhab, A.; Naji, H. Lattice Boltzmann simulation of convective heat transfer from heated blocks in a horizontal channel. *Numer. Heat Transf. A Appl.* **2009**, *56*, 422–443.
22. Amaya-Ventura, G.; Rodríguez-Romo, S. 2D Lattice Boltzmann Simulation of Chemical Reactions within Rayleigh–Bénard and Poiseuille–Bénard Convection Systems. In Proceedings of the International Conference on Numerical Analysis and Applied Mathematics 2011 (ICNAAM 2011), Halkidiki, Greece, 19–25 September 2011; pp. 1798–1801.
23. Chen, S.Y.; Doolen, G.D. Lattice Boltzmann method for fluid flows. *Annu. Rev. Fluid Mech.* **1998**, *30*, 329–364.
24. He, X.Y.; Chen, S.Y.; Doolen, G.D. A novel thermal model for the lattice Boltzmann method in incompressible limit. *J. Comput. Phys.* **1998**, *146*, 282–300.
25. Zhang, L.; Wang, M. Modeling of electrokinetic reactive transport in micropore using a coupled lattice Boltzmann method. *J. Geophys. Res. Solid Earth* **2015**, *120*, 2877–2890.
26. Yamamoto, K.; He, X.Y.; Doolen, G.D. Simulation of combustion field with lattice Boltzmann method. *J. Stat. Phys.* **2002**, *107*, 367–383.
27. Chiavazzo, E.; Karlin, I.V.; Gorban, A.N.; Boulouchos, K. Coupling of the model reduction technique with the lattice Boltzmann method for combustion simulations. *Combust. Flame* **2010**, *157*, 1833–1849.
28. Chiavazzo, E.; Karlin, I.V.; Gorban, A.N.; Boulouchos, K. Efficient simulations of detailed combustion fields via the lattice Boltzmann method. *Int. J. Numer. Methods Heat Fluid Flow* **2011**, *21*, 494–517.
29. Charru, F. *Hydrodynamic Instabilities*, 1st ed.; Cambridge University Press: Cambridge, UK, 2011.
30. Levenspiel, O. *Ingeniería de las Reacciones Químicas*; Reverté: Barcelona, Spain, 1987. (In Spanish)

Extension of the Improved Bounce-Back Scheme for Electrokinetic Flow in the Lattice Boltzmann Method

Qing Chen, Hongping Zhou, Xuesong Jiang, Linyun Xu, Qing Li and Yu Ru

Abstract: In this paper, an improved bounce-back boundary treatment for fluid systems in the lattice Boltzmann method [Yin, X.; Zhang J. *J. Comput. Phys.* **2012**, *231*, 4295–4303] is extended to handle the electrokinetic flows with complex boundary shapes and conditions. Several numerical simulations are performed to validate the electric boundary treatment. Simulations are presented to demonstrate the accuracy and capability of this method in dealing with complex surface potential situations, and simulated results are compared with analytical predictions with excellent agreement. This method could be useful for electrokinetic simulations with complex boundaries, and can also be readily extended to other phenomena and processes.

Reprinted from *Entropy*. Cite as: Chen, Q.; Zhou, H.; Jiang, X.; Xu, L.; Li, Q.; Ru, Y. Extension of the Improved Bounce-Back Scheme for Electrokinetic Flow in the Lattice Boltzmann Method. *Entropy* **2015**, *17*, 7406–7419.

1. Introduction

With growing interest in bio-Micro Electro Mechanical Systems (MEMS) and bio-Nano Electro Mechanical Systems (NEMS) applications and fuel cell technologies, electrokinetic flows have become one of the most important non-mechanical techniques in the application of microfluidics and nanofluidics [1,2]. Electro-osmotic flow (EOF) is a promising approach to drive the microfluidics under an external electric field, such as sample injection, chemical reaction, species separation and energy supply [3,4]. Due to these important applications, EOF in microchannels has received interesting attention [5–11].

Electro-osmotic flow (EOF) is a basic electrokinetic phenomenon, where an electrical double layer (EDL) is formed due to the interaction between an electrolyte solution and a dielectric surface [12]. From the macroscopic point of view, the EOFs are governed by the Navier–Stokes (NS) equations for fluid flow and the Poisson–Boltzmann equation for the electrical potential. Many studies have been carried out on electro-osmotic flow in microchannel. The lattice Boltzmann method (LBM) has been generally accepted as a useful simulation method for complex flows [13–15]. The LBM approach is advantageous in dealing with complex boundary geometries [16,17] and could be potentially more efficient with advanced computational technologies. Because of its distinctive advantages over conventional numerical methods, the lattice Boltzmann method has introduced to simulate electro-osmotic flow in micro devices. Warren [18] made the first attempt to apply the LBM to solve the Navier–Stokes equations for the solution, while the conservation equation for each ionic species and the Poisson equation for the electrical potential were solved via the "moment propagation" method. He and Li [19] proposed a lattice Boltzmann scheme for analyzing the electrochemical processes in an electrolyte based on a locally electrically neutral assumption. With a multiple-component LBM model, this scheme has also been utilized to study the electrohydrodynamic drop deformation in an electric field [20]. The electrokinetic flows in microchannels is simulated by the lattice Boltzmann method with one-dimensional linearized solution of the Poisson–Boltzmann equation [5,7,8]. Melchionna and Succi [21] solved the nonlinear Poisson–Boltzmann equation by an efficient multi-grid technique and then predicted the flow behavior using a lattice Boltzmann scheme. The multi-grid technique has great efficiency to solve the nonlinear Poisson–Boltzmann equation; however it has rarely been extended for complex geometries [22]. Recently, The LBM has been applied to study the mixing enhancement in heterogeneously charged microchannels [23–28] and the roughness and cavitation effects in electro-osmotic microfluidics [29,30].

As with other numerical methods, boundary conditions play crucial roles for the simulation validity and stability. However, unlike the tremendous efforts in developing accurate boundary treatments for LBM models for fluid flows and convection-diffusion systems [31–36], boundary methods for LBM models of electric field have not been addressed adequately. Typical electric field LBM simulations are performed in regular domains with flat boundaries aligned along the lattice grid lines. Several studies have considered rough surfaces [26,30]; however, the rough surfaces were actually modeled as flat, stair-like patches. Recently, Yin and Zhang [32] developed an improved bounce-back method for fluid flows, which is discussed in [37–39].

In this paper, an improved bounce-back method for fluid flows [32] is extended to simulate the electrokinetic flows with arbitrary boundary shapes. Numerical simulations demonstrate that the boundary treatments have accurately represented the spatial geometry as well as the surface potential. An example calculation is also performed to illustrate the application of our boundary treatment for electrokinetic studies. This study could be useful for LBM simulations of electric fields in systems with complex surface geometry and surface conditions.

2. Macroscopic Governing Equations for EOF

For incompressible EOF in microfluidic channel, the governing equations including the continuity equation and the momentum equation can be described as follows:

$$\nabla \cdot \mathbf{u} = 0 \tag{1}$$

$$\rho \frac{\partial \mathbf{u}}{\partial t} + \rho \mathbf{u} \cdot \nabla \mathbf{u} = -\nabla P + \rho v \nabla^2 \mathbf{u} + \mathbf{F} \tag{2}$$

where $\mathbf{u}$ is the velocity vector, ρ is the density of solution, P is the pressure, v is the kinetic viscosity of the flow, $\mathbf{F}$ represents the external force and is given as:

$$\mathbf{F} = \rho_e \mathbf{E} \tag{3}$$

where ρ_e is the net charge density, and $\mathbf{E}$ is the external electric field.

The drive force of the EOF is indicated by the body force term ($\rho_e \mathbf{E}$) in the momentum equation and is caused by the action of the induced electrical field on the net charge density in the EDL region. EDL theory [40] related the electrostatic potential and the ion distribution in the bulk solution can be well approximated by the Poisson equation as follows:

$$\nabla^2 \psi = -\frac{\rho_e}{\varepsilon \varepsilon_0} \tag{4}$$

where ψ is the electrical potential, ε and ε_0 are the dimensionless dielectric constant and permittivity of vacuum, respectively.

For the flows over a non-conducting stationary surface, the ion distribution can be well approximated by the Boltzmann distribution:

$$n_i = n_{i,\infty} \sinh\left(-\frac{z_i e \psi}{k_b T}\right) \tag{5}$$

where n_i is the ionic number concentration of the i-th species, $n_{i,\infty}$ is the ion concentration in the bulk solution, z_i is the valence of type-i ions, e is the charge of an electron, k_b is the Boltzmann constant and T is the absolute temperature. For a symmetric electrolyte ($z_i = z$ and $n_{i,\infty} = n_\infty$) considered in the present study, the net charge density is given as follows:

$$\rho_e = -2 n_\infty z e \sinh\left(\frac{z e \psi}{k_b T}\right) \tag{6}$$

Combining Equations (4) and (6) yields the nonlinear Poisson–Boltzmann distribution for the EDL potential in the dilute electrolyte solution:

$$\nabla^2\psi = \frac{2n_\infty ze}{\varepsilon\varepsilon_0}\sinh\left(\frac{ze\psi}{k_bT}\right) \tag{7}$$

For the surfaces with low surface electric potentials, the Debye–Hückel approximation ($\sinh\left(\frac{ze\psi}{k_bT}\right) \approx \frac{ze\psi}{k_bT}$) can be applied and the Poisson–Boltzmann equation can be linearized to:

$$\nabla^2\psi = \frac{2n_\infty z^2e^2}{\varepsilon\varepsilon_0 k_bT}\psi = \kappa^2\psi \tag{8}$$

where:

$$\kappa = \sqrt{\frac{2n_\infty z^2e^2}{\varepsilon\varepsilon_0 k_bT}} \tag{9}$$

and its reciprocal κ^{-1}, the so-called Debye length, is usually used as a measure of the EDL thickness in Debye–Hückel theory.

3. Numerical Method

The numerical method adopted in this work requires the solution of the Navier–Stokes equations for fluid flow and the Poisson–Boltzmann equation for electric potential distribution. A lattice structure D2Q9 with complex boundary conditions is proposed to solve the governing equations using two LBM model, corresponding to the fluid flow and electric potential, respectively. It is necessary to introduce the LB evolution equations and the boundary treatments in this section.

3.1. Lattice Boltzmann Model for the NS Equations

The evolution equation corresponding to the NS equations with external force is given as:

$$f_\alpha(\mathbf{x}+\mathbf{e}_\alpha\Delta t, t+\Delta t) - f_\alpha(\mathbf{x},t) = -\frac{1}{\tau_f}\left[f_\alpha(\mathbf{x},t) - f_\alpha^{eq}(\mathbf{x},t)\right] + \Delta t F_\alpha \tag{10}$$

where $f_\alpha(\mathbf{x},t)$ is the distribution function for the flow fields at location $\mathbf{x}$ and time t and the subscript α indicates the lattice direction. Δt is the time step. The parameter τ_f is the relaxation time for the density distribution function. F_α is the forcing term corresponding to the applied external electric field.

In order to recover the correct NS equation, the density equilibrium distribution in this work can be typically expressed as:

$$f_{\alpha}^{eq} = \omega_{\alpha}\rho\left[1 + 3\frac{\mathbf{e}_{\alpha} \cdot \mathbf{u}}{c^2} + 9\frac{(\mathbf{e}_{\alpha} \cdot \mathbf{u})^2}{2c^4} - \frac{3\mathbf{u}^2}{2c^2}\right] \tag{11}$$

where c is the lattice speed defined as $\Delta x/\Delta t$, Δx the lattice grid size. For the D2Q9 lattice model, the lattice weight factors depends on the length of the corresponding lattice vector, and is given by $\omega_0 = 4/9$, $\omega_{1-4} = 1/9$, $\omega_{5-8} = 1/36$.

The relaxation time for fluid flow is related with the fluid viscosity by:

$$\tau_f = \frac{3v}{c\Delta x} + 0.5 \tag{12}$$

The forcing term caused by the interaction of the EDL field with the externally applied electrical field is incorporated into the discrete Boltzmann equation by following the method described:

$$F_{\alpha} = \left(1 - \frac{1}{2\tau_f}\right)\omega_{\alpha}\left[\frac{3(\mathbf{e}_{\alpha} - \mathbf{u})}{c^2} + \frac{9(\mathbf{e}_{\alpha} \cdot \mathbf{u})}{c^4}\mathbf{e}_{\alpha}\right] \cdot \mathbf{F} \tag{13}$$

The Chapman–Enskog expansion can be used to recover the macroscopic NS equations, and Macroscopic quantities such as the density and fluid velocity can then be evaluated from the distribution functions as:

$$\rho = \sum_{\alpha} f_{\alpha} \tag{14}$$

$$\rho\mathbf{u} = \sum_{\alpha} \mathbf{e}_{\alpha} f_{\alpha} + \frac{\Delta t}{2}\mathbf{F} \tag{15}$$

3.2. Lattice Boltzmann Model for Poisson–Boltzmann Equation

Here, to solve the Poisson–Boltzmann equation, we adopt an LBM algorithm proposed by Oulaid *et al.* [41] because of its good numerical accuracy. The Poisson–Boltzmann equation can be considered as a convection-diffusion equation at the steady state. A lattice distribution function g_{α} is introduced, and its evolution is described by the following lattice Boltzmann equation:

$$g_{\alpha}(\mathbf{x} + \mathbf{e}_{\alpha}\Delta t, t + \Delta t) - g_{\alpha}(\mathbf{x}, t) = -\frac{1}{\tau_g}\left[g_{\alpha}(\mathbf{x}, t) - g_{\alpha}^{eq}(\mathbf{x}, t)\right] + \Delta t G_{\alpha} + \frac{\Delta t^2}{2}\overline{D}_{\alpha} G_{\alpha} \tag{16}$$

where $g_\alpha(\mathbf{x}, t)$ is the distribution function for electric potential. τ_g is the relaxation time for the electric potential distribution function. G_α is related to the net charge term in Equation (4) by:

$$G_\alpha = -\omega_\alpha \chi \frac{\rho_e}{\varepsilon \varepsilon_0} \tag{17}$$

and the operator $\overline{D}_\alpha = \partial_t + \theta \mathbf{e}_\alpha \cdot \nabla$, with $\theta \in [0, 1]$ as a parameter for different schemes. Both the minimum and maximum values of θ (0 and 1) have been tested with diffusion and convection-diffusion systems, and no significant influence on the solution accuracy is found. In this work we use a forward scheme for the temporal derivative with $\theta = 1$ for simplicity:

$$\overline{D}_\alpha G_\alpha = \frac{G_\alpha(\mathbf{x}, t) - G_\alpha(\mathbf{x} - \mathbf{e}_\alpha \Delta t, t - \Delta t)}{\Delta t} \tag{18}$$

The electric potential ψ can be calculated from the distribution functions by:

$$\psi = \sum_\alpha g_\alpha \tag{19}$$

and the equilibrium distribution g_α^{eq} of electrical potential evolution function is:

$$g_\alpha^{eq} = \omega_\alpha \psi \tag{20}$$

Following the spirit for the fluid flow, the following differential equation through the Chapman–Enskog analysis can be derived:

$$\frac{\partial \psi}{\partial t} = \chi \nabla^2 \psi + \chi \frac{\rho_e}{\varepsilon \varepsilon_0} \tag{21}$$

and the solution to the Poisson Equation (4) can be obtained at the steady state of the simulation when the partial differential term on the left-hand side approaches zero. The potential diffusivity in Equation (21) is defined as:

$$\chi = \frac{(2\tau_g - 1)\, c \Delta x}{6} \tag{22}$$

3.3. Boundary Conditions

As with any other numerical methods, correct and accurate boundary treatments play a crucial role in LBM simulation. Many useful schemes for boundary condition have been developed for solving different physical problems. To model the fluid-solid interaction on the complex geometries, the mid-point bounce-back scheme [30] is used for the flow field. As shown in Figure 1, the link between the fluid node $\mathbf{x}_f$ and the solid node $\mathbf{x}_s$ intersect the physical boundary at $\mathbf{x}_b$. The fraction of the intersected link in the solid domain region is $\Delta = |\mathbf{x}_s - \mathbf{x}_b| / |\mathbf{x}_s - \mathbf{x}_f|$.

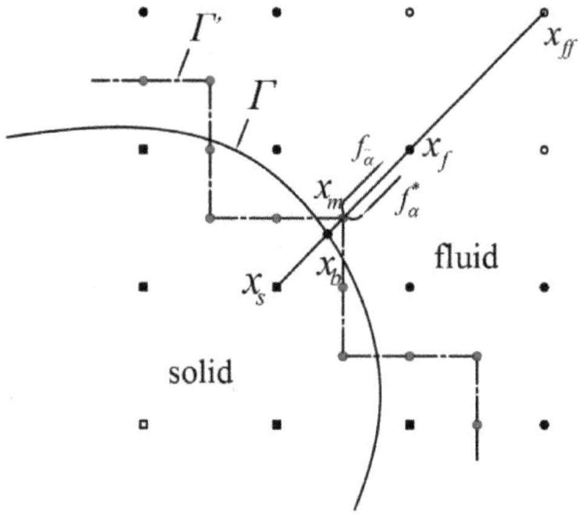

Figure 1. Schematic illustrations for the mid-point bounce-back scheme.

As we known, the evolution equations consist of two computational steps:
Collision:

$$\varphi_\alpha^*(\mathbf{x}, t) - \varphi_\alpha(\mathbf{x}, t) = -\frac{1}{\tau}\left[\varphi_\alpha(\mathbf{x}, t) - \varphi_\alpha^{eq}(\mathbf{x}, t)\right] \tag{23a}$$

Streaming:

$$\varphi_\alpha(\mathbf{x} + \mathbf{e}_\alpha\delta_t, t + \delta_t) = \varphi_\alpha^*(\mathbf{x}, t) \tag{23b}$$

with $\varphi_\alpha = f_\alpha$ or g_α. When implementing the boundary conditions with LBM, the difficulty is how to finish the collision and streaming steps at the boundaries. In Figure 1, after the collision step at the fluid node $\mathbf{x}_f$, the distribution function f_α^* leave $\mathbf{x}_f$, and is then assumed to be bounce-back at the midpoint $\mathbf{x}_m$ in the reversed direction and with a modified magnitude as $f_{\bar{\alpha}}$:

$$f_{\bar{\alpha}} = f_\alpha^* - \frac{2\rho\omega_\alpha}{c_s^2}\mathbf{u}_m \cdot e_\alpha \tag{24}$$

where $e_{\bar{\alpha}} = -e_\alpha$, $\mathbf{u}_m$ is the midpoint velocity at the point $\mathbf{x}_m$ to be determined.

For $\Delta \leq 1/2$, the midpoint $\mathbf{x}_m$ locates between $\mathbf{x}_b$ and $\mathbf{x}_f$, and the midpoint velocity $\mathbf{u}_m$ can be readily obtained with a linear interpolation:

$$\mathbf{u}_m = \frac{\frac{1}{2}\mathbf{u}_b + (\frac{1}{2} - \Delta)\mathbf{u}_f}{1 - \Delta} \tag{25}$$

where $\mathbf{u}_b$ is the imposed boundary velocity at the intersection point $\mathbf{x}_b$, and $\mathbf{u}_f$ is the flow velocity calculated at the fluid node $\mathbf{x}_f$. For $\Delta > 1/2$, the midpoint $\mathbf{x}_m$ is in the solid domain and therefore an extrapolation is needed to obtain velocity $\mathbf{u}_m$:

$$\mathbf{u}_m = \frac{\frac{3}{2}\mathbf{u}_b - (\Delta - \frac{1}{2})\mathbf{u}_{ff}}{2 - \Delta} \quad (26)$$

where $\mathbf{u}_{ff}$ is the velocity at the fluid point $\mathbf{x}_{ff}$.

Following the above velocity boundary treatment, here, we use the electric potential ψ_m at the midpoint to calculate the electric potential distribution function at the bounce-back nodes:

$$g_{\bar{\alpha}}(x_f, t + \delta t) = -g_{\alpha}^{+}(x_f, t) + 2\omega_\alpha \psi_m \quad (27)$$

with the midpoint electric potential ψ_m obtained via:

$$\psi_m = \begin{cases} \frac{\frac{1}{2}\psi_b + (\frac{1}{2} - \Delta)\psi_f}{1-\Delta} & \Delta \le 1/2 \\ \frac{\frac{3}{2}\psi_b - (\Delta - \frac{1}{2})\psi_{ff}}{2-\Delta} & \Delta > 1/2 \end{cases} \quad (28)$$

where ψ_b is the imposed boundary electric potential at the intersection point $\mathbf{x}_b$, ψ_f is the electric potential calculated at the fluid node $\mathbf{x}_f$, and ψ_{ff} is the electric potential at the fluid point $\mathbf{x}_{ff}$.

4. Validation and Discussions

4.1. Electric Potential with Flat Surface

First, we apply the improved bounce-back scheme to calculate the potential profile between two parallel plates, both of constant potential, immersed in an electrolyte solution. Near the charged surfaces, ions in the electrolyte solution will be redistributed and the electric diffuse layer will be established. The ion charge density can be related to the local potential via the Boltzmann distribution. For surfaces with low surface potentials, the Debye–Huckel approximation can be applied and the Poisson–Boltzmann equation can be solved by Equation (8). The solution of this linearized Poisson–Boltzmann equation between two identical plates of a separation H is:

$$\psi(x) = \psi_0 \frac{\cosh(\kappa x)}{\cosh(\kappa H/2)} \quad (29)$$

where x is the transverse location across the gap with $x = 0$ at the centerline. We use $\psi_0 = 1$, $\kappa = 0.02$, and different channel heigh $H = 16, 32, 64, 128$ in our simulation. The calculated potential profile is plotted in Figure 2. The symbols are results from LBM simulations, and the black lines are theoretical solutions predicted from Equation (29). Excellent agreement can be seen with different channel height between the simulation results and theory.

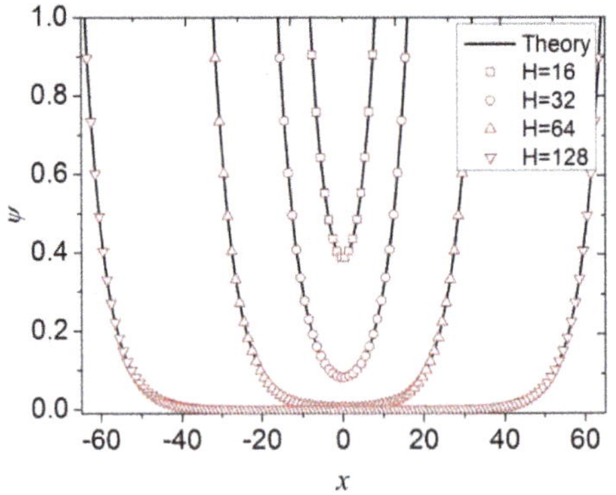

Figure 2. Electric potential distributions from our LBM simulation (symbols) and the analytical solution (black lines) with different height between two identically charged plates in an electrolyte solution.

For a more quantitative analysis, the simulations with channel height $H = 16$ and different offset (Δ=0.2, Δ=0.5, and Δ=0.7) are implemented by the classical bounce-back treatment and the improved bounce-back treatment, as shown in Figure 3. When the offset Δ=0.5, the results from different treatment are identical. When the offset Δ=0.2 or Δ=0.7, for the classical bounce-back methods, these different offset values have no influence on the electric potential distribution. However, the improved bounce-back method can correctly follow the theoretical solution.

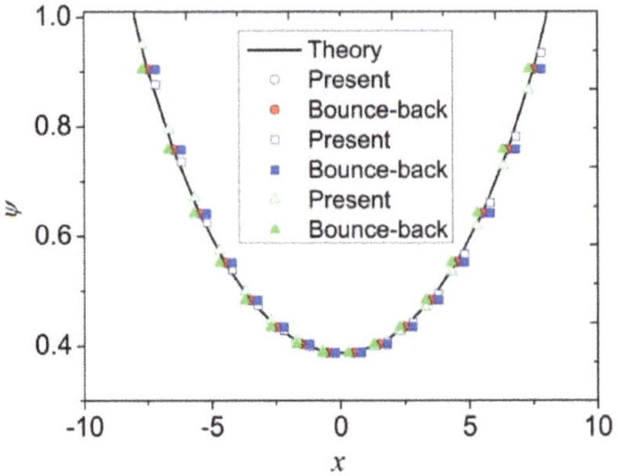

Figure 3. Electric potential distributions from present treatment, classical Bounce-Back treatment and the analytical solution with different offset (red symbol Δ=0.5; blue symbol Δ=0.7; green symbol Δ=0.2).

As for typical LBM boundary models, the numerical accuracy is studied. We choose different channel heigh $H = 16,\ 32,\ 64,\ 128$ and calculate the error between the LBM results and theoretical solutions. The errors are plotted in Figure 4, and linear fitting are conducted in the logarithmic graph. The fitting slope is usually considered as the accuracy order of a numerical model. As show in Figure 5, the accuracy order is 1.97, about 2, indicating a second-order for this system by the improved bounce-back treatment.

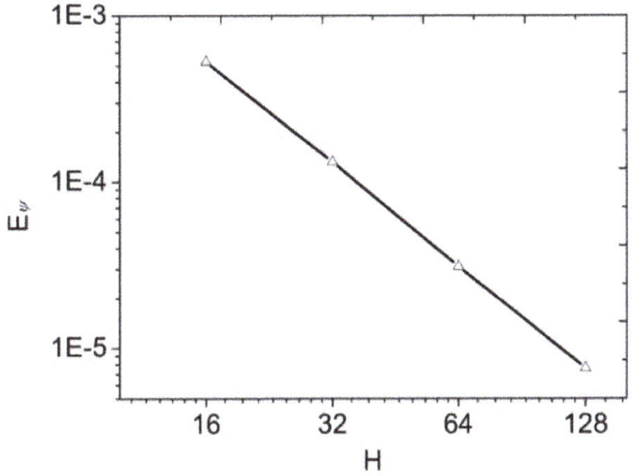

Figure 4. The error for electric potential between two identically charged plates in an electrolyte solution.

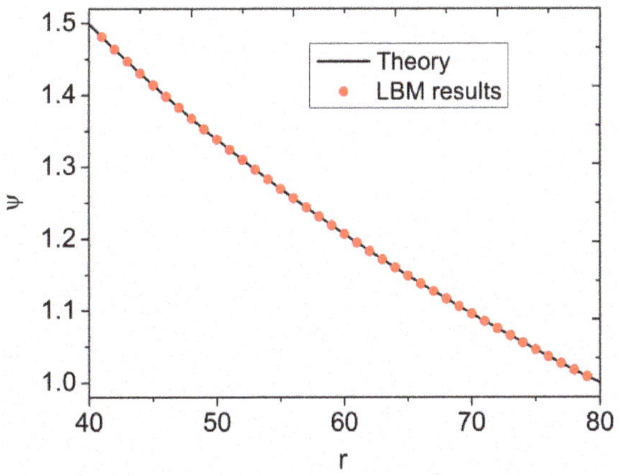

Figure 5. Electric potential distributions between two coaxial circular surfaces.

4.2. Electric Potential with Complex Geometry

Next, to examine the performance of our method for more complex boundary shapes and conditions, we consider the electric field between two coaxial circular surfaces with inner and outer radii R_{in} and R_{out}, respectively. With no net charge, the general solution is given as:

$$\psi(r) = C_1 \ln r + C_2 \tag{30}$$

where the constants C_1 and C_2 can be determined by the boundary conditions on the surfaces. For the Dirichlet boundary conditions, the electric potential on both surfaces are $\psi(R_{in}) = \psi_1$ and $\psi(R_{out}) = \psi_2$.

The corresponding exact solution for this case is:

$$\psi(r) = \psi_1 + (\psi_2 - \psi_1)\frac{\ln(r/R_{in})}{\ln(R_{out}/R_{in})} \tag{31}$$

In the simulation, we set $R_{in} = 40$, R_{out}=80, ψ_1=1.5 and ψ_2=1. The domain size is 201 $\times$ 201 and the surfaces are put at the center of the domain. Simulation result for this system is plotted in Figure 3. We also find that the results agree well with analytical solution.

4.3. Application in Electro-Osmotic Flows

In this section, a charged spherical particle immersed in an electrolyte solution is considered with a side length of 2 μm. A 100 $\times$ 100 $\times$ 100 uniform grid is used and the particle center (x_c, y_c, z_c) locates at the center of the cubic domain. Detailed numerical results on electro-osmotic flow and the effects of variation of ionic concentration, the sphere radius, external electric field and electric potential on velocity profile are presented. The numerical results are also compared with analytical solutions. In the simulation, the Poisson–Boltzmann equation is solved to obtain a steady solution firstly. And then the Navier–Stokes equations with the external force term is solved. We select a symmetric solution with $z : z = 1 : 1$ (for example, KCl, NaCl, *etc.*) and assume the solution has similar physical properties. The parameters are the ionic molar concentration $c_\infty = 0.01M$, $n_\infty = c_\infty N_A$, where N_A is Avogadro's number, the dielectric constant of the solution $\varepsilon\varepsilon_0 = 6.95 \times 10^{-10}\mathrm{C}^2/Jm$, the temperature $T = 273\mathrm{K}$, the density $\rho = 1.0 \times 10^3\mathrm{kg}/m^3$, and the electric potential with ψ_0 as a constant. The external electric field is only applied in x-direction, *i.e.*, $\mathbf{E} = (E_x, 0, 0)$. The dimensionless relaxation time τ_f and τ_g are set to be 1.0. Periodic boundary conditions are applied in all the three directions, and hence the simulated system actually represents a cubic array of spheres uniformly distribution in space.

The algorithm and boundary treatment described in previous sections have been used to simulate the electric flow with curved boundary. For the purpose of validation, the solution of the Poisson–Boltzmann equation around a spherical particle with thin EDL layers is given as:

$$\psi(r) = \psi_0 \frac{R}{r} e^{-\kappa(r-R)} \tag{32}$$

where r is the distance to the spherical center. We use $\psi_0 = 10$ mV and $E_x = 500\text{V}/m$ in our simulation. The particle has a radius of R = 0.6 μm.

Figure 6a shows the electric potential as a function of the distance to the particle center. The red circles are from our LBM calculation and the black curve is the theory solution according to Equation (32). Good agreement can be observed between them. The potential distribution at $y = y_c$ is presented in Figure 6b. The distribution appears circularly symmetric and isotropic, and this is confirmed by the fact that all the simulated ψ~r data points fall approximately on a single curve in Figure 6a.

When an external electric field is applied, an electric force **F** will be generated in the electrolyte solution near the surface due to non-zero charge in that region, and the electrostatic force can thus induce fluid flows along the electric field direction. This phenomenon is called the electro-osmosis. Figure 6c displays the velocity component u in the $y = y_c$ with different locations x = 1.0 μm, x = 1.5 μm, and x = 2.0 μm. Only the upper half ($z \geq zc$ = 1.0 μm) is shown for these symmetric curves. At x = 1.0 μm (black solid line), the velocity increases from 0 at the surface to a plateau value near the top boundary. This is similar to the typical plug-like velocity profile of electro-osmotic flows in straight channels, since the electric force only exists in the thin EDL near the surface. Away from this particular location, the cross-sectional area for the flow passage increases, and therefore the flow velocity decreases due to the mass conservation. The electro-osmotic flow streamlines in the $y = y_c$ plane are plotted in Figure 6d. The red arrows indicate the velocity magnitude and direction. The flow pattern is symmetric about both $x = x_c$ and $z = z_c$ due to the symmetric system geometry and the creeping electro-osmotic flow. These simulation results demonstrate that our method for electro-kinetic flows is useful.

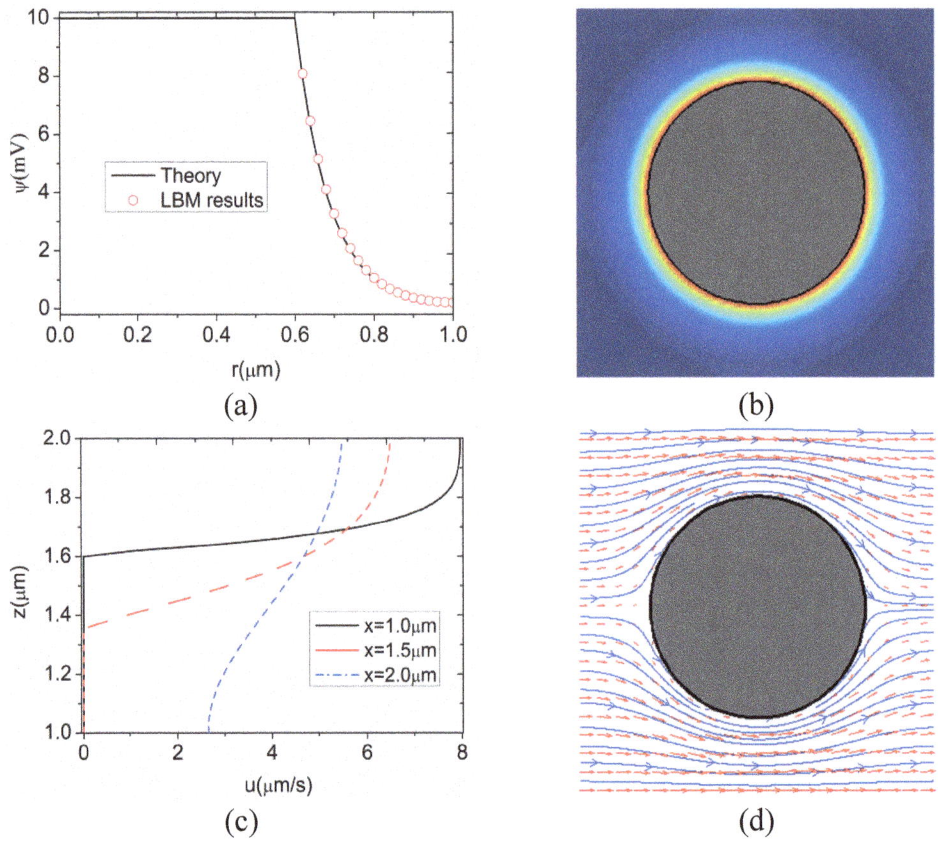

Figure 6. The electric potential distribution (**a**) and (**b**) and electro-osmotic flow (**c**) and (**d**) around the spherical particle in the y = y_c plane.

5. Conclusions

In this paper, we have extended the improved bounce-back boundary treatment for LBM flow simulations to electric field simulations. Several simulations have also been performed to examine our boundary methods in term of ability to deal with complex boundary situations. An example simulation of electro-osmotic flow with a charge sphere particle immersed in an electrolyte solution has also been presented. Comparisons with theoretical predictions show excellent agreement for all simulations, and our method therefore could be useful for future electrokinetic simulations with complex boundary geometries. Furthermore, the boundary treatment in this work can be applied to LBM simulations for other processes and phenomena that can be described by similar differential equations.

Acknowledgments: This work was supported by the National "Twelfth Five-Year" Science and Technology Support Plan (2014BAD08B04) and the Priority Academic Program Development of Jiangsu Higher Education Institutions (PAPD).

Author Contributions: Q.C. and Q.L. produced the model and wrote the codes. H.Z., X.J., L.X. and Y.R. implemented the simulations and obtained the results. All authors have read and approved the final manuscript.

Conflicts of Interest: The authors declare no conflict of interest.

References

1. Stone, H.A.; Stroock, A.D.; Ajdari, A. Engineering flows in small devices: Microfluidics toward a lab-on-a-chip. *Annu. Rev. Fluid. Mech.* **2004**, *36*, 381–411.
2. Darguji, H.; Yang, P.D.; Majumdar, A. Ion transport in nanofluidic channels. *Nano Lett.* **2004**, *4*, 137–142.
3. Ho, C.M.; Tai, Y.C. Micro-electro-mechanical-systems (MEMS) and fluid flow. *Annu. Rev. Fluid. Mech.* **1998**, *30*, 579–612.
4. Wang, C.Y. Fundamental models for fuel cell engineering. *Chem. Rev.* **2004**, *104*, 4727–4766.
5. Sinton, D.; Li, D.Q. Electroosmotic velocity profiles in microchannels. *Colloids Surf. A Physicochem. Eng. Asp.* **2003**, *222*, 273–283.
6. Li, B.; Kwok, D.Y. Lattice Boltzmann model of microfluidics with high Reynolds numbers in the presence of external forces. *Langmuir* **2003**, *19*, 3041–3048.
7. Li, B.; Kwok, D.Y. Electrokinetic microfluidic phenomena by a lattice Boltzmann model using a modified Poisson–Boltzmann equation with an excluded volume effect. *J. Chem. Phys.* **2004**, *120*, 947–953.
8. Tian, F.Z.; Li, B.M.; Kwok, D.Y. Tradeoff between mixing and transport for electroosmotic flow in heterogeneous microchannels with non-uniform surface potentials. *Langmuir* **2005**, *21*, 1126–1131.
9. Chakraborty, S. Augmentation of peristaltic microflows through electro-osmotic mechanisms. *J. Phys. D Appl. Phys.* **2006**, *39*.
10. Benzi, R.; Biferale, L.; Sbragaglia, M.; Succi, S.; Toschi, F. Mesoscopic two-phase model for describing apparent slip in micro-channel flows. *Europhys. Lett.* **2006**, *74*.
11. Sbragaglia, M.; Benzi, R.; Biferale, L.; Succi, S.; Toschi, F. Surface roughness-hydrophobicity coupling in microchannel and nanochannel flows. *Phys. Rev. E* **2006**, *97*, 204503.
12. Karniadakis, G.; Beskok, A.; Aluru, N. *Microflows and nanoflows*; Springer: New York, NY, USA, 2005.
13. Chen, S.; Doolen, G.D. Lattice Boltzmann method for fluid flows. *Annu. Rev. Fluid Mech.* **1998**, *30*, 329–364.
14. Succi, S. *The Lattice Boltzmann Equation for Fluid Dynamics and Beyond*; Oxford University Press: Oxford, UK, 2001.
15. Zhang, J.F. Lattice Boltzmann method for microfluidics: models and applications. *Microfluid Nanofluid* **2011**, *10*.

16. Benzi, R.; Succi, S.; Vergassola, M. The Lattice Boltzmann equation-theory and applications. *Phys. Rep.* **1992,** *222,* 145–197.
17. De Rosis, A.; Ubertini, S.; Ubertini, F. A comparison between the interpolated bounce-back scheme and the immersed boundary method to treat solid boundary conditions for laminar flows in the lattice Boltzmann framework. *J. Sci. Comput.* **2014,** *61,* 477–489.
18. Warren, P.B. Electroviscous transport problems via lattice-Boltzmann. *Int. J. Mod. Phys. C* **1997,** *8,* 889–898.
19. He, X.Y.; Li, N. Lattice Boltzmann simulation of electrochemical systems. *Comput. Phys. Commun.* **2000,** *129,* 158–166.
20. Zhang, J.; Kwok, D.Y. A 2D lattice Boltzmann study on electrohydrodynamic drop deformation with the leaky dielectric theory. *J. Comput. Phys.* **2005,** *206,* 150–161.
21. Melchionna, S.; Succi, S. Electrorheology in nanopores via lattice Boltzmann simulation. *J. Chem. Phys.* **2004,** *120,* 4492–4497.
22. Wu, J.; Srinivasan, V.; Xu, J.; Wang, C.Y. Newton–Krylov-Multigrid algorithms for battery simulation. *J. Electrochem. Soc.* **2002,** *149.*
23. Wang, J.; Wang, M.; Li, Z. Lattice Boltzmann simulations of mixing enhancement by the electro-osmotic flow in microchannels. *Mod. Phys. Lett. B* **2005,** *19,* 1515–1518.
24. Wang, J.; Wang, M.; Li, Z. Lattice Poisson–Boltzmann simulations of electro-osmotic flows in microchannels. *J. Colloid Interface Sci.* **2006,** *296,* 729–736.
25. Tang, G.H.; Li, Z.; Wang, J.K.; He, Y.L.; Tao, W.Q. Electroosmotic flow and mixing in microchannels with the lattice Boltzmann method. *J. Appl. Phys.* **2006,** *100,* 094908.
26. Tang, G.H.; He, Y.L.; Tao, W.Q. Numerical analysis of mixing enhancement for micro-electroosmotic flow. *J. Appl. Phys.* **2010,** *107,* 104906.
27. Chai, Z.; Guo, Z.; Shi, B. Study of electro-osmotic flows in microchannels packed with variable porosity media via lattice Boltzmann method. *J. Appl. Phys.* **2007,** *101,* 014913.
28. Wang, M.; Wang, J.; Chen, S.; Pan, N. Electrokinetic pumping effects of charged porous media in microchannels using the lattice Poisson–Boltzmann method. *J. Colloid Interface Sci.* **2006,** *304,* 246–253.
29. Chai, Z.; Guo, Z.; Zheng, L.; Shi, B. Lattice Boltzmann simulation of surface roughness effect on gaseous flow in a microchannel. *J. Appl. Phys.* **2008,** *104,* 014902.
30. Wang, M.; Wang, J.; Chen, S. Roughness and cavitations effects on electro-osmotic flows in rough microchannels using the lattice Poisson–Boltzmann methods. *J. Comput. Phys.* **2007,** *226,* 836–851.
31. Ladd, A.J.C. Numerical simulations of particulate suspensions via a discretized Boltzmann equation. Numerical simulations of particulate suspensions via a discretized Boltzmann equation. Part 1. Theoretical foundation. *J. Fluid Mech.* **1994,** *271,* 285–309.
32. Yin, X.W.; Zhang, J.F. An improved bounce-back scheme for complex boundary conditions in lattice Boltzmann method. *J. Comput. Phys.* **2012,** *231,* 4295–4303.
33. Chen, Q.; Zhang, X.B.; Zhang, J.F. Improved treatments for general boundary conditions in the lattice Boltzmann method for convection-diffusion and heat transfer processes. *Phys. Rev. E* **2013,** *88,* 033304.

34. Chen, Q.; Zhang, X.B.; Zhang, J.F. Numerical simulation of Neumann boundary condition in the thermal lattice Boltzmann model. *Int. J. Mod. Phys. C* **2014**, *25*.
35. Li, L.; Mei, R.W.; Klausner, J.F. Boundary conditions for thermal lattice Boltzmann equation method. *J. Comput. Phys.* **2013**, *237*, 366–395.
36. Li, L.; Chen, C.; Mei, R.W.; Klausner, J.F. Conjugate heat and mass transfer in the lattice Boltzmann equation method. *Phys. Rev. E* **2014**, *89*, 043308.
37. Filippova, O.; Hänel, D. Lattice Boltzmann simulation of gas-particle flow in filters. *Comput. Fluids* **1997**, *26*, 697–712.
38. Bouzidi, M.; Firdaouss, M.; Lallemand, P. Momentum transfer of a Boltzmann-lattice fluid with boundaries. *Phys. Fluids* **2001**, *13*, 3452–3459.
39. Yu, D.Z.; Mei, R.W.; Luo, L.-S.; Shyy, W. Viscous flow computations with the method of lattice Boltzmann equation. *Prog. Aerosp. Sci.* **2003**, *39*, 329–367.
40. Oulaid, O.; Chen, Q.; Zhang, J.F. Accurate boundary treatments for lattice Boltzmann simulations of electric fields and electro-kinetic applications. *J. Phys. A Math. Theor.* **2013**, *46*.
41. Honig, B.; Nicholls, A. classical electrostatics in biology and chemistry. *Science* **1995**, *268*, 1144–1149.

A Truncation Scheme for the BBGKY2 Equation

Gregor Chliamovitch, Orestis Malaspinas and Bastien Chopard

Abstract: In recent years, the maximum entropy principle has been applied to a wide range of different fields, often successfully. While these works are usually focussed on cross-disciplinary applications, the point of this letter is instead to reconsider a fundamental point of kinetic theory. Namely, we shall re-examine the *Stosszahlansatz* leading to the irreversible Boltzmann equation at the light of the MaxEnt principle. We assert that this way of thinking allows to move one step further than the factorization hypothesis and provides a coherent—though implicit—closure scheme for the two-particle distribution function. Such higher-order dependences are believed to open the way to a deeper understanding of fluctuating phenomena.

Reprinted from *Entropy*. Cite as: Chliamovitch, G.; Malaspinas, O.; Chopard, B. A Truncation Scheme for the BBGKY2 Equation. *Entropy* **2015**, *17*, 7522–7529.

1. Introduction

While the formulation of equilibrium statistical mechanics in terms of the maximum entropy ("MaxEnt") principle goes back to Jaynes' seminal work [1,2] in the 50s, the last decade has seen a spectacular revival of this approach. In particular, the MaxEnt-based characterization of complex systems presented in [3] paved the way to applications in a variety of fields ranging from linguistics to biology [4–8]. First focussed on equilibrium situations, these works soon turned their attention to non-equilibrium properties as well [9–12].

However, it seems that these authors paid comparatively little attention to more "fundamental" issues. In the present letter, we would like to reconsider some aspects of the kinetic theory of gases at the light of the MaxEnt philosophy. More precisely, kinetic theory relies heavily on the so-called *Stosszahlansatz* which asserts that, before colliding, particles are uncorrelated. While this bold assumption can be motivated physically, it is not completely clear how it should be generalized when considering higher-order distribution functions. Our point is that if one considers the *Stosszahlansatz* as a heuristically motivated assumption, it generalizes naturally to higher-order cases—even though this raises extra mathematical challenges!

We start with a brief reminder on MaxEnt distributions as well as on the BBGKY hierarchy leading to kinetic equations, and bridge both in the last two sections.

2. Maximum Entropy Distributions

The Shannon entropy $H(X) = -\int dx P(x) \ln P(x)$ has all properties one would expect from an uncertainty measure [13,14]. In other terms, among a set of distributions, the least biased guess an observer can make is the one having the largest entropy while still satisfying available observational constraints. Assume for instance we try to maximize $H(X)$ under the constraint $\langle O \rangle = \int dx O(x) P(x) = \mu$. Introducing a multiplier for the constraint and another for the probabilistic normalization, one has to compute

$$\frac{\partial}{\partial P}\left(\int dx P(x) \ln \frac{1}{P(x)} + \lambda_0 \int dx P(x) + \lambda \int dx O(x) P(x)\right) = \int dx \left(\ln \frac{1}{P(x)} - 1 + \lambda_0 + \lambda O(x)\right). \quad (1)$$

Letting this expression vanish yields

$$P(x) = \exp\left(-1 + \lambda_0 + \lambda O(x)\right), \quad (2)$$

where the multipliers have to be determined to match the constraints. This result extends straightforwardly to the case of several constraints, namely $\langle O_k \rangle = \int dx O_k(x) P(x) = \mu_k$ for $k = 1, 2, \ldots, K$. Then

$$P(x) = \frac{1}{Z} \exp\left(\sum_{k=1}^{K} \lambda_k O_k(x)\right). \quad (3)$$

In what follows we shall be concerned primarily with constraints on marginals instead of averages. An appropriate use of δ functions allows to generalize the previous result. As an example consider the case of a quadrivariate variable $\mathbf{x} = (w, x, y, z)$ the marginal $P_{123}(a, b, c)$ of which is assumed to be known. Putting $O(\mathbf{x}) = \delta(w, a)\delta(x, b)\delta(y, c)$ we can write

$$\begin{aligned}
\langle O \rangle &= \sum_{\mathbf{x}} O(\mathbf{x}) P(\mathbf{x}) \\
&= \sum_{w,x,y} \delta(w, a)\delta(x, b)\delta(y, c) \sum_{z} P(\mathbf{x}) \\
&= \sum_{w,x,y} \delta(w, a)\delta(x, b)\delta(y, c) P_{123}(w, x, y) \\
&= P_{123}(a, b, c). \quad (4)
\end{aligned}$$

Using the result (3) derived for the averages one gets

$$P(\mathbf{x}) = \frac{1}{Z} \exp\left(\sum_{a,b,c} \lambda(a, b, c) O(\mathbf{x})\right) = \frac{1}{Z} \exp \lambda(w, x, y), \quad (5)$$

λ now denoting a *function*. This result can be extended to any number of marginals of any order. If for instance besides P_{123} the marginals P_{124} and P_{34} are given we get

$$P(\mathbf{x}) = \frac{1}{Z} \exp\left(\lambda_1(w,x,y) + \lambda_2(w,x,z) + \lambda_3(y,z)\right) \tag{6}$$

for functions λ_1, λ_2 and λ_3 to be determined. Unfortunately, this determination is difficult except in the case of univariate constraints. Then $\lambda_1(w) = \ln P_1(w)$, *etc.*, obviously solves the problem, so that the corresponding MaxEnt distribution is the factorized distribution. When turning our attention to the applicability of this result to the realm of kinetic theory, it will appear that in that context the problem can be slightly simplified due to the structure of reduced distributions.

3. The BBGKY Hierarchy

Let us consider N particles of mass m, the coordinates of which in phase space are their positions $\mathbf{q}_i$ and momenta $\mathbf{p}_i$. It will be convenient to define a condensed notation $x_i = (\mathbf{q}_i, \mathbf{p}_i)$. Let $f_N(x_1, ..., x_N, t)$ denote the joint distribution function characterizing the system, which obeys Liouville's equation

$$\frac{df_N}{dt} = \frac{\partial f_N}{\partial t} + \sum_{i=1}^{N} \frac{\mathbf{p}_i}{m} \frac{\partial f_N}{\partial \mathbf{q}_i} + \sum_{i=1}^{N} \mathbf{F}_i \frac{\partial f_N}{\partial \mathbf{p}_i} = 0, \tag{7}$$

where $\mathbf{F}_i$ denotes the force exerted on particle i. We shall restrict ourselves to the case without external force and where particles interact pairwise through some radial potential $V(|\mathbf{q}_i - \mathbf{q}_j|) = V_{ij}$. Then $\mathbf{F}_i = -\sum_{j \neq i} \frac{\partial V_{ij}}{\partial \mathbf{q}_i}$.

Reminding that f_N itself is normalized to $N!$, we now introduce the reduced s-particle distribution $f_s(x_1, ..., x_s, t) = \frac{N!}{(N-s)!} \int dx_{s+1} ... dx_N f_N(x_1, ..., x_N, t)$. The standard result [15] is that by integrating Liouville's equation, one obtains a dynamical equation for f_s given by the so-called BBGKY hierarchy (from the non-chronological list of its co-discoverers' names : Bogoliubov, Born, Green, Kirkwood, Yvon) :

$$\frac{\partial f_s}{\partial t} + \sum_{i=1}^{s} \frac{\mathbf{p}_i}{m} \frac{\partial f_s}{\partial \mathbf{q}_i} - \sum_{i=1}^{s} \sum_{j \neq i}^{s} \frac{\partial V_{ij}}{\partial \mathbf{q}_i} \frac{\partial f_s}{\partial \mathbf{p}_i} - \int dx_{s+1} \sum_{i=1}^{s} \frac{\partial V_{i,s+1}}{\partial \mathbf{q}_i} \frac{\partial f_{s+1}}{\partial \mathbf{p}_i} = 0. \tag{8}$$

This expression forms a hierarchy since the dynamics for f_s is expressed in terms of the higher-order distribution f_{s+1}. Of course each equation can be deduced from its higher-order precursor by integration, at the cost of an information loss. In what follows we shall denote the s-th equation of the hierarchy as BBGKY-s. BBGKY1 and BBGKY2, in which we are primarily interested here, are

$$\frac{\partial f_1}{\partial t} + \frac{\mathbf{p}_1}{m}\frac{\partial f_1}{\partial \mathbf{q}_1} = \int dx_2 \frac{\partial V_{12}}{\partial \mathbf{q}_1}\frac{\partial f_2}{\partial \mathbf{p}_1} \tag{9}$$

and

$$\frac{\partial f_2}{\partial t} + \frac{\mathbf{p}_1}{m}\frac{\partial f_2}{\partial \mathbf{q}_1} + \frac{\mathbf{p}_2}{m}\frac{\partial f_2}{\partial \mathbf{q}_2} - \frac{\partial V_{12}}{\partial \mathbf{q}_1}\left(\frac{\partial}{\partial \mathbf{p}_1} - \frac{\partial}{\partial \mathbf{p}_2}\right) f_2 = \int dx_3 \left(\frac{\partial V_{13}}{\partial \mathbf{q}_1}\frac{\partial f_3}{\partial \mathbf{p}_1} + \frac{\partial V_{23}}{\partial \mathbf{q}_2}\frac{\partial f_3}{\partial \mathbf{p}_2}\right), \tag{10}$$

where of course $f_1 = f_1(\mathbf{p}_1, \mathbf{q}_1, t)$, $f_2 = f_2(\mathbf{p}_1, \mathbf{q}_1, \mathbf{p}_2, \mathbf{q}_2, t)$ and $f_3 = f_3(\mathbf{p}_1, \mathbf{q}_1, \mathbf{p}_2, \mathbf{q}_2, \mathbf{p}_3, \mathbf{q}_3, t)$. The purpose of this paper is to investigate the second of these equations. As stressed above, we shall not try to express BBGKY1 and BBGKY2 as a set of coupled equations relating f_1 and f_2, since such an approach would not "fit" nicely in the spirit of the BBGKY approach. Instead, we shall manage to truncate BBGKY2 in order to obtain a single, self-standing equation for f_2.

4. The *Stosszahlansatz* for BBGKY2

The procedure leading from the BBGKY1 equation to a consistent kinetic equation for f_1 is standard: the *Stosszahlansatz* asserts that before colliding two particles are uncorrelated, *i.e.*, f_2 factorizes as $f_2(x_1, x_2) = f_1(x_1)f_1(x_2)$. This allows us to express the collision integral in terms of f_1, so that BBGKY1 becomes a closed equation for f_1. Since this factorization hypothesis may be supported from a physical standpoint, it is tempting to use this ansatz in the collision term for BBGKY2 as well. But this raises an issue: if BBGKY2 can be cast into an equation relating a streaming term expressed in terms of f_2 to a collision term expressed in terms of f_1, then this equation is obviously not consistent by itself and has to be supplemented, so as to obtain a system of coupled equations.

Our point is that this issue vanishes if the *Stosszahlansatz* is reconsidered as a heuristic ansatz instead of a physically-grounded assumption. We propose to reformulate it as follows: since the exact codependence of particles entering the collision range is unknown, one has to make a reasonable guess on it, and the MaxEnt distribution steps out at this point. The MaxEnt guess for f_2, compatible with the univariate distribution appearing in the streaming term, is the factorized one, but on the contrary the guess for f_3, compatible with the f_2 appearing in the left-hand side, will be quite different from a factorized distribution (as exemplified by (6)).

Let us now see to what extent the result (6) obtained for the MaxEnt distribution may be particularized to our current purpose. We showed that, given bivariate marginals, the MaxEnt estimate for $f_3(x_1, x_2, x_3)$ was given by

$$f_3^{ME}(x_1, x_2, x_3) = \frac{1}{Z} \exp\left(\lambda_1(x_1, x_2) + \lambda_2(x_1, x_3) + \lambda_3(x_2, x_3)\right) \tag{11}$$

for some λ_1, λ_2 and λ_3. The point is that these marginals are the same for each pair by definition of the reduced distribution f_2, and accordingly all three λs are the same. Absorbing the normalization, one is therefore allowed to write that

$$f_3^{ME}(x_1, x_2, x_3) = g(x_1, x_2) g(x_1, x_3) g(x_2, x_3) \tag{12}$$

for a function g that is nevertheless unknown, except for the fact that it has to satisfy the marginal constraint

$$g(x, y) \int dz g(x, z) g(y, z) = f_2(x, y). \tag{13}$$

5. The Collision Term

Using this ansatz we now proceed to write down the kinetic equation for f_2. All through, we shall retain the usual assumptions of kinetic theory, leading us to neglect triple collisions: the streaming term for the two-particle distribution characterizing particles "1" and "2" will thus be altered by (1) binary collisions between "1"and another particle, "2" being spectator, and (2) binary collisions between "2"and another particle, "1" being spectator.

The binary interaction is defined as occurring when two particles meet in a ball B of radius R. Defining ternary interactions is more subtle since, inasmuch as the interaction potential is the same whatever the order of the interaction, it seems artificial to introduce a specific cutoff. We shall therefore define the range of triple collisions as the lenticular overlap of balls $B_R^{(1)}$ and $B_R^{(2)}$ characterizing the domain of interaction with "1" and "2" respectively. Neglecting triple collisions thus amounts to assuming that $|\mathbf{q}_1 - \mathbf{q}_2| > 2R$.

We first compute the contribution of collisions of "2" with "3", "1" being left aside. Let us recall that the collision term is given by

$$\left(\frac{\partial f_2}{\partial t}\right)_{coll} = \int dx_3 \left(\frac{\partial V_{13}}{\partial \mathbf{q}_1}\frac{\partial f_3}{\partial \mathbf{p}_1} + \frac{\partial V_{23}}{\partial \mathbf{q}_2}\frac{\partial f_3}{\partial \mathbf{p}_2}\right). \tag{14}$$

In the usual derivation of the Boltzmann equation from the BBGKY hierarchy, the right-hand side of BBGKY1 is transformed using BBGKY2. Similarly, we can transform $(\partial_t f_2)_{coll}$ using BBGKY3, namely

$$\frac{\partial f_3}{\partial t} + \frac{\mathbf{p}_1}{m}\frac{\partial f_3}{\partial \mathbf{q}_1} + \frac{\mathbf{p}_2}{m}\frac{\partial f_3}{\partial \mathbf{q}_2} + \frac{\mathbf{p}_3}{m}\frac{\partial f_3}{\partial \mathbf{q}_3}$$
$$- \frac{\partial V_{12}}{\partial \mathbf{q}_1}\left(\frac{\partial}{\partial \mathbf{p}_1} - \frac{\partial}{\partial \mathbf{p}_2}\right) f_3 - \frac{\partial V_{13}}{\partial \mathbf{q}_1}\left(\frac{\partial}{\partial \mathbf{p}_1} - \frac{\partial}{\partial \mathbf{p}_3}\right) f_3 - \frac{\partial V_{23}}{\partial \mathbf{q}_2}\left(\frac{\partial}{\partial \mathbf{p}_2} - \frac{\partial}{\partial \mathbf{p}_3}\right) f_3 = \left(\frac{\partial f_3}{\partial t}\right)_{coll} \tag{15}$$

(we do not make explicit the collision term $(\partial_t f_3)_{coll}$ since we shall cancel it soon anyway). Under usual dimensional assumptions, we can write $\partial_t f_3 \approx 0$ and $(\partial_t f_3)_{coll} \approx 0$, so that, substituting in the collision term, $(\partial_t f_2)_{coll}$ is rewritten as

$$\begin{aligned}\left(\frac{\partial f_2}{\partial t}\right)_{coll} &= \int dx_3 \left(\frac{\mathbf{p}_1}{m}\frac{\partial f_3}{\partial \mathbf{q}_1} + \frac{\mathbf{p}_2}{m}\frac{\partial f_3}{\partial \mathbf{q}_2} + \frac{\mathbf{p}_3}{m}\frac{\partial f_3}{\partial \mathbf{q}_3} - \frac{\partial V_{12}}{\partial \mathbf{q}_1}\left(\frac{\partial}{\partial \mathbf{p}_1} - \frac{\partial}{\partial \mathbf{p}_2}\right) f_3 + \left(\frac{\partial V_{13}}{\partial \mathbf{q}_1} + \frac{\partial V_{23}}{\partial \mathbf{q}_2}\right)\frac{\partial f_3}{\partial \mathbf{p}_3}\right) \\ &= \int dx_3 \left(\frac{\mathbf{p}_2}{m}\frac{\partial f_3}{\partial \mathbf{q}_2} + \frac{\mathbf{p}_3}{m}\frac{\partial f_3}{\partial \mathbf{q}_3}\right)\end{aligned} \tag{16}$$

(the last term vanishes due to the boundary condition $f_3(|\mathbf{p}| \to \infty) = 0$, the penultimate since "1" and "2" are supposed far apart from each other and the first because f_3 depends but weakly on $\mathbf{q}_1$). More precisely,

$$\left(\frac{\partial f_2}{\partial t}\right)_{coll} = \int_{\mathbf{q}_3 \in B_R^{(2)}} d\mathbf{q}_3 d\mathbf{p}_3 \left(\frac{\mathbf{p}_2}{m}\frac{\partial f_3}{\partial \mathbf{q}_2} + \frac{\mathbf{p}_3}{m}\frac{\partial f_3}{\partial \mathbf{q}_3}\right). \tag{17}$$

The following is standard [15]. We introduce the relative coordinate $\mathbf{r}_{23} = \mathbf{q}_3 - \mathbf{q}_2$ and use Gauss' theorem in order to rewrite $(\partial_t f_2)_{coll}$ as a surface integral, so that

$$\begin{aligned}\left(\frac{\partial f_2}{\partial t}\right)_{coll} &= \int_{\mathbf{r}_{23} \in B_R} d\mathbf{r}_{23} d\mathbf{p}_3 \frac{\mathbf{p}_3 - \mathbf{p}_2}{m}\frac{\partial}{\partial \mathbf{r}_{23}} f_3(\mathbf{q}_1, \mathbf{p}_1, \mathbf{q}_2, \mathbf{p}_2, \mathbf{q}_3, \mathbf{p}_3, t) \\ &= \int_{S_R} d\mathbf{p}_3 d\Sigma \cdot \frac{\mathbf{p}_3 - \mathbf{p}_2}{m} f_3(\mathbf{q}_1, \mathbf{p}_1, \mathbf{q}_2, \mathbf{p}_2, \mathbf{q}_3, \mathbf{p}_3, t) \\ &= \int_{S_R^- \cup S_R^+} d\mathbf{p}_3 d\Sigma \cdot \frac{\mathbf{p}_3 - \mathbf{p}_2}{m} f_3(\mathbf{q}_1, \mathbf{p}_1, \mathbf{q}_2, \mathbf{p}_2, \mathbf{q}_3, \mathbf{p}_3, t),\end{aligned} \tag{18}$$

where $d\Sigma$ denotes the surface element of the sphere S_R such that $|\mathbf{r}_{23}| = R$. The southern hemisphere is interpreted as the contribution of oncoming collisions since $(\mathbf{p}_3 - \mathbf{p}_2) \cdot d\Sigma < 0$, while the northern one is the contribution of ending collisions since $(\mathbf{p}_3 - \mathbf{p}_2) \cdot d\Sigma > 0$.

Orienting the polar axis along $\mathbf{p}_3 - \mathbf{p}_2$, we have $d\Sigma \cdot (\mathbf{p}_3 - \mathbf{p}_2) = |\mathbf{p}_3 - \mathbf{p}_2| R^2 \sin\theta \cos\theta d\theta d\phi$. This can be re-expressed in terms of the surface element of the azimuthal plane such that $\theta = \pi/2$. Letting r denote the radial component on the plane, we obviously have $r = R\sin\theta$, whence $dr = \pm R\cos\theta d\theta$ (depending on θ

being lesser or larger than $\pi/2$) and $d\Sigma \cdot (\mathbf{p}_3 - \mathbf{p}_2) = \pm|\mathbf{p}_3 - \mathbf{p}_2|d\omega$. The collision term can thus be rewritten as (approximating $\mathbf{q}_3 \approx \mathbf{q}_2$)

$$\left(\frac{\partial f_2}{\partial t}\right)_{coll} = \int_{after} d\mathbf{p}_3 d\omega \frac{|\mathbf{p}_3 - \mathbf{p}_2|}{m} f_3(\mathbf{q}_1, \mathbf{p}_1, \mathbf{q}_2, \mathbf{p}_2, \mathbf{q}_2, \mathbf{p}_3, t)$$
$$- \int_{before} d\mathbf{p}_3 d\omega \frac{|\mathbf{p}_3 - \mathbf{p}_2|}{m} f_3(\mathbf{q}_1, \mathbf{p}_1, \mathbf{q}_2, \mathbf{p}_2, \mathbf{q}_2, \mathbf{p}_3, t). \quad (19)$$

Now comes the ansatz. Before the collision we obviously have

$$f_3(\mathbf{q}_1, \mathbf{p}_1, \mathbf{q}_2, \mathbf{p}_2, \mathbf{q}_2, \mathbf{p}_3, t) = g(\mathbf{q}_1, \mathbf{p}_1, \mathbf{q}_2, \mathbf{p}_2, t) g(\mathbf{q}_1, \mathbf{p}_1, \mathbf{q}_2, \mathbf{p}_3, t) g(\mathbf{q}_2, \mathbf{p}_2, \mathbf{q}_2, \mathbf{p}_3, t). \quad (20)$$

The ansatz may be extended after the collision using the fact that, by Liouville's equation, $f_3(\mathbf{q}_1, \mathbf{p}_1, \mathbf{q}_2, \mathbf{p}_2, \mathbf{q}_2, \mathbf{p}_3, t) = f_3(\mathbf{q}_1^{-\tau}, \mathbf{p}_1, \mathbf{q}_2^{-\tau}, \mathbf{p}_2', \mathbf{q}_2^{-\tau}, \mathbf{p}_3', t - \tau)$, where τ is the retardation such that at $t - \tau$ the particles are entering the collision range with momenta $\mathbf{p}_2'$, $\mathbf{p}_3'$. Since $\mathbf{q}_i^{-\tau} \approx \mathbf{q}_i$ and $t \approx t - \tau$, and since $\mathbf{p}_2'$, $\mathbf{p}_3'$ are pre-collisional momenta, the ansatz may be introduced in the first integral as well with arguments $\mathbf{p}_2'$, $\mathbf{p}_3'$. We are therefore eventually led to the following Boltzmann-like form for the BBGKY2 equation:

$$\frac{\partial f_2}{\partial t} + \frac{\mathbf{p}_1}{m}\frac{\partial f_2}{\partial \mathbf{q}_1} + \frac{\mathbf{p}_2}{m}\frac{\partial f_2}{\partial \mathbf{q}_2} - \frac{\partial V_{12}}{\partial \mathbf{q}_1}\left(\frac{\partial}{\partial \mathbf{p}_1} - \frac{\partial}{\partial \mathbf{p}_2}\right) f_2$$
$$= \int d\mathbf{p}_3 d\omega \frac{|\mathbf{p}_3 - \mathbf{p}_2|}{m} (g(\mathbf{q}_1, \mathbf{p}_1, \mathbf{q}_2, \mathbf{p}_2') g(\mathbf{q}_1, \mathbf{p}_1, \mathbf{q}_2, \mathbf{p}_3') g(\mathbf{q}_2, \mathbf{p}_2', \mathbf{q}_2, \mathbf{p}_3')$$
$$- g(\mathbf{q}_1, \mathbf{p}_1, \mathbf{q}_2, \mathbf{p}_2) g(\mathbf{q}_1, \mathbf{p}_1, \mathbf{q}_2, \mathbf{p}_3) g(\mathbf{q}_2, \mathbf{p}_2, \mathbf{q}_2, \mathbf{p}_3))$$
$$+ (1 \leftrightarrow 2). \quad (21)$$

The last term accounts for the contribution of collisions undergone by particle "1".

6. Final Remarks

As promised, equation (21) is coherent for f_2 since g can—in principle—be solved in terms of f_2. Unfortunately, we are not aware of any readily available solution to the integral equation (13). It is interesting to note that kinetic theory provides a motivation for studying the mathematical object (12), which proves surprisingly involved in spite of its deceptive apparent simplicity. In the authors' opinion, it might turn sound to tackle the problem from a linearized vantage point, considering situations where the particles are almost uncorrelated, that is where g is almost factorizable. Such an approach would also be in line with usual methods of kinetic theory [16].

This current implicit form of the collision term bears a close ressemblance with the one appearing in the standard Boltzmann equation. This ressemblance might however turn deceptive since g is likely to be a complicated functional of f_2. We hope that the present letter can prompt further work on this collision term, which might eventually lead to a form susceptible of a hydrodynamical treatment. Our hope is that such a treatment could lead to a deeper understanding of fluctuating phenomena, for which the first-order theory provides only lacunary insights. In particular, it seems reasonable to expect that (21) can be cast in the lattice-based formalism which proved so successful for the usual Boltzmann equation.

Acknowledgments: The authors would like to acknowledge funding from the European Union Seventh Framework Programme, under grant agreement 317534 (Sophocles).

Author Contributions: G.C., O.M. and B.C. performed the research; B.C. supervised the project; G.C. wrote the manuscript. All authors have read and approved the final manuscript.

Conflicts of Interest: The authors declare no conflict of interest.

Bibliography

1. Jaynes, E.T. Information Theory and Statistical Mechanics. *Phys. Rev.* **1957**, *106*, doi:10.1103/PhysRev.106.620.
2. Jaynes, E.T. Information Theory and Statistical Mechanics. II. *Phys. Rev.* **1957**, *108*, doi:10.1103/PhysRev.108.171.
3. Schneidman, E.; Still, S.; Berry, M.J.; Bialek, W. Network Information and Connected Correlations. *Phys. Rev. Lett.* **2003**, *91*, 238701.
4. Schneidman, E.; Berry, M.J.; Segev, R.; Bialek, W. Weak Pairwise Correlations Imply Strongly Correlated Network States in a Neural Population. *Nature* **2006**, *440*, 1007–1012.
5. Stephens, G.J.; Bialek, W. Statistical Mechanics of Letters in Words. *Phys. Rev. E* **2010**, *81*, 066119.
6. Mora, T.; Bialek, W. Are Biological Systems Poised at Criticality ? *J. Stat. Phys.* **2011**, *144*, 268–302.
7. Bialek, W.; Cavagna, A.; Giardina, I.; Mora, T.; Silvestri, E.; Viale, M.; Walczak, A.M. Statistical Mechanics for Natural Flocks of Birds. *Proc. Natl. Acad. Sci. USA* **2012**, *109*, 4786–4791.
8. Stephens, G.J.; Mora, T.; Tkačik, G.; Bialek, W. Statistical Thermodynamics of Natural Images. *Phys. Rev. Lett.* **2013**, *110*, 018701.
9. Van der Straeten, E. Maximum Entropy Estimation of Transition Probabilities of Reversible Markov Chains. *Entropy* **2009**, *11*, 867–887.
10. Marre, O.; El Boustani, S.; Frégnac, Y.; Destexhe, A. Prediction of Spatiotemporal Patterns of Neural Activity from Pairwise Correlations. *Phys. Rev. Lett.* **2009**, *102*, 138101.
11. Cavagna, A.; Giardina, I.; Ginelli, F.; Mora, T.; Piovani, D.; Tavarone, R.; Walczak, A.M. Dynamical Maximum Entropy Approach to Flocking. *Phys. Rev. E* **2014**, *89*, 042707.
12. Chliamovitch, G.; Dupuis, A.; Golub, A.; Chopard, B. Improving Predictability of Time Series Using Maximum Entropy Methods. *Europhys. Lett.* **2015**, *110*, doi:10.1209/0295-5075/110/10003.
13. Shannon, C.E. A Mathematical Theory of Communication. *Bell Syst. Tech. J.* **1948**, *27*, 379–423.

14. Khinchin, A.Y. *Mathematical Foundations of Information Theory*; Dover: Mineola, NY, USA, 1957.
15. Kreuzer, H.J. *Nonequilibrium Thermodynamics and its Statistical Foundations*; Oxford University Press: Oxford, UK, 1984.
16. Liboff, R.L. *Kinetic Theory*; Springer: New York, NY, USA, 2003.

From Lattice Boltzmann Method to Lattice Boltzmann Flux Solver

Yan Wang, Liming Yang and Chang Shu

Abstract: Based on the lattice Boltzmann method (LBM), the lattice Boltzmann flux solver (LBFS), which combines the advantages of conventional Navier–Stokes solvers and lattice Boltzmann solvers, was proposed recently. Specifically, LBFS applies the finite volume method to solve the macroscopic governing equations which provide solutions for macroscopic flow variables at cell centers. In the meantime, numerical fluxes at each cell interface are evaluated by local reconstruction of LBM solution. In other words, in LBFS, LBM is only locally applied at the cell interface for one streaming step. This is quite different from the conventional LBM, which is globally applied in the whole flow domain. This paper shows three different versions of LBFS respectively for isothermal, thermal and compressible flows and their relationships with the standard LBM. In particular, the performance of isothermal LBFS in terms of accuracy, efficiency and stability is investigated by comparing it with the standard LBM. The thermal LBFS is simplified by using the D2Q4 lattice velocity model and its performance is examined by its application to simulate natural convection with high Rayleigh numbers. It is demonstrated that the compressible LBFS can be effectively used to simulate both inviscid and viscous flows by incorporating non-equilibrium effects into the process for inviscid flux reconstruction. Several numerical examples, including lid-driven cavity flow, natural convection in a square cavity at Rayleigh numbers of 10^7 and 10^8 and transonic flow around a staggered-biplane configuration, are tested on structured or unstructured grids to examine the performance of three LBFS versions. Good agreements have been achieved with the published data, which validates the capability of LBFS in simulating a variety of flow problems.

Reprinted from *Entropy*. Cite as: Wang, Y.; Yang, L.; Shu, C. From Lattice Boltzmann Method to Lattice Boltzmann Flux Solver. *Entropy* **2015**, *17*, 7713–7735.

1. Introduction

Since its earliest appearance in 1988 [1], the lattice Boltzmann equation (LBE)-based method [2–11] has been developed into an effective and efficient solver for simulating a variety of complex fluid flow problems, such as isothermal and thermal flows [9,10,12], multi-phase and multi-component flows [8,13–15], compressible flows [16–18] and micro non-equilibrium flows [19–21]. To effectively study these problems, a large number of LBE-based methods have been continuously proposed and refined [9–23].

Historically, the LBM was developed from the lattice gas cellular automata (LGCA) method, aiming to remove its statistical noise and limitation to use Boolean numbers. It was later proven that the solid foundation of LBM roots in gas kinetic theory and the Chapman–Enskog expansion analysis, through which both continuous and lattice Boltzmann equation can recover the Navier–Stokes equations. This confers the LBM with an appealing kinetic nature. Besides, LBM has very simple numerical algorithms with algebraic manipulations. In particular, two simple steps of streaming and collision are involved in its solution process. The streaming process is linear, which moves particles with different distribution functions to neighboring points, but effectively considers non-linear physics in fluid dynamics. The collision process takes place locally in either lattice velocity space or macroscopic moment space according to different collision models [2,4,22]. Both single relaxation time (SRT) model (lattice Bhatnagar–Gross–Krook (LBGK) model) [4] and multiple-relaxation-time (MRT) model [22] can be applied. It seems that the MRT model eliminates some defects of the LBGK model and enriches the connotation of LBM, although the computational effort is also increased. With these distinguishing features, the LBM has emerged as an alternative powerful tool in many complex fluid flow problems [13,20,23–25]. However, the LBM also suffers from some drawbacks. Due to lattice regularity, the standard LBM usually restricts its applications to uniform grids, which hinders its direct application in certain problems with curved boundaries. In addition, as compared with conventional Navier–Stokes solvers, LBM usually needs more virtual storage to store both the distribution functions and flow variables. Furthermore, the time step in LBM is coupled with mesh spacing, which presents a great challenge for applications on multi-block and adaptive grids. Moreover, LBM is mostly applied to simulate incompressible flows since the equilibrium distribution function is obtained from low Mach number approximation.

To remove the drawbacks of standard LBM, some efforts [5–7,21] have been made to directly solve the discrete velocity Boltzmann equation (DVBE) with well-established numerical approaches. This way eliminates the coupling issue between the mesh spacing and time step in standard LBM and can be effectively applied on non-uniform grids. However, it loses the primary advantages of LBM such as simple implementation and algebraic operation. It may also involve a large numerical dissipation and encounter numerical instability [26,27].

Recently, from Chapman–Enskog (C-E) analysis, several versions of lattice Boltzmann flux solver (LBFS) [8–11] have been developed based on the standard LBM for simulating isothermal, thermal, multiphase and compressible flows. LBFS is a finite volume solver for direct update of the macroscopic flow variables at cell centers. The key idea of the LBFS is to evaluate numerical fluxes at each cell interface by local application of LBM solution for one streaming step. In other words, in LBFS, LBM is applied locally and independently at each interface, and the streaming time step is nothing to do with the real time step. Instead, it is only applied to calculate the relaxation parameter used in the process of solution reconstruction by LBM. For a control cell with four cell interfaces of a two-dimensional case, the local LBM solution at different interface is reconstructed independently and the streaming time step could be different. This gives us a great flexibility for application of the solver to non-uniform meshes and complex geometry. This feature effectively removes the drawback of coupling issue between the mesh spacing and time step in the conventional LBM. In addition, the LBFS does not track the evolution of distribution function and the dependent variables are the macroscopic flow variable. As a consequence, the required virtual memory is reduced substantially, and the physical boundary conditions can be implemented directly without converting to those for distribution functions. Moreover, as compared with conventional incompressible Navier–Stokes solvers, such as the well-known semi-implicit method for pressure linked equations (SIMPLE) method and its variants [28], the LBFS overcomes their drawbacks of tedious discretization of the second order derivatives, the requirement of staggered grids for preventing pressure oscillations and slow convergence on fine grids [29]. Indeed, the LBFS combines the advantages of lattice Boltzmann solver (simplicity and kinetic nature) with those of Navier–Stokes solver by the finite volume method (FVM) (geometric flexibility and easy implementation of boundary conditions). It shows the progressive development from LBM to LBFS.

Since LBFS is a new solver, its performance has not been fully investigated and further improvements and simplifications for different flow problems may be required. Particularly, there is a lack of a systematic investigation on its accuracy, efficiency, stability and capability in simulating isothermal flows, thermal flows at high Reynolds/Rayleigh numbers, and compressible flows. Motived by this, the LBFS will be refined and examined comprehensively in this paper. Firstly, numerical simulations of two-dimensional lid-driven cavity flows [30–34] are considered. A detailed comparison of its performances with the standard LBM will be carried out. Secondly, the thermal LBFS will be simplified in this work by using the D2Q4 model and then applied to simulate natural convection at two high Rayleigh numbers of $Ra = 10^7$ and 10^8. The obtained results on different grids will be compared with those by MRT-based LBM [35]. Thirdly, the compressible LBFS will be improved by incorporating non-equilibrium effects into the process for inviscid flux reconstruction and introducing a switch function to control numerical diffusion. Its performance will be examined by simulating transonic inviscid and viscous flows around a staggered-biplane configuration on unstructured grids. The obtained results will be compared with the published data.

2. Lattice Boltzmann Method

For a general case, we consider incompressible thermal fluid flows and the LBE with BGK approximation can be written as [34,36]:

$$f_\alpha(\boldsymbol{r}+\boldsymbol{e}_\alpha\delta_t, t+\delta_t) = f_\alpha(\boldsymbol{r},t) + \frac{f_\alpha^{eq}(\boldsymbol{r},t) - f_\alpha(\boldsymbol{r},t)}{\tau_v}, \alpha = 0,1,...,N, \tag{1}$$

$$g_\alpha(\boldsymbol{r}+\boldsymbol{e}_\alpha\delta_t, t+\delta_t) = g_\alpha(\boldsymbol{r},t) + \frac{g_\alpha^{eq}(\boldsymbol{r},t) - g_\alpha(\boldsymbol{r},t)}{\tau_c}, \alpha = 0,1,...,M, \tag{2}$$

where f_α and g_α are density distribution functions (DDF) and temperature distribution functions (TDF) respectively; f_α^{eq} and g_α^{eq} are equilibrium states of f_α and g_α; τ_v and τ_c are relaxation parameters; δ_t is the streaming time; M and N are the total number of discrete particles for f_α and g_α. The macroscopic fluid properties of density ρ, velocity $\boldsymbol{u}$ and temperature T are evaluated from lattice moments of f_α and g_α:

$$\rho = \sum_{\alpha=0}^{N} f_\alpha,\ \rho\boldsymbol{u} = \sum_{\alpha=0}^{N} f_\alpha \boldsymbol{e}_\alpha,\ T = \sum_{\alpha=0}^{M} g_\alpha \tag{3}$$

The equilibrium DDF f_α^{eq} is given by [4]:

$$f_\alpha^{eq}(\boldsymbol{r},t) = \rho w_\alpha \left[1 + \frac{\boldsymbol{e}_\alpha \cdot \boldsymbol{u}}{c_s^2} + \frac{(\boldsymbol{e}_\alpha \cdot \boldsymbol{u})^2 - (c_s|\boldsymbol{u}|)^2}{2c_s^4}\right] \tag{4}$$

The D2Q9 model, whose lattice components are written as $\boldsymbol{e}_\alpha = (0,0), (\pm1,0)$, $(0,\pm1)$ and $(\pm1,\pm1)$, is commonly applied for f_α^{eq}. The weights w_α and the sound speed c_s in Equation (4) are $w_0 = 4/9$, $w_{1\sim4} = 1/9$, $w_{5\sim8} = 1/36$ and $c_s = 1/\sqrt{3}$. The equilibrium TDF g_α^{eq} is given by [37]:

$$g_\alpha^{eq}(\boldsymbol{r},t) = \frac{T}{4}\left[1 + 2\boldsymbol{e}_\alpha \cdot \boldsymbol{u}\right] \tag{5}$$

The D2Q4 lattice velocity model, whose lattice components are written as $\boldsymbol{e}_\alpha = (\pm1,\pm1)$, is usually used for g_α^{eq}. τ_v and τ_c are respectively calculated by the dynamic viscosity μ and thermal diffusivity κ:

$$\mu = \left(\tau_v - \frac{1}{2}\right)\rho c_s^2 \delta_t \tag{6a}$$

$$\kappa = \frac{1}{2}\left(\tau_c - \frac{1}{2}\right)\delta_t \tag{6b}$$

Note that Equations (1) and (2) are applied to simulate thermal flows. If isothermal flows are considered, we can just apply Equation (1). For the application of LBM, we have to apply Equations (1) and (2) in the whole computational domain and track the evolution of distribution functions f_α and g_α. In addition, physical boundary conditions for macroscopic variables are converted to those for f_α and g_α. For no-slip condition, bounce-back scheme can be easily applied. In some cases, accurate implementation of other boundary conditions in the LBM may not be as straightforward as in Navier–Stokes solvers.

3. Chapman–Enskog Analysis

As a mesoscopic method with microscopic particle distribution functions, the LBM described in Section 2 has been well applied to study weakly compressible fluid flows in incompressible limit. Conventionally, such flows are governed and solved by the macroscopic conservation laws of mass, momentum and energy. This indicates that the solution of a physical flow problem can be either obtained by applying the mesoscopic LBM or the macroscopic Navier–Stokes solver. It is interesting to note that these two intrinsically different numerical methods, one from microscopic statistical physics and the other from macroscopic conservation laws, are both able to study the same fluid flow problem. From this point of view, it can be inferred that these two methods should have some connections. Indeed, it has been proven that the lattice Boltzmann equation is able to recover the Navier–Stokes equation through the multi-scale Chapman–Enskog expansion analysis, which is briefly introduced below.

A multi-scale expansion of the DDF, TDF, the temporal derivative and the spatial derivative can be respectively given by:

$$f_\alpha = f_\alpha^{(0)} + \varepsilon f_\alpha^{(1)} + \varepsilon^2 f_\alpha^{(2)} \tag{7a}$$

$$g_\alpha = g_\alpha^{(0)} + \varepsilon g_\alpha^{(1)} + \varepsilon^2 g_\alpha^{(2)} \tag{7b}$$

$$\frac{\partial}{\partial t} = \varepsilon \frac{\partial}{\partial t_0} + \varepsilon^2 \frac{\partial}{\partial t_1} \tag{7c}$$

$$\nabla_r = \varepsilon \nabla_{r1} \tag{7d}$$

where ε is a small parameter proportional to the Knudsen number. Applying the second order Taylor-series expansion to Equations (1) and (2) gives:

$$(\frac{\partial}{\partial t} + e_\alpha \cdot \nabla) f_\alpha + \frac{\delta_t}{2}(\frac{\partial}{\partial t} + e_\alpha \cdot \nabla)^2 f_\alpha + \frac{1}{\tau \delta_t}(f_\alpha - f_\alpha^{eq}) + O(\delta_t^2) = 0, \tag{8}$$

$$(\frac{\partial}{\partial t} + e_\alpha \cdot \nabla) g_\alpha + \frac{\delta_t}{2}(\frac{\partial}{\partial t} + e_\alpha \cdot \nabla)^2 g_\alpha + \frac{1}{\tau \delta_t}(g_\alpha - g_\alpha^{eq}) + O(\delta_t^2) = 0. \tag{9}$$

The macroscopic Navier–Stokes equations for conservation laws can be recovered by substituting Equation (7) into Equations (8) and (9) [9,10]:

$$\frac{\partial \mathbf{W}}{\partial t} + \nabla \cdot \mathbf{F} = 0 \tag{10}$$

where:

$$\mathbf{W} = \begin{Bmatrix} \rho \\ \rho \boldsymbol{u} \\ T \end{Bmatrix} \text{ and } \mathbf{F} = \begin{Bmatrix} \boldsymbol{P} \\ \boldsymbol{\Pi} \\ \boldsymbol{Q} \end{Bmatrix} \tag{11}$$

In Equation (11), $\boldsymbol{P}$, $\boldsymbol{\Pi}$ and $\boldsymbol{Q}$ are respectively mass, momentum and energy fluxes and can be given by lattice summations of DDFs and TDFs [9,10]:

$$\boldsymbol{P} = \rho \boldsymbol{u} = \sum_{\alpha=0}^{N} f_{\alpha}^{eq} \boldsymbol{e}_{\alpha} \tag{12a}$$

$$\boldsymbol{\Pi} = \rho u \otimes u + p\mathbf{I} - \mu \left(\nabla u + (\nabla u)^{T} \right) = \sum_{\alpha=0}^{N} e_{\alpha} e_{\alpha} \hat{f}_{\alpha} \tag{12b}$$

$$\boldsymbol{Q} = \boldsymbol{u}T - \kappa \nabla T = \sum_{\alpha=0}^{N} e_{\alpha} \hat{g}_{\alpha} \tag{12c}$$

where:

$$\hat{f}_{\alpha} = f_{\alpha}^{eq} + \left(1 - \frac{1}{2\tau_{v}}\right) f_{\alpha}^{neq} \text{ and } \hat{g}_{\alpha} = g_{\alpha}^{eq} + \left(1 - \frac{1}{2\tau_{c}}\right) g_{\alpha}^{neq} \tag{13}$$

$$f_{\alpha}^{neq} = -\tau_{v} \delta_{t} \left(\frac{\partial}{\partial t} + \boldsymbol{e}_{\alpha} \cdot \nabla \right) f_{\alpha}^{eq} \tag{14a}$$

$$g_{\alpha}^{neq} = -\tau_{c} \delta_{t} \left(\frac{\partial}{\partial t} + \boldsymbol{e}_{\alpha} \cdot \nabla \right) g_{\alpha}^{eq} \tag{14b}$$

In the conventional LBM, the Chapman–Enskog (C-E) analysis is usually used to verify that the flow variables obtained by LBM can satisfy the macroscopic Navier–Stokes equations. In some interesting works, this C-E analysis is also applied to build a hybrid solver by combing the LBM with the FVM scheme for Navier–Stokes equations [37–39]. In our recent work [9,10], it was found that the relationship between flow variables/fluxes and particle distribution functions given in Equation (12) can be applied to build a new solver named LBFS with a better performance, which effectively combines the advantages of Navier–Stokes solver and lattice Boltzmann solver and at the same time removes some of their drawbacks. The reliability and flexibility of the LBFS have been demonstrated in [8–10].

4. LBFS for Isothermal and Thermal Incompressible Flows

Based on the LBM given by Equations (1) and (2), the LBFS [9] has been proposed by solving Equation (10) directly. Unlike conventional Navier–Stokes solver, which applies well-established schemes to discretize Equation (10), the LBFS is a finite volume solver and reconstructs its fluxes locally with Equation (12) derived from Chapman–Enskog analysis. It may be pointed out that Equation (12) for flux reconstruction can be applied with different lattice velocity models. In our previous work [9], the D2Q9 thermal model for temperature field is applied, which is obviously complex and inefficient as compared with the D2Q4 model [34]. In this work, this drawback will be removed by applying the D2Q4 model. The details are given below.

4.1. Finite Volume Discretization

A finite volume discretization of Equation (10) over a control volume Ω_i gives the following formulation:

$$\frac{d\mathbf{W}_i}{dt} = -\frac{1}{dV_i}\sum_k \mathbf{R}_k dl_k,\ \mathbf{R}_k = (\boldsymbol{n} \cdot \mathbf{F})_k \tag{15}$$

where dV_i is the area of the control cell Ω_i, dl_k is the length of the k-th control surface enclosed Ω_i and $\boldsymbol{n} = (n_x, n_y)$ is the unit normal vector on the k-th control surface. With the D2Q9 and D2Q4 models, the flux $\mathbf{R}_k$ at each cell interface can be written as follows:

$$\mathbf{R}_k = \begin{pmatrix} n_x\left(f_1^{eq} - f_3^{eq} + f_5^{eq} - f_6^{eq} - f_7^{eq} + f_8^{eq}\right) + n_y\left(f_2^{eq} - f_4^{eq} + f_5^{eq} + f_6^{eq} - f_7^{eq} - f_8^{eq}\right) \\ n_x\left(\hat{f}_1+\hat{f}_3+\hat{f}_5+\hat{f}_6+\hat{f}_7+\hat{f}_8\right) + n_y\left(\hat{f}_5 - \hat{f}_6+\hat{f}_7 - \hat{f}_8\right) \\ n_x\left(\hat{f}_5 - \hat{f}_6+\hat{f}_7 - \hat{f}_8\right) + n_y\left(\hat{f}_2+\hat{f}_4+\hat{f}_5+\hat{f}_6+\hat{f}_7+\hat{f}_8\right) \\ n_x\left(\hat{g}_1 - \hat{g}_3\right) + n_y\left(\hat{g}_2 - \hat{g}_4\right) \end{pmatrix}_k \tag{16}$$

As can be seen, the formulation of the energy flux reconstructed by the simplified D2Q4 lattice velocity model is much simpler than that given by the original D2Q9 model [9]. In Equation (16), the unknowns are $\hat{f}_\alpha$ and $\hat{g}_\alpha$ defined in Equation (13), which include both equilibrium and non-equilibrium density and temperature distribution functions f_α^{eq}, f_α^{neq}, g_α^{eq} and g_α^{neq}. All these unknowns can be obtained through local reconstruction of the LBE solution without tracking the evolution of the DDF and TDF.

4.2. Local Reconstruction of Fluxes at Each Interface

Figure 1 shows an interface between two adjacent control cells for local reconstruction of fluxes [10], in which the D2Q9 and D2Q4 lattice velocity models respectively for DDF and TDF are embedded. The non-equilibrium parts for DDF and

TDF are given by Equation (14). At each cell interface, discretization of Equation (14) with the second-order Taylor-series expansion gives the following equations [9]:

$$f_\alpha^{neq}(\boldsymbol{r},t) = -\tau_\nu \left[f_\alpha^{eq}(\boldsymbol{r},t) - f_\alpha^{eq}(\boldsymbol{r} - \boldsymbol{e}_\alpha \delta_t, t - \delta_t) \right] \tag{17a}$$

$$g_\alpha^{neq}(\boldsymbol{r},t) = -\tau_c \left[g_\alpha^{eq}(\boldsymbol{r},t) - g_\alpha^{eq}(\boldsymbol{r} - \boldsymbol{e}_\alpha \delta_t, t - \delta_t) \right] \tag{17b}$$

where $\boldsymbol{r}$ represents the location of the cell interface. Equation (17) shows that both f_α^{neq} and g_α^{neq} can be approximated from f_α^{eq} and g_α^{eq} at the interface and its surrounding points $\boldsymbol{r} - \boldsymbol{e}_\alpha \delta_t$. Following the convention in LBM, f_α^{eq} and g_α^{eq} at the position and time $(\boldsymbol{r} - \boldsymbol{e}_\alpha \delta_t, t - \delta_t)$ are computed by using Equations (4) and (5). The involved flow quantities of density ρ, velocity $\boldsymbol{u}$ and temperature T at $(\boldsymbol{r} - \boldsymbol{e}_\alpha \delta_t, t - \delta_t)$ can be obtained through interpolations:

$$\psi(\boldsymbol{r} - \boldsymbol{e}_\alpha \delta_t, t - \delta_t) = \begin{cases} \psi(\boldsymbol{r}_i) + (\boldsymbol{r} - \boldsymbol{e}_\alpha \delta_t - \boldsymbol{r}_i)\nabla\psi(\boldsymbol{r}_i), & \text{when } \boldsymbol{r} - \boldsymbol{e}_\alpha \delta_t \text{ is in } \Omega_i \\ \psi(\boldsymbol{r}_{i+1}) + (\boldsymbol{r} - \boldsymbol{e}_\alpha \delta_t - \boldsymbol{r}_{i+1})\nabla\psi(\boldsymbol{r}_{i+1}), & \text{when } \boldsymbol{r} - \boldsymbol{e}_\alpha \delta_t \text{ is in } \Omega_{i+1} \end{cases} \tag{18}$$

where $\boldsymbol{r}_i$ and $\boldsymbol{r}_{i+1}$ are the locations of the two cell centers and ψ represents any of the flow variables. After interpolation, $f_\alpha^{eq}(\boldsymbol{r} - \boldsymbol{e}_\alpha \delta_t, t - \delta_t)$ and $g_\alpha^{eq}(\boldsymbol{r} - \boldsymbol{e}_\alpha \delta_t, t - \delta_t)$ can be obtained.

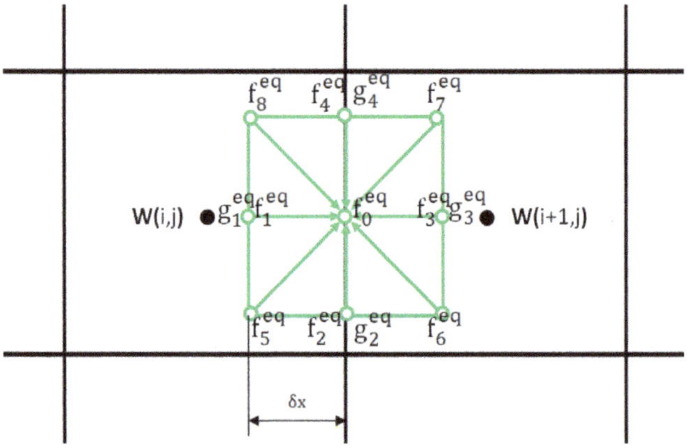

Figure 1. Evaluation of flux at an interface between two control cells.

Similarly, $f_\alpha^{eq}(\boldsymbol{r},t)$ and $g_\alpha^{eq}(\boldsymbol{r},t)$ can be calculated by Equations (4) and (5) if macroscopic flow variables are known. So, the challenging issue is to evaluate ρ, $\boldsymbol{u}$ and T at the cell interface $(\boldsymbol{r},t)$. Previous studies [9,10] have shown that they can be reconstructed locally by streaming the particle from the neighboring point $(\boldsymbol{r} - \boldsymbol{e}_\alpha \delta_t, t - \delta_t)$ to the cell interface. As a result, we have:

$$\rho(\boldsymbol{r}, t) = \sum_{\alpha=0}^{N} f_\alpha^{eq}(\boldsymbol{r} - \boldsymbol{e}_\alpha \delta_t, t - \delta_t), \tag{19a}$$

$$\rho(\boldsymbol{r}, t)\boldsymbol{u}(\boldsymbol{r}, t) = \sum_{\alpha=0}^{N} f_\alpha^{eq}(\boldsymbol{r} - \boldsymbol{e}_\alpha \delta_t, t - \delta_t)\boldsymbol{e}_\alpha \tag{19b}$$

$$T(\boldsymbol{r}, t) = \sum_{\alpha=0}^{N} g_\alpha^{eq}(\boldsymbol{r} - \boldsymbol{e}_\alpha \delta_t, t - \delta_t) \tag{19c}$$

With the flow quantities obtained from Equation (19), $f_\alpha^{eq}(\boldsymbol{r}, t)$ and $g_\alpha^{eq}(\boldsymbol{r}, t)$ can be computed by using Equations (4) and (5). Once f_α^{eq}, f_α^{neq}, g_α^{eq} and g_α^{neq} are obtained, the fluxes at each cell interface can be evaluated numerically. Then Equation (15) can be solved by conventional Runge–Kutta scheme.

5. LBFS for Compressible Flows

5.1. Navier–Stokes Equations Discretized by FVM and Compressible Lattice Boltzmann Model

In Section 4, the local LBM solution is reconstructed by using the conventional lattice Boltzmann model, which is limited to incompressible flows. Thus, LBFS in Section 4 can only be applied to simulate incompressible flows. To simulate compressible flows by LBFS, we need to use the compressible lattice Boltzmann model to reconstruct the solution at the cell interface. However, the existing compressible lattice Boltzmann models are very complicated and inefficient for 2D and 3D cases. To simplify the solution process and make LBFS be applicable for simulation of compressible flows, we apply the 1D compressible lattice Boltzmann model along the normal direction of cell interface to evaluate the inviscid flux, and the viscous flux is still approximated by conventional finite difference schemes. This process is equivalent to developing a Riemann solver by 1D compressible lattice Boltzmann model.

In LBFS, the Navier–Stokes equations are discretized by FVM and the fluxes at the cell interface are evaluated by local reconstruction of LBM solutions [11]. Thanks to the application of FVM, it is convenient for LBFS to apply on arbitrary meshes. The compressible Navier–Stokes equations discretized by FVM can be written as:

$$\frac{d\mathbf{W}_i}{dt} = -\frac{1}{dV_i}\sum_{k=1}^{N_f} (\mathbf{F}_{ck} - \mathbf{F}_{vk})\, dl_k \tag{20}$$

where i is the index of a control volume, dV_i and N_f represent the volume and the number of interfaces of the control volume i, respectively. $\mathbf{W}$, $\mathbf{F}_c$ and $\mathbf{F}_v$ are the conservative variables at the cell center and the inviscid and viscous fluxes at the cell interface given by:

$$\mathbf{W} = \begin{bmatrix} \rho \\ \rho u \\ \rho v \\ \rho E \end{bmatrix}, \mathbf{F}_c = \begin{bmatrix} \rho U_n \\ \rho u U_n + n_x p \\ \rho v U_n + n_y p \\ (\rho E + p) U_n \end{bmatrix}, \mathbf{F}_v = \begin{bmatrix} 0 \\ n_x \tau_{xx} + n_y \tau_{xy} \\ n_x \tau_{yx} + n_y \tau_{yy} \\ n_x \Theta_x + n_y \Theta_y \end{bmatrix} \tag{21}$$

Here, ρ and p are the density and pressure of mean flow, respectively. (u, v) and (n_x, n_y) denote the velocity vector and unit normal vector on the control surface in the Cartesian coordinate system, respectively. E is the total energy of mean flow. U_n represents the normal velocity. Furthermore, τ_{ij} denotes the components of viscous stress tensor and Θ_i represents the term describing the work of viscous stress and the heat conduction in the fluid.

To compute $\mathbf{F}_c$ by LBFS, the compressible lattice Boltzmann model is required. In this work, the non-free parameter D1Q4 model [11] is utilized. The configuration of this model is shown in Figure 2, and it is given by:

$$\begin{aligned} g_1 &= \frac{\rho\left(-d_1 d_2^2 - d_2^2 u + d_1 u^2 + d_1 c^2 + u^3 + 3uc^2\right)}{2d_1\left(d_1^2 - d_2^2\right)} \\ g_2 &= \frac{\rho\left(-d_1 d_2^2 + d_2^2 u + d_1 u^2 + d_1 c^2 - u^3 - 3uc^2\right)}{2d_1\left(d_1^2 - d_2^2\right)} \\ g_3 &= \frac{\rho\left(d_1^2 d_2 + d_1^2 u - d_2 u^2 - d_2 c^2 - u^3 - 3uc^2\right)}{2d_2\left(d_1^2 - d_2^2\right)} \\ g_4 &= \frac{\rho\left(d_1^2 d_2 - d_1^2 u - d_2 u^2 - d_2 c^2 + u^3 + 3uc^2\right)}{2d_2\left(d_1^2 - d_2^2\right)} \end{aligned} \tag{22}$$

$$\begin{aligned} d_1 &= \sqrt{u^2 + 3c^2 - \sqrt{4u^2c^2 + 6c^4}} \\ d_2 &= \sqrt{u^2 + 3c^2 + \sqrt{4u^2c^2 + 6c^4}} \end{aligned} \tag{23}$$

where c represents the peculiar velocity of particles defined as $c = \sqrt{p/\rho}$. Note that in Equation (22), g is the equilibrium distribution function f^{eq}. As D1Q4 model is a one-dimensional model, it needs to be applied along the normal direction of cell interface for multi-dimensional problems [11]. That is, the velocity u in Equations (22) and (23) should be replaced by normal velocity U_n when multi-dimensional problems are considered.

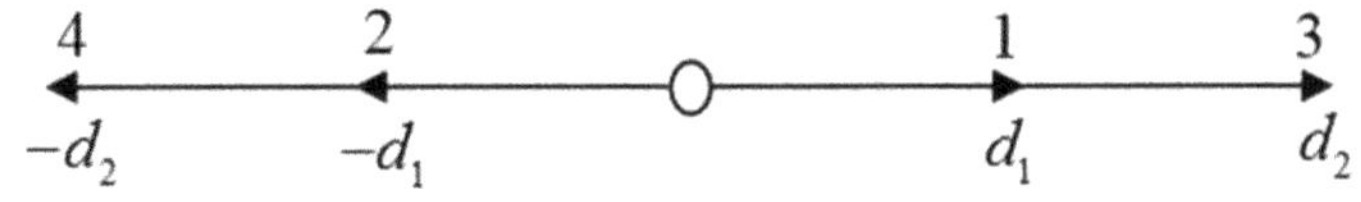

Figure 2. Configuration of non-free parameter D1Q4 model.

Suppose that the cell interface is located at $\mathbf{r} = 0$. For simplicity, a local-coordinate system with x-axis pointing to the normal direction and y-axis pointing to the tangential direction of the cell interface is used. As a result, the inviscid flux at the cell interface in the local-coordinate system can be written as:

$$\mathbf{F}_c^* = \begin{bmatrix} \rho U_n & \rho U_n U_n + p & \rho U_n U_\tau & (\rho E + p) U_n \end{bmatrix}^T = \mathbf{F}_c^n + \mathbf{F}_c^t \quad (24a)$$

$$\mathbf{F}_c^n = \begin{bmatrix} \rho U_n & \rho U_n U_n + p & 0 & \left(\rho\left(\frac{1}{2}U_n^2 + e\right) + p\right) U_n \end{bmatrix}^T \quad (24b)$$

$$\mathbf{F}_c^t = \begin{bmatrix} 0 & 0 & \rho U_n U_\tau & \frac{1}{2}\rho U_\tau^2 U_n \end{bmatrix}^T \quad (24c)$$

Here, $\mathbf{F}_c^n$ and $\mathbf{F}_c^t$ are respectively the flux attributed to the normal velocity and tangential velocity, e is the potential energy of mean flow. From Chapman–Enskog analysis, and with the use of non-free parameter D1Q4 model, $\mathbf{F}_c^n$ can be computed by:

$$\mathbf{F}_c^n = \sum_{i=1}^{4} \xi_i \boldsymbol{\varphi}_i f_i(0,t) \quad (25)$$

where:

$$f_i(0,t) = f^{eq}(0,t) + f^{neq}(0,t) = g_i(0,t) - \tau_0\left[g_i(0,t) - g_i(-\xi_i\delta t, t - \delta t)\right], \quad (26)$$

ξ_i is the particle velocity in the i-direction, *i.e.*, $\xi_1 = d_1$, $\xi_2 = -d_1$, $\xi_3 = d_2$ and $\xi_4 = -d_2$. $\boldsymbol{\varphi}_i$ stands for the moments:

$$\boldsymbol{\varphi}_i = \left(1, \xi_i, 0, \frac{1}{2}\xi_i^2 + e_p\right)^T \quad (27)$$

Here, e_p is the potential energy of particles; $f_i(0,t)$ is the distribution function at the cell interface. $g_i(0,t)$ and $g_i(-\xi_i\delta t, t-\delta t)$ are respectively the equilibrium distribution function at the cell interface and the surrounding point of the cell interface. The non-equilibrium part in Equation (26) is viewed as the numerical dissipation since only the inviscid flux is computed by LBFS. τ_0 is the dimensionless collision time, which can be regarded as the weight of the numerical dissipation [40,41]. Substituting Equation (26) into Equation (25), we have the final expression of $\mathbf{F}_c^n$ as follows:

$$\mathbf{F}_c^n = (1-\tau_0)\,\mathbf{F}_c^{n,I} + \tau_0\mathbf{F}_c^{n,II} \tag{28}$$

where:

$$\mathbf{F}_c^{n,I} = \sum_{i=1}^{4} \xi_i \boldsymbol{\varphi}_a g_i\,(0,t) \text{ and } \mathbf{F}_c^{n,II} = \sum_{i=1}^{4} \xi_i \boldsymbol{\varphi}_a g_i\,(-\xi_i \delta t, t-\delta t) \tag{29}$$

To evaluate the flux attributed to the tangential velocity, one of the feasible ways can be expressed as:

$$\mathbf{F}_c^t(3) = \sum_{i=1,3} \xi_i f_i\,(0,t)\,U_\tau^L + \sum_{i=2,4} \xi_i f_i\,(0,t)\,U_\tau^R \tag{30a}$$

$$\mathbf{F}_c^t(4) = \frac{1}{2}\left[\sum_{i=1,3} \xi_i f_i\,(0,t)\left(U_\tau^L\right)^2 + \sum_{i=2,4} \xi_i f_i\,(0,t)\left(U_\tau^R\right)^2\right] \tag{30b}$$

where, U_τ^L and U_τ^R are the tangential velocity at the left and right side of cell interface, respectively. From $\mathbf{F}_c^*$, the actual inviscid flux $\mathbf{F}_c$ in Equation (20) in the Cartesian coordinate system can be obtained by a transformation [41]:

$$\mathbf{F}_c = \left[\begin{array}{cccc} F_c^*(1) & F_c^*(2)n_x - F_c^*(3)n_y & F_c^*(2)n_y + F_c^*(3)n_x & F_c^*(4) \end{array}\right] \tag{31}$$

Equation (31) forms the LBFS for inviscid compressible flows. It can be treated as a Riemann solver developed by 1D compressible lattice Boltzmann model. As for the viscous flux $\mathbf{F}_v$, it is still approximated by conventional finite difference schemes in this work.

5.2. Evaluation of Inviscid Flux by LBFS

To compute $\mathbf{F}_c^{n,I}$, the equilibrium distribution function at the cell interface should be computed in advance. As the equilibrium distribution function is the function of conservative variables, we just need to calculate the conservative variables at the cell interface first. The conservative variables at the cell interface in the local-coordinate system can be expressed as:

$$\mathbf{W}^* = \left[\begin{array}{cccc} \rho & \rho U_n & \rho U_\tau & \rho E \end{array}\right]^T = \mathbf{W}^n + \mathbf{W}^t \tag{32a}$$

$$\mathbf{W}^n = \left[\begin{array}{cccc} \rho & \rho U_n & 0 & \rho\left(\frac{1}{2}U_n^2 + e\right) \end{array}\right]^T \tag{32b}$$

$$\mathbf{W}^t = \left[\begin{array}{cccc} 0 & 0 & \rho U_\tau & \frac{1}{2}\rho U_\tau^2 \end{array}\right]^T \tag{32c}$$

Like Equation (19), the conservative variables at the cell interface attributed to the normal velocity can be evaluated by:

$$\mathbf{W}^n = \sum_{i=1}^{4} \varphi_i g_i\left(-\xi_i \delta t, t - \delta t\right) \tag{33}$$

It is assumed that a local Riemann problem is formed at the cell interface. Then, the equilibrium distribution function $g_i\left(-\xi_i \delta t, t - \delta t\right)$ can be given by:

$$g_i\left(-\xi_i \delta t, t - \delta t\right) = \begin{cases} g_i^L & \text{if } i = 1,3 \\ g_i^R & \text{if } i = 2,4 \end{cases} \tag{34}$$

where g_i^L and g_i^R are the equilibrium distribution functions at the left and right sides of cell interface as shown in Figure 3. In addition, the conservative variables attributed to the tangential velocity can be approximated by:

$$\mathbf{W}^t(3) = \sum_{i=1,3} g_i\left(-\xi_i \delta t, t - \delta t\right) U_\tau^L + \sum_{i=2,4} g_i\left(-\xi_i \delta t, t - \delta t\right) U_\tau^R \tag{35a}$$

$$\mathbf{W}^t(4) = \frac{1}{2}\left[\sum_{i=1,3} g_i\left(-\xi_i \delta t, t - \delta t\right)\left(U_\tau^L\right)^2 + \sum_{i=2,4} g_i\left(-\xi_i \delta t, t - \delta t\right)\left(U_\tau^R\right)^2\right] \tag{35b}$$

Once the conservative variables at the cell interface $\mathbf{W}^*$ are obtained, the flux attributed to the equilibrium distribution function at the cell interface $\mathbf{F}_c^{n,I}$ can be calculated by substituting the above conservative variables directly into the expression of inviscid flux. As for the flux $\mathbf{F}_c^{n,II}$, it can be obtained by substituting Equation (34) into Equation (29) directly.

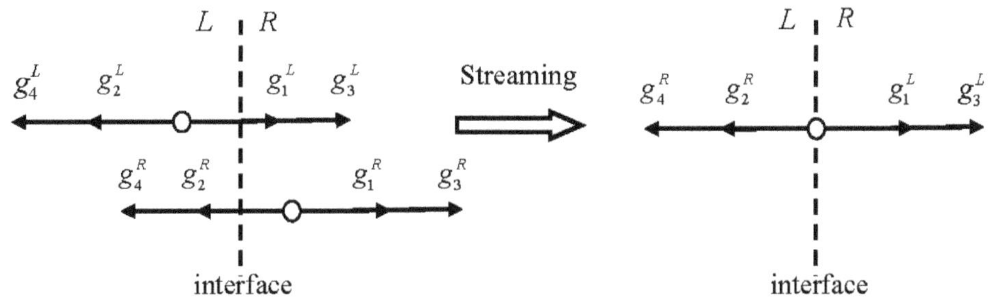

Figure 3. Streaming process of D1Q4 model at the cell interface.

6. Numerical Examples and Discussion

In this section, several benchmark cases will be studied to examine the performance of newly-developed LBFS. In particular, lid-driven cavity flows and steady natural convection in a square cavity at high Rayleigh numbers of 10^7 and 10^8 will be simulated on non-uniform grids by the incompressible LBFS. Transonic flows around a staggered-biplane configuration will be studied on unstructured grids by using the compressible LBFS.

6.1. Lid-Driven Cavity Flows

As a benchmark problem, the classical lid-driven cavity flow has been studied extensively by many researchers [30–34]. This problem involves several geometrical and flow parameters: the length of the cavity L, the velocity of the lid U and the density and dynamic viscosity of the fluid ρ and μ. With these parameters, the Reynolds number is defined as $\text{Re} = \rho UL/\mu$. In present study, this problem will be solved by using the LBFS and the standard LBM to compare their accuracy, stability and efficiency in detail.

At first, the accuracy of the LBFS and LBM are examined by comparing the pressure and velocity profiles for lid-driven cavity flows at Re = 5000. In the computation, a non-uniform grid of 101×101 for the LBFS and uniform grid of 301×301 for LBM are applied. Note that much finer grid is required for the LBM to maintain numerical stability at this Reynolds number. Figure 4 compares the pressure and velocity profiles obtained by these two methods. It can be seen that excellent agreements have been achieved. Note that the LBFS uses about one-ninth of the total grid points adopted by the LBM, which shows its advantage and capability in applying non-uniform grids. It may also be noted that both LBFS and LBM have the second order of accuracy and quantitatively similar solutions will be obtained when the same grids are applied [8–10].

After that, the pure stability of the LBFS and LBM is investigated without considering numerical accuracy. This test is conducted by simulating the lid-driven cavity flows at Reynolds numbers from 100 to 5000. The minimum grids to get stable solution required respectively by the LBFS and LBM are recorded and shown in Table 1. It can be seen that, with the increase of Reynolds number, the LBM requires more grid points to maintain stability while the LBFS only needs a grid size of 4×4 for all cases considered. This feature indicates that the LBFS is more stable than LBM. In addition, the stability of these two methods can be further examined by comparing numerical solutions for pressure. Figure 5 compares the pressure contours of lid-driven cavity flows at Reynolds number of 5000. A grid size of 101×101 is applied by the LBFS and 301×301 is applied by the LBM. It is obvious that the results of LBM for pressure field have substantial unphysical oscillations at the top left, top right and bottom right corners while those of the LBFS are smooth all over the flow domain. This phenomenon indicates that the stability of LBFS is superior to that of the LBM.

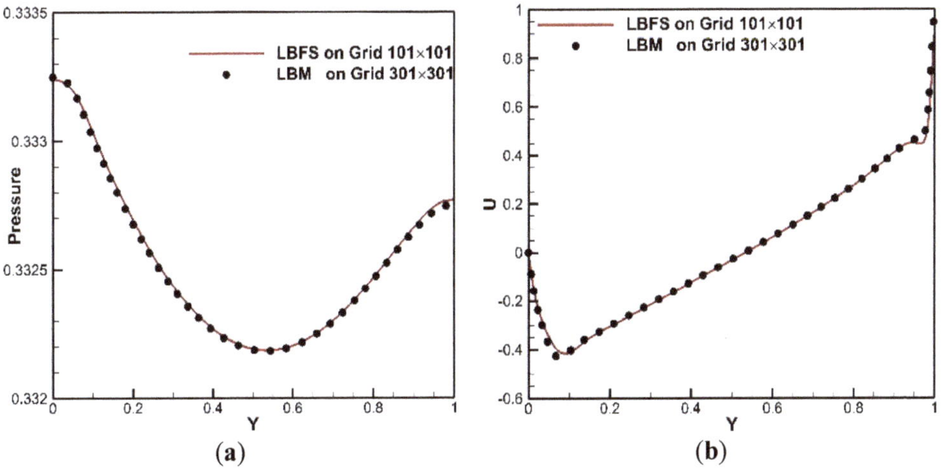

Figure 4. Comparison of pressure and velocity profiles on the vertical centerline for lid-driven cavity flows at Re = 5000. (**a**) Pressure; (**b**) Velocity.

Table 1. Minimum required mesh resolution of the lattice Boltzmann flux solver (LBFS) and lattice Boltzmann method (LBM) for lid-driven cavity flows at different Reynolds numbers.

Re	100	1000	5000	7500
LBGK	5 × 5	39 × 39	219 × 219	341 × 341
LBFS	4 × 4	4 × 4	4 × 4	4 × 4

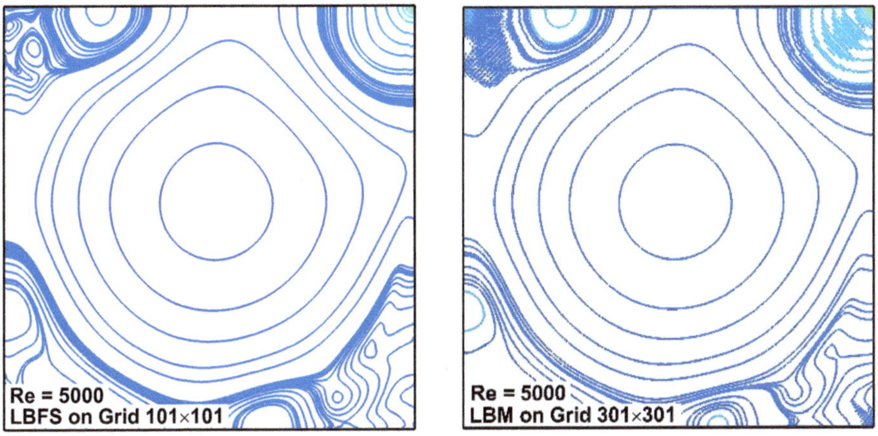

(**a**) LBFS (**b**) LBM

Figure 5. Comparison of pressure contours for lid-driven cavity flows at Re = 5000. (**a**) lattice Boltzmann flux solver (LBFS); (**b**) lattice Boltzmann method (LBM).

Moreover, the efficiency of the LBFS and LBM are investigated by considering two cases. The first one is their efficiency on the same uniform grid of 301×301 by simulating the lid-driven cavity flow at Re = 5000. The other case is to examine the computational effort of the LBFS on non-uniform grids when its results agree well with those of the LBM. Table 2 compares the CPU time consumed by each solver in these two cases. As can be seen, when applied on the same grid, the LBFS takes about 3.8 times the computational time and 47% of the virtual memory that are required by the LBM. This is mainly attributed to the fact that, in the LBFS, interpolations of physical quantities are performed at each cell for flux reconstruction, which degrades its efficiency. On the other hand, when the solution of the LBFS compares well with that of LBM as shown in Figure 4, the LBFS only needs a non-uniform grid of 101×101. As compared with those of LBM, only about 16.6% of computational time and 5.3% of virtual memory are consumed by the LBFS. This indicates that the LBFS needs less computational resources and can be more efficient for applications on non-uniform grids.

Table 2. Comparison of performance of the lattice Boltzmann flux solver (LBFS) and lattice Boltzmann method (LBM) on different grids for 2D lid-driven cavity flows at Re = 5000.

Object Type	Method	Min.	Max.	Avg.	Std. Dev.
	V2SFCA	5	67	25.53	13.95
	EV2SFCA	5	118	27.71	23.88
Population	V2SFCA	5	63	23.20	15.08
	EV2SFCA	5	70	25.49	17.54

6.2. Natural Convection in a Square Cavity at High Rayleigh Number

Natural convection in a square cavity is a benchmark case for validating various numerical methods [35,37,42]. For instance, Peng *et al.* [42] simulated this problem at Rayleigh numbers from $Ra = 10^3$ to 10^6 to validate their simplified SRT-based thermal LB model. Guo *et al.* [37] studied this problem to examine their D2Q4 thermal LB model. Contrino *et al.* [35] validated their MRT-based thermal LB model by simulating this flow at high Rayleigh numbers of 10^7 and 10^8. Recently, with SRT-based D2Q9 thermal LB model, the LBFS [9] has also been validated through its application to simulate this problem from $Ra = 10^3$ to 10^6. It is noticed that, to study this problem, all versions of LBM restrict their computations on uniform grids. However, non-uniform grids may be more accurate and efficient to capture thin boundary layers, especially for flows at high Rayleigh numbers. This provides a good chance to examine the performance of the LBFS with D2Q4 thermal LB model.

The flow pattern of this problem is characterized by two normalized parameters: the Prandtl number Pr and the Rayleigh number Ra, which can be respectively defined as:

$$Pr = \frac{\nu}{\chi}, \ Ra = \frac{g\beta \cdot \Delta T \cdot L^3}{\nu \cdot \chi} = \frac{V_c^2 \cdot L^2}{\nu \cdot \chi} \tag{36}$$

where L is the length of the square cavity, ΔT is the temperature difference between the hot and cold walls and $V_c = \sqrt{g\beta L \cdot \Delta T}$ is the characteristic thermal velocity. The flow parameters are set as follows: $L = 1$, $V_c = 0.1$, $Pr = 0.71$. Two high Rayleigh numbers of $Ra = 10^7$ and 10^8 are considered. Non-uniform grids with different grid sizes (201×201 and 301×301 for $Ra = 10^7$; 301×301 and 401×401 for $Ra = 10^8$) are applied. To quantify the result, the mean Nusselt number $Nu_{1/2}$ along the line of $x = L/2$ is computed and compared:

$$Nu_{1/2} = \frac{L}{\chi \cdot \Delta T} \cdot \frac{1}{L} \oint_{x=L/2} \left(uT - \chi \frac{\partial T}{\partial x} \right) dl \tag{37}$$

Tables 3 and 4 show the maximum absolute value of the stream-function $|\varphi|_{\max}$ and its position, the mean Nusselt number $Nu_{1/2}$ along the vertical centerline, the maximum u-velocity $U_{\max}$ along $x = L/2$ and its vertical position, the maximum v-velocity $V_{\max}$ along $y = L/2$ and its horizontal position. The numerical results of Contrino *et al.* [35] obtained by the MRT-LBM and those of Quere [43] obtained by a high order pseudo-spectral method are also included for comparison. With the increase of the grid size, the present solution is closer to the benchmark solution of Quere [43] for all cases considered. In particular, the relative error of $|\varphi|_{\max}$ between the present solution on the finest grid and those of Quere [43] is within 0.26% and that of $Nu_{1/2}$ is within 0.08%. This indicates that both the strength of the flow field represented by the stream-function and the heat transfer rate represented by the mean Nusselt number are well predicted by the LBFS. Figures 6 and 7 show the streamlines and isotherms at Rayleigh numbers of 10^7 and 10^8 respectively. At $Ra = 10^7$, a large clock-wise recirculation is formed and attached to all walls. Both the flow and temperature boundary layers close to the hot and cold walls are very thin. The temperature at the same height of the cavity is almost a constant near central area. As Ra is increased to 10^8, the boundary layer separates near the bottom left and top right region. Figure 8 shows the u-velocity along the vertical centerline and v-velocity along the horizontal centerline. As can be seen, the velocity component v is almost zero in a large zone except for those near the wall. This indicates that vertical convection in the central area can be very weak and heat conduction dominates this region. This is consistent with the observation that the temperature on the same latitude is almost a constant in the central area of the cavity.

Table 3. Comparison of representative quantities for natural convection in a square cavity at $Ra = 10^7$.

Grid Size	Present		Contrino *et al.* [35]			Quere [43]
	201^2	301^2	379^2	1019^2	1531^2	
$\lvert\varphi\rvert_{max}$	30.165	30.164	30.349	30.310	30.185	30.165
x	0.0868	0.0857	0.0848	0.0856	0.0857	0.86
y	0.5545	0.5559	0.5578	0.5562	0.5559	0.556
$Nu_{1/2}$	16.550	16.543	16.526	16.523	16.523	16.52
U_{max}	148.17	148.84	148.48	148.57	148.58	148.59
y	0.8788	0.8789	0.8794	0.8793	0.8793	0.879
V_{max}	699.19	699.91	699.11	699.27	699.31	699.18
x	0.0204	0.0216	0.0214	0.0213	0.0213	0.021

Table 4. Comparison of representative quantities for natural convection in a square cavity at $Ra = 10^8$.

Grid Size	Present		Contrino *et al.* [35]			Quere [43]
	301	401^2	379^2	1019^2	1531^2	
$\lvert\varphi\rvert_{max}$	53.955	53.893	54.870	54.106	53.953	53.85
x	0.0482	0.4760	0.0469	0.0478	0.0480	0.048
y	0.5536	0.5528	0.5594	0.5545	0.5533	0.553
$Nu_{1/2}$	30.353	30.301	30.257	30.229	30.227	30.225
U_{max}	316.07	323.65	315.08	320.74	321.37	321.9
x	0.9267	0.9288	0.9239	0.9273	0.9276	0.928
V_{max}	2221.1	2222.9	2221.4	2222.1	2222.3	2222
y	0.0118	0.1192	0.0121	0.0120	0.0120	0.012

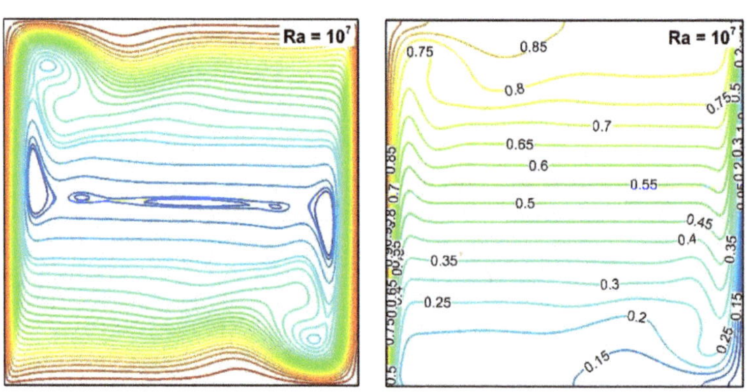

(a) **(b)**

Figure 6. Streamlines and isotherms for natural convection at $Ra = 10^7$. (**a**) Streamlines; (**b**) Isotherms.

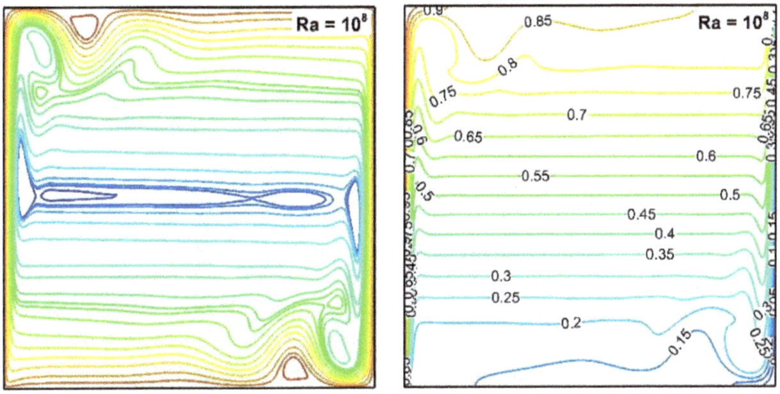

(a) (b)

Figure 7. Streamlines and isotherms for natural convection at $Ra = 10^8$. (**a**) Streamlines; (**b**) Isotherms.

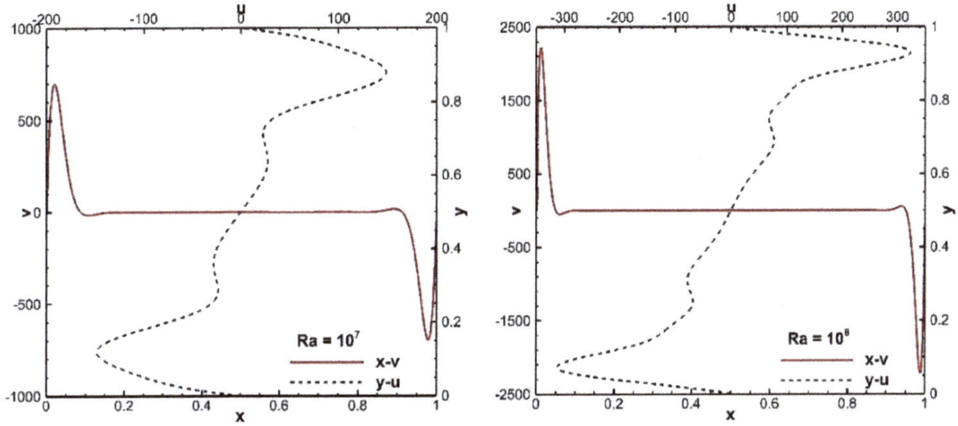

(a) (b)

Figure 8. Illustration of the velocity profiles at the centerlines for natural convection at $Ra = 10^7$ and 10^8. (**a**) $Ra = 10^7$; (**b**) $Ra = 10^8$.

6.3. Inviscid and Viscous Transonic Flows Around a Staggered-Biplane Configuration

To validate the compressible LBFS for simulation of flows with complex geometry, the inviscid and viscous transonic flows around a staggered-biplane configuration are simulated. This test example is taken from the work of Jawahar and Kamath [44]. It comprises two NACA0012 airfoils, staggered by half a chord length in the pitchwise as well as chordwise directions. At first, the inviscid flow with the free-stream Mach number of 0.7 and the angle of attack of 0 degree is simulated. In the test, the unstructured grid containing 256 points on each airfoil and 36,727 triangular cells in the computational domain is utilized, and its partial view is shown in Figure 9a. The pressure contours obtained from present scheme are shown in Figure 9b. It can be seen from the figure that, a strong normal shock is formed between two airfoils and near the trailing edge of bottom airfoil, which is in line with those observed by Jawahar and Kamath [44] and Lerat and Wu [45]. Figure 10 shows the pressure coefficient distribution on the airfoil surface computed by present scheme. Also displayed in this figure are the results of Jawahar and Kamath [44] and Lerat and Wu [45]. Clearly, the results of current scheme agree well with the published data.

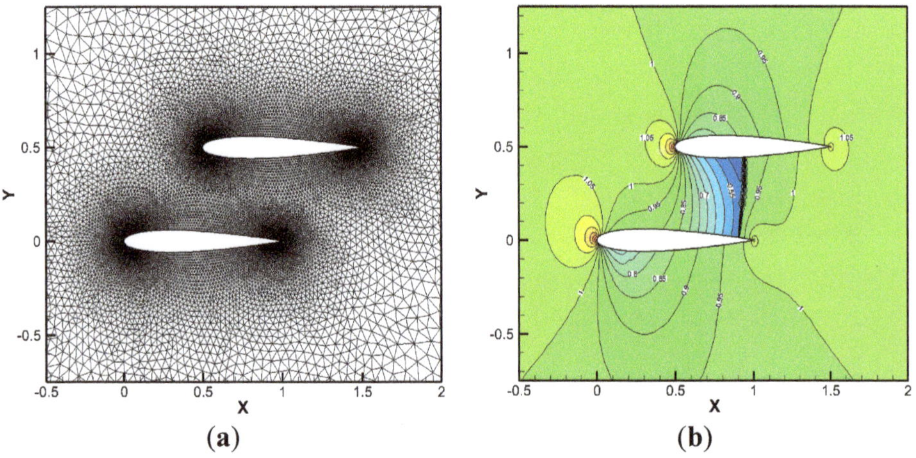

Figure 9. Partial view of computational mesh and pressure contours for inviscid biplane configuration. (**a**) Computational mesh; (**b**) Pressure contours.

In addition, the viscous flow with the free-stream Mach number of 0.8, the Reynolds number of 500 and the angle of attack of 10 degree is simulated. In the simulation, the unstructured grid containing 512 points on each airfoil and 65,861 cells in the computational domain is used, and its partial view is shown in Figure 11a. Figure 11b shows the streamline pattern obtained from present scheme. From this figure, it can be observed that the separation region on the upper surface of top airfoil reveals two vortices. This observation is consistent with the results reported in [44]. As pointed out by Jawahar and Kamath [44], the secondary vortex is introduced by the bottom airfoil. The comparison of pressure coefficient and skin friction coefficient distributions on the airfoil surface obtained by present scheme with those reported in [44] is shown in Figure 12. Once again, the results obtained by present scheme compare well with the reference data.

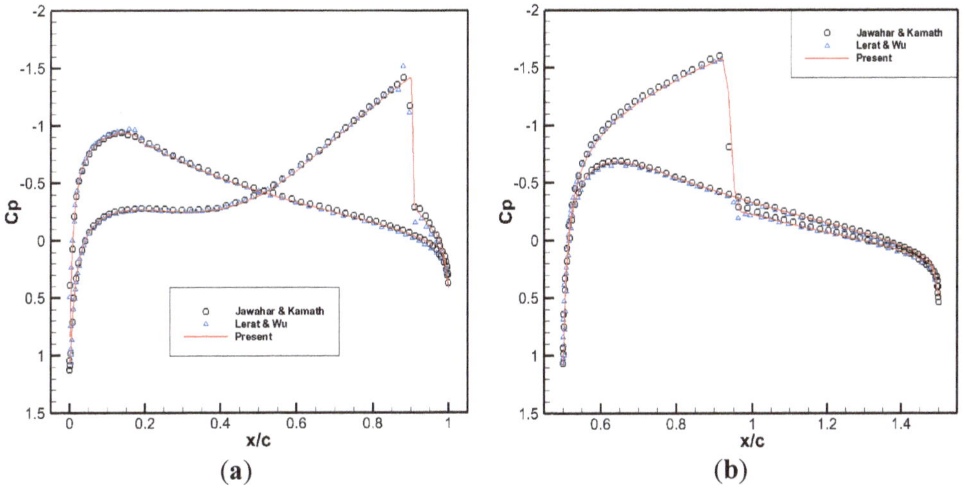

Figure 10. Comparison of pressure coefficient distribution on the surface of bottom and top airfoils for inviscid biplane configuration. (**a**) Bottom airfoil; (**b**) Top airfoil.

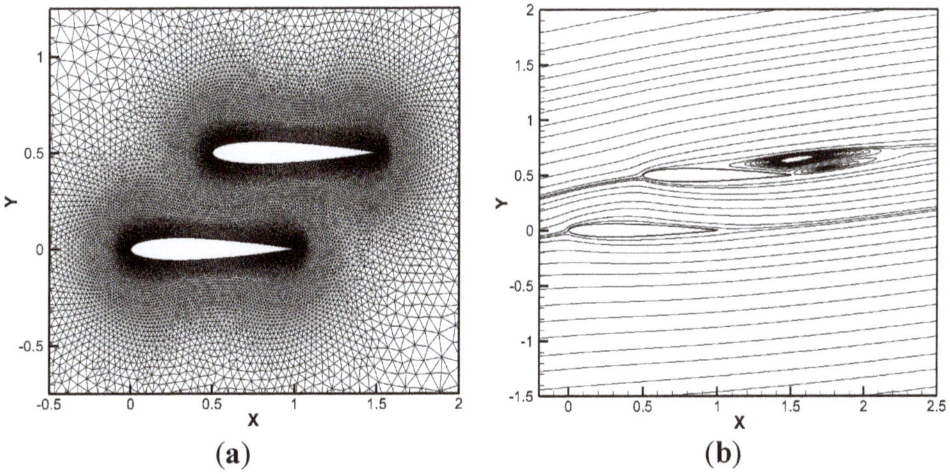

Figure 11. Partial view of computational mesh and streamline pattern for viscous biplane configuration. (**a**) Computational mesh; (**b**) Streamline pattern.

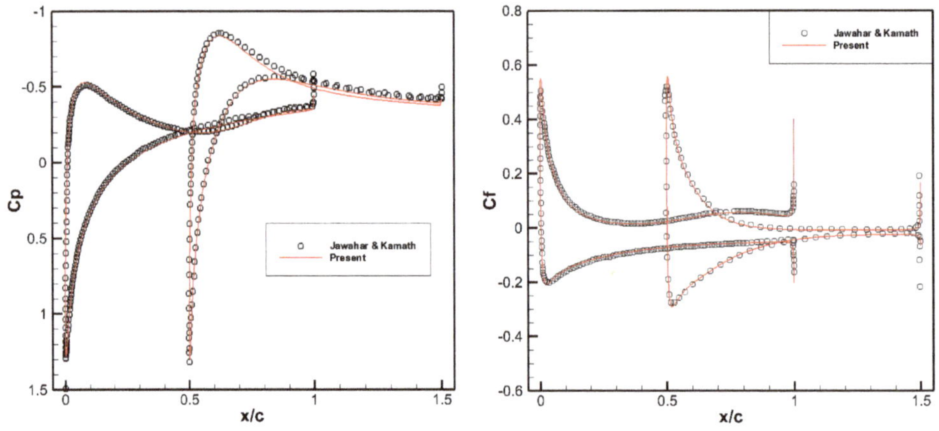

(a) (b)

Figure 12. Comparison of pressure coefficient and skin friction coefficient distributions for viscous biplane configuration. (**a**) Pressure coefficient distribution; (**b**) Skin friction distribution.

7. Conclusions

As a finite-volume solver, the LBFS directly updates macroscopic flow variables at cell centers by solving macroscopic governing equations. Its fluxes are reconstructed locally at each interface through lattice moments of particle distribution functions, in which the relationships obtained from the Chapman–Enskog theory are applied. During local reconstruction, the LBM is applied locally in one streaming time step, which is different from the global application of the conventional LBM. As a consequence, the LBFS is able to combine the advantages of the finite volume method and the LBM.

In this work, the historic development from the LBM to the LBFS is briefly introduced and their relationships with the macroscopic conservation laws are also described through the multi-scale Chapman–Enksog analysis. The major contribution of this work is to refine and examine three different versions of the LBFS, proposed respectively for isothermal, thermal and compressible flows. In particular, the accuracy, stability and efficiency of the isothermal LBFS are compared with the LBM in detail. The LBFS for temperature field is simplified by the D2Q4 model, which reduces computational effort as compared with that by D2Q9 model. The LBFS for compressible flows is improved by incorporating non-equilibrium effects into the process for inviscid flux reconstruction, in which numerical dissipation can be controlled through a switch function.

Several benchmark problems, including lid-driven cavity flows, natural convection in a square cavity at high Rayleigh numbers of 10^7 and 10^8 and transonic flows around a staggered-biplane configuration, have been carried out to examine the solvers. Numerical results show that the LBFS is able to obtain comparable solutions with much less non-uniform grid points and its efficiency can be greatly improved. It is also shown that LBFS is much more stable than LBM and does not generate unphysical pressure oscillations for lid-driven cavity flows. With the application of one-dimensional compressible lattice Boltzmann model, the LBFS can be effectively applied for simulation of compressible flows on unstructured grids.

Author Contributions: The authors contributed equally to the research and writing of this article. All authors have read and approved the final manuscript.

Conflicts of Interest: The authors declare no conflict of interest.

References

1. McNamara, G.R.; Zanetti, G. Use of the Boltzmann Equation to Simulate Lattice-Gas Automata. *Phys. Rev. Lett.* **1988**, *61*, 2332–2335.
2. Higuera, F.J.; Succi, S.; Benzi, R. Lattice gas-dynamics with enhanced collisions. *Europhys. Lett.* **1989**, *9*, 345–349.
3. Chen, S.Y.; Chen, H.D.; Martinez, D.; Matthaeus, W. Lattice Boltzmann model for simulation of magnetohydrodynamics. *Phys. Rev. Lett.* **1991**, *67*, 3776–3779.
4. Qian, Y.H.; D'Humières, D.; Lallemand, P. Lattice BGK models for Navier–Stokes equation. *Europhys. Lett.* **1992**, *17*, 479–484.
5. Mei, R.; Shyy, W. On the Finite Difference-Based Lattice Boltzmann Method in Curvilinear Coordinates. *J. Comput. Phys.* **1998**, *143*, 426–448.
6. Hejranfar, K.; Ezzatneshan, E. Implementation of a high-order compact finite-difference lattice Boltzmann method in generalized curvilinear coordinates. *J. Comput. Phys.* **2014**, *267*, 28–49.
7. Zarghami, A.; Ubertini, S.; Succi, S. Finite-volume lattice Boltzmann modeling of thermal transport in nanofluids. *Comput. Fluids* **2013**, *77*, 56–65.
8. Wang, Y.; Shu, C.; Huang, H.B.; Teo, C.J. Multiphase lattice Boltzmann flux solver for incompressible multiphase flows with large density ratio. *J. Comput. Phys.* **2015**, *280*, 404–423.
9. Wang, Y.; Shu, C.; Teo, C.J. Thermal lattice Boltzmann flux solver and its application for simulation of incompressible thermal flows. *Comput. Fluids* **2014**, *94*, 98–111.
10. Shu, C.; Wang, Y.; Teo, C.J.; Wu, J. Development of Lattice Boltzmann Flux Solver for Simulation of Incompressible Flows. *Adv. Appl. Math. Mech.* **2014**, *6*, 436–460.
11. Yang, L.M.; Shu, C.; Wu, J. A moment conservation-based non-free parameter compressible lattice Boltzmann model and its application for flux evaluation at cell interface. *Comput. Fluids* **2013**, *79*, 190–199.
12. Guo, Z.; Shi, B.; Wang, N. Lattice BGK Model for Incompressible Navier–Stokes Equation. *J. Comput. Phys.* **2000**, *165*, 288–306.
13. Shan, X.; Chen, H. Lattice Boltzmann model for simulating flows with multiple phases and components. *Phys. Rev. E* **1993**, *47*, 1815–1819.

14. Lee, T.; Lin, C.-L. A stable discretization of the lattice Boltzmann equation for simulation of incompressible two-phase flows at high density ratio. *J. Comput. Phys.* **2005**, *206*, 16–47.
15. He, X.; Chen, S.; Zhang, R. A Lattice Boltzmann Scheme for Incompressible Multiphase Flow and Its Application in Simulation of Rayleigh–Taylor Instability. *J. Comput. Phys.* **1999**, *152*, 642–663.
16. Li, Q.; He, Y.L.; Wang, Y.; Tao, W.Q. Coupled double-distribution-function lattice Boltzmann method for the compressible Navier–Stokes equations. *Phys. Rev. E* **2007**, *76*, 056705.
17. Chen, F.; Xu, A.; Zhang, G.; Li, Y.; Succi, S. Multiple-relaxation-time lattice Boltzmann approach to compressible flows with flexible specific-heat ratio and Prandtl number. *Europhys. Lett.* **2010**, *90*.
18. He, Y.-L.; Liu, Q.; Li, Q. Three-dimensional finite-difference lattice Boltzmann model and its application to inviscid compressible flows with shock waves. *Physica A* **2013**, *392*, 4884–4896.
19. Nie, X.; Doolen, G.D.; Chen, S. Lattice-Boltzmann Simulations of Fluid Flows in MEMS. *J. Stat. Phys.* **2002**, *107*, 279–289.
20. Lim, C.Y.; Shu, C.; Niu, X.D.; Chew, Y.T. Application of lattice Boltzmann method to simulate microchannel flows. *Phys. Fluids* **2002**, *14*, 2299–2308.
21. Meng, J.; Zhang, Y. Accuracy analysis of high-order lattice Boltzmann models for rarefied gas flows. *J. Comput. Phys.* **2011**, *230*, 835–849.
22. Lallemand, P.; Luo, L.-S. Theory of the lattice Boltzmann method: Dispersion, dissipation, isotropy, Galilean invariance, and stability. *Phys. Rev. E* **2000**, *61*, 6546–6562.
23. Niu, X.D.; Shu, C.; Chew, Y.T. A lattice Boltzmann BGK model for simulation of micro flows. *Europhys. Lett.* **2004**, *67*, 600–606.
24. Yuan, H.-Z.; Niu, X.-D.; Shu, S.; Li, M.; Yamaguchi, H. A momentum exchange-based immersed boundary-lattice Boltzmann method for simulating a flexible filament in an incompressible flow. *Comput. Math. Appl.* **2014**, *67*, 1039–1056.
25. Yuan, H.-Z.; Shu, S.; Niu, X.-D.; Li, M.; Hu, Y. A Numerical Study of Jet Propulsion of an Oblate Jelly fish Using a Momentum Exchange-Based Immersed Boundary-Lattice Boltzmann Method. *Adv. Appl. Math. Mech.* **2014**, *6*, 307–326.
26. Ubertini, S.; Succi, S. Recent advances of Lattice Boltzmann techniques on unstructured grids. *Prog. Comput. Fluid Dyn. Int. J.* **2005**, *5*, 85–96.
27. Rossi, N.; Ubertini, S.; Bella, G.; Succi, S. Unstructured lattice Boltzmann method in three dimensions. *Int. J. Numer. Methods Fluids* **2005**, *49*, 619–633.
28. Patankar, S.V.; Spalding, D.B. A calculation procedure for heat, mass and momentum transfer in three-dimensional parabolic flows. *Int. J. Heat Mass Transf.* **1972**, *15*, 1787–1806.
29. Sun, D.L.; Qu, Z.G.; He, Y.L.; Tao, W.Q. An efficient segregated algorithm for incompressible fluid flow and heat transfer problems-IDEAL (Inner Doubly Iterative Efficient Algorithm for Linked Equations) Part I: Mathematical formulation and solution procedure. *Numer. Heat Transf. B* **2008**, *53*.

30. Bruneau, C.-H.; Saad, M. The 2D lid-driven cavity problem revisited. *Comput. Fluids* **2006**, *35*, 326–348.
31. Kalita, J.C.; Gupta, M.M. A streamfunction–velocity approach for 2D transient incompressible viscous flows. *Int. J. Numer. Methods Fluids* **2010**, *62*, 237–266.
32. Lin, L.-S.; Chang, H.-W.; Lin, C.-A. Multi relaxation time lattice Boltzmann simulations of transition in deep 2D lid driven cavity using GPU. *Comput. Fluids* **2013**, *80*, 381–387.
33. Zhuo, C.; Zhong, C.; Cao, J. Filter-matrix lattice Boltzmann simulation of lid-driven deep-cavity flows, Part I—Steady flows. *Comput. Math. Appl.* **2013**, *65*, 1863–1882.
34. Zhuo, C.; Zhong, C.; Cao, J. Filter-matrix lattice Boltzmann simulation of lid-driven deep-cavity flows, Part II—Flow bifurcation. *Comput. Math. Appl.* **2013**, *65*, 1883–1893.
35. Contrino, D.; Lallemand, P.; Asinari, P.; Luo, L.-S. Lattice-Boltzmann simulations of the thermally driven 2D square cavity at high Rayleigh numbers. *J. Comput. Phys.* **2014**, *275*, 257–272.
36. Luan, H.-B.; Xu, H.; Chen, L.; Sun, D.-L.; He, Y.-L.; Tao, W.-Q. Evaluation of the coupling scheme of FVM and LBM for fluid flows around complex geometries. *Int. J. Heat Mass Transf.* **2011**, *54*, 1975–1985.
37. Guo, Z.; Shi, B.; Zheng, C. A coupled lattice BGK model for the Boussinesq equations. *Int. J. Numer. Methods Fluids* **2002**, *39*, 325–342.
38. Chen, L.; He, Y.-L.; Kang, Q.; Tao, W.-Q. Coupled numerical approach combining finite volume and lattice Boltzmann methods for multi-scale multi-physicochemical processes. *J. Comput. Phys.* **2013**, *255*, 83–105.
39. Chen, L.; Luan, H.; Feng, Y.; Song, C.; He, Y.-L.; Tao, W.-Q. Coupling between finite volume method and lattice Boltzmann method and its application to fluid flow and mass transport in proton exchange membrane fuel cell. *Int. J. Heat Mass Transf.* **2012**, *55*, 3834–3848.
40. Yang, L.M.; Shu, C.; Wu, J.; Zhao, N.; Lu, Z.L. Circular function-based gas-kinetic scheme for simulation of inviscid compressible flows. *J. Comput. Phys.* **2013**, *255*, 540–557.
41. Yang, L.M.; Shu, C.; Wu, J. A three-dimensional explicit sphere function-based gas-kinetic flux solver for simulation of inviscid compressible flows. *J. Comput. Phys.* **2015**, *295*, 322–339.
42. Peng, Y.; Shu, C.; Chew, Y.T. Simplified thermal lattice Boltzmann model for incompressible thermal flows. *Phys. Rev. E* **2003**, *68*, 026701.
43. Le Quéré, P. Accurate solutions to the square thermally driven cavity at high Rayleigh number. *Comput. Fluids* **1991**, *20*, 29–41.
44. Jawahar, P.; Kamath, H. A High-Resolution Procedure for Euler and Navier–Stokes Computations on Unstructured Grids. *J. Comput. Phys.* **2000**, *164*, 165–203.
45. Lerat, A.; Wu, Z.N. Stable Conservative Multidomain Treatments for Implicit Euler Solvers. *J. Comput. Phys.* **1996**, *123*, 45–64.

Entropy-Assisted Computing of Low-Dissipative Systems

Ilya V. Karlin, Fabian Bösch, Shyam S. Chikatamarla and Sauro Succi

Abstract: Entropy feedback is reviewed and highlighted as the guiding principle to reach extremely low dissipation. This principle is illustrated through turbulent flow simulations using the entropic lattice Boltzmann scheme.

Reprinted from *Entropy*. Cite as: Karlin, I.V.; Bösch, F.; Chikatamarla, S.S.; Succi, S. Entropy-Assisted Computing of Low-Dissipative Systems. *Entropy* **2015**, *17*, 8099–8110.

1. Introduction

The operation of many natural and engineered systems depends crucially on their ability to function at very low dissipation rates, the lower most often the better [1]. Zero-dissipation, however, is an ideal limit which could only be reached if these systems could operate at virtually infinite processing speed. Hence, a very general question arises: how low can one keep dissipation in a given thermodynamic system?

Here we show that the ability of a specific class of fluid-kinetic systems [2–4] to function at a very low dissipation is dramatically enhanced by enforcing the second principle of thermodynamics in the form of an entropic feedback [5]. Through concrete examples of turbulent flows, we highlight how entropy-assisted simulation maintains the system at low viscosity, through a highly orchestrated and self-consistent interplay between local enhancement and reduction of the dissipation. Balancing of these dissipation fluctuations leads to a spatial distribution of the average effective viscosity which keeps the simulation "alive and well". We envisage the entropy-assisted computing procedure to offer a general paradigm for the computer simulation of a wide class of low-dissipative complex phenomena, such as classical and quantum turbulence and wave propagation in active media.

2. Survival below Minimum Dissipation Threshold

The second principle of thermodynamics stands out as one of the most general and inescapable laws of physics, with profound bearings on the time evolution of virtually all natural systems [1]. In its essence, it states that any natural system is driven towards a state of maximum entropy (equilibrium), characterized by a maximum number of microscopic configurations. However, it says little about a most relevant question: how long does it take for a given system to reach its equilibrium state? This question, the heart of non-equilibrium thermodynamics, is all but academic, since most natural phenomena, life in the first place, depend on the time the system is able to borrow from temporary elusion of the second principle [1]. The rate of decay to equilibrium is measured by transport coefficients, such as kinematic viscosity, and can change widely from system to system, from seconds in an ordinary gas, to years and centuries in glassy materials.

The typical form of kinematic viscosity is given by $\nu \sim v_T^2 \tau$, where $v_T = \sqrt{k_B T/m}$ is the thermal speed, τ a typical relaxation time. The kinematic viscosity measures the diffusivity of momentum across the system: such diffusivity results from the competition between kinetic energy, which sustains the free motion of molecules, and potential energy, which controls their interactions (collisions). Kinetic energy drives the system out of equilibrium, while molecular collisions pull it back to a local equilibrium, in which entropy is locally maximized (Boltzmann's H-theorem) [6–8]. Thus, a low-viscous fluid is not one with nearly no collisions, but one where collisions are so frequent and effective that they inhibit any migration of momentum from place to place, which is the source of macroscopic dissipation. From the above argument, it is seen that zero-viscosity is a mirage because it would imply instantaneous relaxation, *i.e.*, $\tau \to 0$. Given that strictly zero-viscosity is a chimera in a real (finite-speed) world, a natural question arises: *what is the minimum dissipation which can be sustained by a given physical system?* While the answer depends on the specific system in mind, here we shall focus on discrete dynamical systems, *i.e.*, featuring a fundamental minimal length scale a and minimal time scale h.

For the case of simulated fluids, for instance, the condition is that the smallest coherent structures (eddies) capable of surviving dissipation be resolved by the discrete grid, *i.e.*, $l_K > a$, where $l_K \sim L/\mathrm{Re}^{3/4}$ is the so-called Kolmogorov's length [9], the smallest active scale in the game and $\mathrm{Re} = UL/\nu$ is the Reynolds number, *i.e.*, the ratio of nonlinear energy transfer to dissipation, for a fluid moving at a macroscopic speed U on a domain of macroscopic size L. As a result, the minimal viscosity is given by $\nu_{\min} = (u/N^{1/3})\nu_l$, where $N = L/a$ is the grid size, $u \equiv U/U_l$, $U_l = a/h$ and $\nu_l = a^2/h$ being the natural lattice speed and lattice viscosity, respectively. Given that $u < 1$ for reasons of numerical stability, we see that the minimal viscosity is always smaller than the lattice viscosity, the ratio of the two decreasing like $1/N^{1/3}$, so that the minimum viscosity can be brought to zero only in the continuum limit $N \to \infty$. Another face of the same chimera. The message is that fluids cannot support viscosity below their minimum bound $\nu_{\min}$. Breaking such constraint leads to two basic scenarios: a mild reaction (*loss of accuracy*), whereby the resolved eddies, $l > a$, still survive, although with a corrupted dynamics, the degree of corruption increasing as they approach a. The second, more dramatic, possibility is *loss of realizability*: the system develops disruptive instabilities, typically in the form of an uncontrolled growth of the smallest eddies. This is nothing short of a survival problem, except that it concerns a discrete dynamical system. From the practical point of view, the art of keeping the system alive and well in the forbidden regime $\nu < \nu_{min}$ is known as turbulence modeling, a topic of utmost practical and conceptual importance. Essentially, the idea is to replace the nominal viscosity with an effective one, representing the effects of unresolved eddies as "random" collisions on the resolved ones. This picture explicitly draws upon an analogy with kinetic theory, where there is a clear scale separation between molecular and hydrodynamic degrees of freedom. Turbulence, on the contrary, features a continuum spectrum of scales, hence the notion of eddy viscosity, although very useful, still resists a rigorous justification.

3. Minimum Viscosity in Discrete Phase-Space-Time

However, a modern formulation of continuum fluid mechanics in a form which explicitly ingrains the discreteness of space-time is known as the lattice Boltzmann equation [4]

$$f_i(x + c_i h; t + h) = f_i' \equiv \left(1 - \frac{h}{2\tau}\right) f_i(x,t) + \left(\frac{h}{2\tau}\right) f_i^{\mathrm{mirr}}(x,t) \quad (1)$$

In the above f_i is the probability of finding a "particle" at position x in the lattice at time t, moving with discrete velocity c_i along b lattice links; f_i^{mirr} is the so-called *mirror state*. In the simplest case [4], it is taken as $f_i^{\mathrm{mirr}} = 2f_i^{\mathrm{eq}} - f_i$, with f_i^{eq} the local equilibrium, which is a universal non-linear function of the local order parameters. For standard fluids, these are the fluid density $n(x,t) = \sum_i f_i(x,t)$ and velocity $u(x,t) = n^{-1} \sum_i c_i f_i(x,t)$. The left-hand side of Equation (1) represents the free-streaming step, while the right-hand side describes the interactions among the discrete populations f_i at each given lattice site. On condition that the lattice obeys proper symmetries, the lattice Boltzmann Equation (1) reproduces fluid dynamics with the viscosity

$$\nu \sim \left(\frac{\tau}{h} - \frac{1}{2}\right) \nu_1 \quad (2)$$

The negative shift, $-1/2$ in Equation (2) is crucial; indeed, if only in principle, it permits to achieve zero viscosity in the limit $\tau \to h/2$, *i.e.*, without sending $\tau \to 0$, unlike in the continuum. This negative shift is the result of the broken time-symmetry, which contributes a *negative* viscosity (sometimes called propagation viscosity) to the overall momentum diffusivity, besides the conventional contribution due to the collisional relaxation. Thus, in a discrete world, the viscosity receives contributions from both dynamical steps of the kinetic description: free streaming and collisions. They carry opposite signs, hence, if only in principle, they can cancel each other, leaving the time step and relaxation time both finite. This property, typically regarded as a very useful numerical artifact, has played a major role in the lattice Boltzmann simulation of a variety of complex flows, and most notably turbulent ones [10].

Amazingly, suitably designed discrete kinetic systems keep describing correct fluid behavior *several orders of magnitude below* the hydrodynamic minimum viscosity bounds mentioned earlier in this paper. How come the minimal viscosity can be eluded by several orders of magnitude?

The key is the second principle in fully discrete setting. Indeed, the lattice Boltzmann Equation (1) is compatible with a discrete-time H-theorem, based on the H-function (negative of the entropy), $H[f] = \sum_i f_i \ln(f_i/w_i)$, where w_i are suitable positive-definite weights. Lattice Boltzmann systems equipped with the H-theorem are known as *entropic* [5] and function on a feedback mechanism, whereby the local relaxation τ is adjusted in space and time, so as to secure the entropic bound, $H[f'] \leq H[f]$, where f and f' are the pre- and post-collisional states, respectively.

The working principle is explained in Figure 1 and amounts to using in Equation (1) the *entropy-supervised mirror state* $f_i^{\mathrm{mirr}} = (1-\alpha)f_i + \alpha f^{\mathrm{eq}}$, where the *stretch* α is found from the isentropic constraint, $H[f^{\mathrm{mirr}}] = H[f]$. This can be interpreted as the *effective viscosity*,

$$\nu_{\mathrm{eff}} \sim \left(\frac{\tau_{\mathrm{eff}}}{h} - \frac{1}{2}\right)\nu_1 \tag{3}$$

with the effective relaxation time, $\tau_{\mathrm{eff}} = 2\tau/\alpha$. The entropy-assisted computation thus *informs* the pre-collision state f about its isentropic mirror f^{mirr} and stipulates the single condition that the second law is respected by the post-collision state f'. Whenever non-equilibrium effects become strong enough to endanger realizability, the entropic constraint adjusts the relaxation time so as to secure compliance with the second principle. This feedback is self-activated "on demand", *i.e.*, only whenever and wherever the need arises. And when the danger is gone, most elegantly, the entropic feedback, leaves the stage unsolicited. The second principle decides by itself: sometimes viscosity is increased ($\nu_{\mathrm{eff}} > \nu$) to smooth out sharp features, sometimes it is reduced ($\nu_{\mathrm{eff}} < \nu$) to sharpen the dying ones. In the most demanding cases, the effective viscosity may even drop negative ($\tau_{\mathrm{eff}} < h/2$) to promote local instabilities and sustain the system against dissipative death. The effective viscosity self-adapts to the actual state of turbulence to literally protect it against defective evolution and disruptive instabilities.

In Figure 2 we illustrate the above by the vorticity field of a flow past a circular cylinder at Re $\sim$ 3300, in which many active scales of motion are visible. The Reynolds number, $\mathrm{Re} = UD/\nu$, is based on the diameter of the cylinder, which is here taken as $D = 30\,a$, while the mean flow velocity is $U = 0.03\,(a/h)$, corresponding to a viscosity $\nu = UD/\mathrm{Re} \sim 2.7 \times 10^{-4}\,(a^2/h)$. With these parameters, the minimum viscosity is $\nu_{\mathrm{min}} = 0.03/30^{1/3} \sim 0.01\,(a^2/h)$, so that $\nu/\nu_{\mathrm{min}} \sim 1/40$. The flow structures in Figure 2a are colored with the effective viscosity, normalized as $R = (\nu_{\mathrm{eff}} - \nu)/\nu$. The high quality of resolution of the flow structures (vortex tubes, tangles *etc.*) is maintained by a concerted action of dampers ($R > 0$) and promoters ($R < 0$). The tiniest structures would not be able to survive unless the effective viscosity is enabled to go negative from time to time, in order to compensate for over-dissipation and "regenerate" small scale structures otherwise doomed by over-damping. Also to be noted (Figure 2b) is the *spottiness* of the effective viscosity, with a highly fine-grained mixing of dampers and promoters.

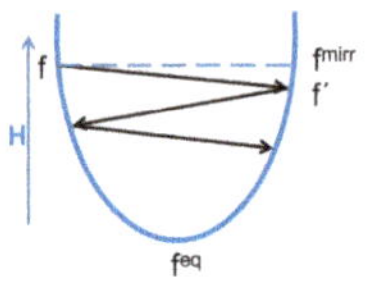

Figure 1. Entropy-assisted computing. The initial state f is over-relaxed to the state f' with the entropy function H value strictly below the value at the entropy mirror state f^{mirr}. The zigzag trajectory of over-relaxations eventually ends up at the bottom of the well—at the equilibrium f^{eq}.

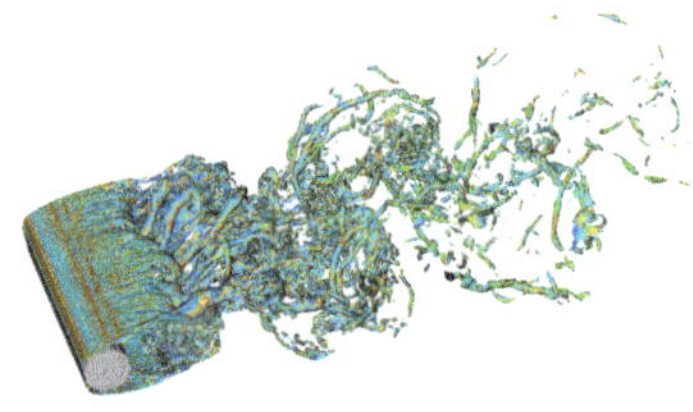

(a)

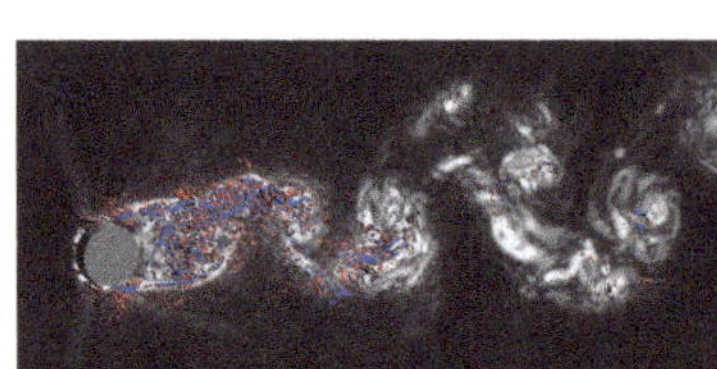

(b)

Figure 2. (**a**) Turbulent flow generated by a round cylinder. Snapshot of the vorticity iso-surfaces are shown, colored with the effective viscosity. Blue: $R > 0$ (dampers); Red/Yellow: $R < 0$ (promoters). The interplay between the dampers and promoters along each vortex tube is clearly seen. (**b**) Snapshot of the intertwining of dampers ($R > 0$, blue) and promoters ($R < 0$, red). Essential dampers ($R > 1.5$) and promoters ($R < -0.6$) are shown. The entropic feedback is concentrated in the region behind the obstacle, where the transition to turbulence occurs. Gray background: Vorticity.

All of the above configures a very elegant preemptive scenario which we can take as the hallmark of entropic computing: very attentive "guardian angels". Amazingly, the spatial pattern of the time-averaged effective viscosity shown in Figure 3 resembles indeed a "guardian angel", protecting the system against numerical crisis!

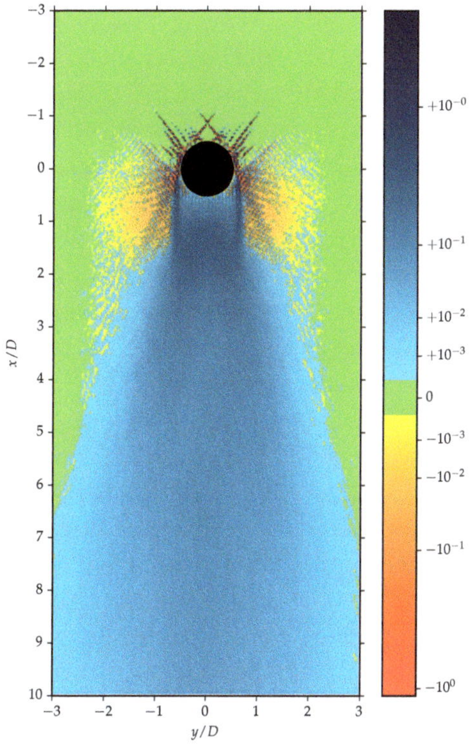

Figure 3. Distribution of the time-averaged normalized effective viscosity $R = (\nu_{\text{eff}} - \nu)/\nu$ at the mid-section of the flow past a round cylinder. Red/Yellow: Promoters ($R < 0$); Blue: Dampers ($R > 0$); Green: Nominal ($R = 0$). While the snapshot in Figure 2 demonstrates a larger variation of the effective viscosity, the time-average picture is much milder: most of the activity (strongest damping neighboring the strongest promotion) is concentrated at and around the twin shear layers, just behind the cylinder. In the rest of the domain, the deviation of the effective viscosity from its nominal value is less than a fraction of a percent.

4. Entropic Lattice Boltzmann Algorithm

Here we describe the essentials of the simulation method used above. In the entropic lattice Boltzmann scheme, populations associated with the discrete velocities

c_i evolve according to (1). The local equilibrium f_i^{eq} was found by minimizing the entropy function $H[f]$. The entropic mirror state $f_i^{\text{mirr}} = (1-\alpha)f_i + \alpha f^{\text{eq}}$ is specified by the stretch α, which is computed as the positive root of the entropy condition: $H[f + \alpha(f^{\text{eq}} - f)] = H[f]$. Whenever the simulation is resolved at a particular grid node x, the stretch α becomes fixed automatically to the value $\alpha = 2$ at that node, and the effective viscosity ν_{eff} (3) reduces to the nominal viscosity ν (2). The stretch was obtained numerically at each grid point using Newton-Raphson method. For the simulation presented above, we used the lattice with $b = 15$ discrete velocities [11]. Apart from an entropy-supported kinetic equation, we require augmenting boundary conditions that are capable of simulating both resolved and under-resolved flows. Existing boundary conditions such as the bounce-back scheme [4], provide reliable results for resolved simulations but with reducing grid sizes and increasing the Reynolds number, the quality of the simulations is lost due to shock-like instabilities generated at the walls. Hence, we used here the recently proposed Tamm–Mott-Smith boundary condition [12] for circumventing these instabilities. Entropy-supported kinetic Equation (1) provides reliable simulations for any choice of lattice and flow velocity as long as the Mach number remains small. Armed with stable boundary condition, the present scheme was extensively tested for various flow setups such as decaying turbulence ($\nu/\nu_{\text{min}} \sim 10^{-4}$) [13], turbulent channel ($\nu/\nu_{\text{min}} \sim 10^{-3}$) [12], grid generated turbulence ($\nu/\nu_{\text{min}} \sim 10^{-3}$), flow past an airfoil ($\nu/\nu_{\text{min}} \sim 10^{-4}$) and others. For the particular simulation of the flow past a circular cylinder presented in Figures 2 and 3, we used a computational domain that is $9D$ long in the span-wise direction, $35D$ along in the stream-wise direction with $10D$ upstream of the cylinder and $25D$ downstream of it; along vertical direction the domain was $21D$ long with cylinder axis in the mid-plane. The cylinder was resolved with the diameter $D = 30$ grid points. Apart from the simulation of turbulent flows, the entropy feedback has significantly improved stability of thermal flows with temperature gradients [14], multiphase flows [15] and other fluid dynamics problems. This gives us strong confidence that entropy-guidance can be extended to other low-dissipative physical systems.

5. ELBM: Questions and Answers

Here we answer some typical questions to the entropic lattice Boltzmann method (ELBM) reviewed above. For the sake of convenience, we rewrite the ELBM equation setting the time step $h = 1$:

$$f_i(x + c_i, t+1) \equiv f_i' = (1-\beta)f_i(x,t) + \beta f_i^{\text{mirr}}(x,t) \tag{4}$$

where $\beta \in [0,1]$, while the entropic mirror state f^{mirr} is

$$f_i^{\text{mirr}} = f_i + \alpha(f_i^{\text{eq}} - f_i) \tag{5}$$

The stretch α is defined by the entropy balance between the pre-collision state f and the mirror state f^{mirr},

$$H(f^{\text{mirr}}) = H(f) \tag{6}$$

A discrete-time H-theorem states: If the non-trivial solution α exists for the entropy balance (6), then the total entropy $\bar{H}(t) = \sum_x H(f(x,t))$ is not increasing, $\bar{H}(t+1) \leq \bar{H}(t)$. Note that the validity of the H-theorem requires not just the equilibrium to be evaluated through the minimization of H but also, and most importantly, the fulfillment of the entropy balance condition (6).

1. *Is the entropic feedback in ELBM a stabilizing technique or a physically sound subgrid-scale model for turbulence?*

   A: The ELBM should be viewed as a built-in subgrid model rather than a mere stabilization technique. Stabilization in ELBM is a by-product of the discrete-time H-theorem. Instead of a mere addition of artificial viscosity, the ELBM allows the effective viscosity to fluctuate around the target value ν. In order to clarify this point, note a few general features of the entropic stretch α.

   - *Over-relaxation:* Thanks to convexity of the entropy function, the solution to Equation (6) always leads to over-relaxation, $\alpha > 1$;
   - *Duality:* Let f be a population vector, and $f(\alpha) \equiv f + \alpha(f^{\text{eq}} - f)$ its entropic mirror state, with the same value of the entropy, $H(f(\alpha)) = H(f)$. If the entropy estimate is applied to $f(\alpha)$ instead of f, then the initial state is recovered in the form $f = f(\alpha) + \alpha^*(f^{\text{eq}} - f(\alpha))$, with another stretch $\alpha^* > 1$ which satisfies a duality relation:

     $$\alpha^*\alpha = \alpha^* + \alpha. \tag{7}$$

     Equation (7) implies that whenever $\alpha \lesseqgtr 2$, the opposite holds for the dual, $\alpha^* \gtreqless 2$.
   - *Hydrodynamic limit:* whenever the simulation is resolved (populations stay close to the local equilibrium), the stretch α tends to the fixed value $\alpha = 2$ (and so does also the dual stretch, $\alpha^* = 2$, according to (7)). Then ELBM self-consistently becomes equivalent to the lattice Bhatnagar–Gross–Krook

(LBGK) equation ($\alpha = 2$) and recovers the Navier-Stokes equations with the kinematic viscosity,

$$\nu = c_s^2 \left(\frac{1}{2\beta} - \frac{1}{2} \right) \tag{8}$$

where c_s is speed of sound (a $O(1)$ lattice-dependent constant).

Note that the above is a direct implication of the built-in H-theorem. Indeed, the resolved simulation, at the kinetic level, is characterized by the fact that all populations are asymptotically close to the local equilibrium. Then, the entropy function becomes well represented by its second-order approximation: at fixed locally conserved fields (density and momentum here), if $\delta f = f - f^{eq}$, $|\delta f / f^{eq}| \ll 1$, then $H(f) \approx H^{eq} + (1/2)\sum_i \delta f_i^2 / f_i^{eq}$. The levels of the entropy are then asymptotically close to the levels of the above quadratic form. It is under such condition that the entropy estimate (6) results in $\alpha = 2$. Note that the standard Chapman–Enskog approximation is valid under precisely the same condition of closeness to the local equilibrium, thereby the viscosity ν is the same for both ELBM and LBGK.

- *Effective viscosity and self-averaging:* The effective viscosity in the above notation reads,

$$\nu_{\text{eff}} = c_s^2 \left(\frac{1}{\alpha\beta} - \frac{1}{2} \right) \tag{9}$$

Depending on the outcome for stretch α, the effective viscosity $\nu_{\text{eff}}(\alpha\beta)$ is larger than the viscosity $\nu \equiv \nu_{\text{eff}}(2\beta)$ if $\alpha < 2$, and it is smaller than ν if $\alpha > 2$. In the first case, the (larger) effective viscosity leads to smoothing the velocity gradient at the given node, while in the second case, the smaller viscosity leads to a sharpening of the velocity gradient. Note that, when $\beta \to 1$ (vanishing viscosity $\nu \to 0$), the effective viscosity (9) can drop to even negative values if $\alpha > 2$. This asymmetry between the over-relaxation being "shorter" ($\alpha < 2$) or "longer" ($\alpha > 2$) than the LBGK over-relaxation $\alpha = 2$ is the crucial implication of the compliance with the H-theorem: even if the effective viscosity becomes negative at some lattice nodes, this does not lead to numerical instability because even in that case the H-theorem (and the proper behavior of the total entropy) remains valid.

Parameterization with the effective viscosity $\nu_{\text{eff}}(\alpha\beta)$ can be seen as an alternative to the parameterization with the over-relaxation α. Let us note that, if a pair $\{\alpha, \alpha^*\}$ is connected by the duality relation (7), then the mean value of the corresponding effective viscosity is equal to the viscosity (8),

$$\frac{\nu_{\text{eff}}(\alpha\beta) + \nu_{\text{eff}}(\alpha^*\beta)}{2} = \nu_{\text{eff}}(2\beta) \equiv \nu \tag{10}$$

The relation (10) is termed self-averaging, and provides important albeit heuristic argument that the averaged-in-time effective viscosity in ELBM simulation is close to the viscosity ν. In other words, we expect that it is only the matter of resolution that the average effective viscosity deviates from ν. This assertion, while not rigorous, is supported by simulation (see Figure 3). The rapid fluctuations of the stretch α around $\alpha = 2$ at a given monitoring point chosen at random in the simulation domain are clearly seen in Figure 4.

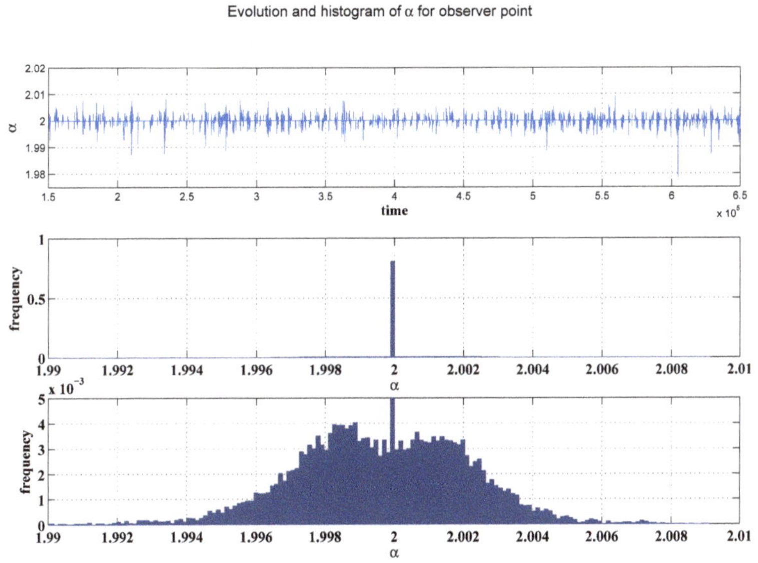

Figure 4. Top: History of the entropic strech α at a monitoring point; Middle: Histogram of α; Bottom: Close-up of the hystogram around the dominant value $\alpha = 2$.

In summary, the ELBGK exploits the self-adaptive mechanism of effective viscosity by choosing automatically the over-relaxation α at each node to guarantee the H-theorem at all sites and all discrete time-steps. When the grid is coarsened, over-relaxation α becomes "smeared" in an interval, $[\alpha_{\min}, \alpha_{\max}]$, with $1 < \alpha_{\min} < 2$, and $\alpha_{\max} > 2$. The self-adapted over-relaxation set up by (6), results in two oppositely directed effects: if $\alpha < 2$, the effective viscosity is larger than ν, and the ELBM will tend to smoothen any flow perturbation. On the other hand, if $\alpha > 2$, the flow perturbation is enhanced (effective viscosity is smaller than ν). In ELBM simulations, these two effects act simultaneously on various nodes, with the net effect combining stabilization (through smoothing, $\alpha < 2$) with the preservation of the resolution (through sharpening, $\alpha > 2$). Note that, as $\beta \to 1$, the effective viscosity can even drop to negative values when $\alpha > 2$. This, however, does not lead to instabilities as the total entropy balance remains under control by the discrete-time H-theorem. This all is very different from a conventional perspective on "eddy viscosity" turbulence modeling, and it is not surprising that ELBM does not reduce to familiar large eddy simulation (LES) models [16].

2. *What is the relation of ELBM to the entropic stabilizing techniques proposed in CFD?*

   A: During the last four decades numerous entropic stabilizing techniques have been proposed in computational fluid dynamics (CFD) (see, e.g., Refs [17–25] and references therein). The idea behind is, roughly speaking, to maintain an appropriate amount of artificial viscosity through the analysis of discretization of the entropy balance (physical or artificial). In this regards, ELBM is based on a different premise: it applies to strictly discrete systems (in velocity-space-time), and the discrete-time H-theorem does not reduce to the estimate of the entropy production (*cf.*, e.g., [26]).

3. *How ELBM performs in comparison to other stabilizing techniques proposed for LBM?*

   A: The closest analog of the conventional stabilization techniques in the LBM setting is perhaps the method of entropic limiters [27–29]. The idea behind is to measure the closeness of the pre-collision state to the corresponding local equilibrium (in the sense of the entropy difference), and to apply equilibration instead of over-relaxation if the difference exceeds a user-defined threshold. This is similar to conventional artificial viscosity stabilization techniques in CFD. Various versions of limiters were considered [27–29]. The authors of [29] claimed that entropic limiters "perform better" than ELBM.

4. *What is the main numerical mechanism promoting stability in ELBM?*

   A: Stability is promoted by the discrete-time H-theorem. Note that the implication of the H-theorem *in the presence of the over-relaxation* allows post-collision distributions to be both closer to the equilibrium than the LBGK outcome ($\alpha < 2$) or further away from the equilibrium ($\alpha > 2$). It must be noted that, in principle, for some pre-collision states, the corresponding entropic mirror state may not exist (and hence no entropy balance is possible). However, this happens beyond the domain of validity of the lattice Boltzmann models, and is of no concern in practice. In particular, pathological cases (no solution for α) occur in none of the simulations referred to in this paper.

5. *Very recently, in Ref. [30], Karlin et al. presented a new entropic stabilizer for LB schemes. How is it different from the ELBM?*

   A: ELBM is based on the discrete-time H-theorem which is imposed in a rather "orthodox" manner through the entropy balance condition (6) for the over-relaxation. A different realization of the entropic control was introduced recently by three of the present authors in [30] (we refer to this as KBC model). The idea is to replace the entropic over-relaxation on all the non-conserved moments as it is done in ELBM by a combination of the standard (unsupervised) over-relaxation of the stresses with the proper equilibration of the rest of the non-conserved moments. More specifically, if we write a moment representation of the populations, $f_i = k_i + s_i + h_i$, where k_i is the contribution of locally conserved fields, s_i are the stresses and h_i are the remaining higher-order moments, then the mirror state for KBC models reads,

$$f_i^{\mathrm{mirr}} = k_i + [2s_i^{\mathrm{eq}} - s_i] + \left[(1-\gamma)h_i + \gamma h_i^{\mathrm{eq}}\right] \tag{11}$$

   where γ is the *entropic stabilizer* which is found by minimizing the entropy in the post-collision state (4) with the mirror state (11):

$$\frac{dH[f'(\gamma)]}{d\gamma} = 0 \tag{12}$$

   The rationale behind is this: The over-relaxation of the stresses in the mirror state is the only formal condition to recover the viscosity ν (8); hence, an optimal post-collision state should minimize the entropy under this constraint. Thus, the KBC post-collision state is a quasi-equilibrium which corresponds to the minimum of the entropy function once all the relevant constraints are applied. Moreover, Equation (12) admits the following approximate solution,

$$\gamma = \frac{1}{\beta} - \left(2 - \frac{1}{\beta}\right)\frac{\langle \delta s | \delta h \rangle}{\langle \delta h | \delta h \rangle}, \tag{13}$$

   with $\langle X|Y\rangle = \sum_{i,j} X_i [\partial^2 H/\partial f_i \partial f_j]_{\mathrm{eq}} Y_j$ the entropic scalar product, and $\delta s_i = s_i - s_i^{\mathrm{eq}}$, $\delta h_i = h_i - h_i^{\mathrm{eq}}$. While (11) lumps together all the higher-order moments in the h-part of the populations, a generalization which makes a distinction within these moments is straightforward: For $h_i = \sum_m h_{mi}$ with m labeling the different higher-order moments (or groups of such moments), we have instead of (11),

$$f_i^{\mathrm{mirr}} = k_i + [2s_i^{\mathrm{eq}} - s_i] + \sum_m \left[(1-\gamma_m)h_i + \gamma_m h_i^{\mathrm{eq}}\right] \tag{14}$$

while the formula (13) generalizes to

$$\gamma_m = \frac{1}{\beta} - \left(2 - \frac{1}{\beta}\right) \sum_n [C^{-1}]_{mn} \langle \delta s | \delta h_n \rangle \tag{15}$$

with C^{-1} the inverse of the correlation matrix $C_{mn} = \langle \delta h_m | \delta h_n \rangle$. While the H-theorem is not directly imposed in the KBC models (unlike the ELBM), simulations of various setups demonstrated they are 'virtually indestructible' (Ref. [31]).

Note that in both ELBM and KBC models (and eventually in *any* lattice Boltzmann model) a statement that "it recovers the viscosity ν" refers *only to a fully resolved simulation*. Validity of the Navier-Stokes equation at small scales for a given simulation is checked independently, for example, by measuring the viscosity in the energy and enstrophy balance equations. For a detailed analysis of these aspects for the KBC models we refer to recent papers [32,33].

6. Conclusion

The second law of thermodynamics provides a parameter-free solution to the problem of controlling the effective turbulent viscosity, so as to tame numerical disruption. Besides turbulence, to which the above findings have an immediate impact on, the general notion of entropy-assisted computing, likely with different realizations of the entropy feedback, is expected to apply to other states of matter characterized by extremely low dissipation, such as superfluids [34] and cosmological fluids [35] near black-hole horizons. It is also of interest to explore whether a similar paradigm might inform the behavior of active matter systems [36]. Finally, one may extrapolate even further and conjecture that entropy-assisted feedback systems, functioning according to the feedback loop discussed above, may be engineered outside the realm of fluid mechanics, typically at the intersection of information, biology and statistical physics [37–39]. In an even broader perspective, we surmise that entropy-assisted procedures might also inspire the design of novel active feedback systems in natural, biological and possibly also medical sciences.

Acknowledgments: Ilya V. Karlin, Fabian Bösch, and Shyam S. Chikatamarla were supported by the European Research Council (ERC) Advanced Grant No. 291094-ELBM. Computational resources at the Swiss National Super Computing Center CSCS were provided under the grant S492.

Author Contributions: Ilya V. Karlin introduced the concept of the entropic lattice Boltzmann method. Fabian Bösch and Shyam S. Chikatamarla wrote the code and run the simulations. Sauro Succi and Ilya V. Karlin developed the concept of the paper and wrote it. All authors have read and approved the final manuscript.

Conflicts of Interest: The authors declare no conflict of interest.

References

1. Nicolis, G.; Prigogine, I. *Self-Organization in Nonequilibrium Systems*; Wiley: Hoboken, NJ, USA, 1977.
2. Frisch, U.; Hasslacher, B.; Pomeau, Y. Lattice-gas automata for the Navier-Stokes equation. *Phys. Rev. Lett.* **1986**, *56*, doi:10.1103/PhysRevLett.56.1505.
3. Higuera, F.J.; Succi, S. Simulating the flow around a circular cylinder with a lattice Boltzmann equation. *Europhys. Lett.* **1989**, *8*, doi:10.1209/0295-5075/8/6/005.
4. Succi, S. *The Lattice Boltzmann Equation for Fluid Dynamics and Beyond*; Clarendon Press: Oxford, UK, 2001.
5. Karlin, I.V.; Ferrante, A.; Öttinger, H.C. Perfect entropy functions of the lattice Boltzmann method. *Europhys. Lett.* **1999**, *47*, 182–188.
6. Boltzmann, L. *Vorlesungen über Gastheorie*; Johann Ambrosius Barth: Leipzig, Germany, 1896. (In German)
7. Lebowitz, J.L. Boltzmann's entropy and time's arrow. *Phys. Today* **1993**, *46*, 32.
8. Lieb, E.H.; Yngvason, J. The Physics and Mathematics of the Second Law of Thermodynamics. *Phys. Rep.* **1999**, *310*, 1–96.
9. Kolmogorov, A.N. The local structure of turbulence in incompressible viscous fluid for very large Reynolds number. *Dokl. Akad. Nauk SSSR* **1941**, *30*, 301–305.
10. Chen, H.; Kandasamy, S.; Orszag, S.; Shock, R.; Succi, S.; Yakhot, V. Extended Boltzmann kinetic equation for turbulent flows. *Science* **2003**, *301*, 633–636.
11. Chikatamarla, S.S.; Ansumali, S.; Karlin, I.V. Entropic lattice Boltzmann models for hydrodynamic in three dimensions. *Phys. Rev. Lett.* **2006**, *97*, 010201.
12. Chikatamarla, S.S.; Karlin, I.V. Entropic lattice Boltzmann method for turbulent flow simulations: Boundary conditions. *Physica A* **2013**, *392*, 1925–1930.
13. Karlin, I.V.; Succi, S.; Chikatamarla, S.S. Comment on "Numerics of the lattice Boltzmann method: Effects of collision models on the lattice Boltzmann simulations". *Phys. Rev. E* **2011**, *84*, 068701.
14. Frapolli, N.; Chikatamarla, S.S.; Karlin, I.V. Multispeed entropic lattice Boltzmann model for thermal flows. *Phys. Rev. E* **2014**, *90*, 043306.
15. Mazloomi M, A.; Chikatamarla, S.S.; Karlin, I.V. Entropic lattice Boltzmann method for multiphase flows. *Phys. Rev. Lett.* **2015**, *114*, 174502.
16. Malaspinas, O.; Deville, M.; Chopard, B. Towards a physical interpretation of the entropic lattice Boltzmann method. *Phys. Rev. E* **2008**, *78*, 066705.
17. Lax, P.D. *Hyperbolic Systems of Conservation Laws and the Mathematical Theory of Shock Waves*; Society for Industrial and Applied Amathematics (SIAM): Philadelphia, PA, USA, 1973.
18. Harten, A.; Hyman, J.M.; Lax, P.D.; Keyfitz, B. On finite-difference approximations and entropy conditions for shocks. *Commun. Pure Appl. Math.* **1976**, *29*, 297–322.
19. Harten, A. On the symmetric form of systems of conservation laws with entropy. *J. Comput. Phys.* **1983**, *49*, 151–164.

20. Hughes, T.J.R.; Franca, L.P.; Mallet, M. A new finite element formulation for computational fluid dynamics: I. Symmetric forms of the compressible Euler and Navier-Stokes equations and the second law of thermodynamics. *Comput. Method Appl. Mech. Eng.* **1986**, *54*, 223–234.
21. Tadmor, E. Entropy stability theory for difference approximations of nonlinear conservation laws and related time-dependent problems. *Acta Numer.* **2003**, *12*, 451–512.
22. Tadmor, E.; Zhong, W.G. Entropy stable approximations of Navier-Stokes equations with no artificial numerical viscosity. *J. Hyper. Differ. Equ.* **2006**, *3*, 529–559.
23. Naterer, G.F.; Camberos, J.A. Entropy and the second law fluid flow and heat transfer simulation. *J. Thermophys. Heat Transf.* **2003**, *17*, 360–371.
24. Naterer, G.F.; Camberos, J.A. *Entropy Based Design and Analysis of Fluids Engineering Systems*; CRC Press: Boca Raton, FL, USA, 2008.
25. Fisher, T.C.; Carpenter, M.H. High-order entropy stable finite difference schemes for nonlinear conservation laws: Finite domains. *J. Comput. Phys.* **2013**, *252*, 518–557.
26. Succi, S.; Karlin, I.V.; Chen, H.D. Colloquium: Role of the *H* theorem in lattice Boltzmann hydrodynamics. *Rev. Mod. Phys.* **2002**, *74*, 1203–1220.
27. Brownlee, R.A.; Gorban, A.N.; Levesley, J. Stabilization of the lattice Boltzmann method using the Ehrenfests' coarse-graining idea. *Phys. Rev. E* **2006**, *74*, 037703.
28. Brownlee, R.A.; Gorban, A.N.; Levesley, J. Nonequilibrium entropy limiters in lattice Boltzmann methods. *Physica A* **2008**, *387*, 385–406.
29. Gorban, A.N.; Packwood, D.J. Enhancement of the stability of lattice Boltzmann methods by dissipation control. *Physica A* **2014**, *414*, 285–299.
30. Karlin, I.V.; Bösch, F.; Chikatamarla, S.S. Gibbs' principle for the lattice-kinetic theory of fluid dynamics. *Phys. Rev. E* **2014**, *90*, 031302.
31. Mattila, K.K.; Hegele, L.A.; Philippi, P.C. Investigation of an entropic stabilizer for the lattice-Boltzmann method. *Phys. Rev. E* **2015**, *91*, 063010.
32. Bösch, F.; Chikatamarla, S.S.; Karlin, I.V. Entropic multi-relaxation lattice Boltzmann scheme for turbulent flows. *Phys. Rev. E* **2015**, *92*, 043309.
33. Bösch, F.; Chikatamarla, S.S.; Karlin, I.V. Entropic Multi-Relaxation Models for Simulation of Fluid Turbulence. Available online: http://arxiv.org/abs/1507.02509 (accessed on 8 December 2015).
34. Finne, A.P.; Araki, T.; Blaauwgeers, R.; Eltsov, V.B.; Kopnin, N.B.; Krusius, M.; Skrbek, L.; Tsubota, M.; Volovik, G.E. An intrinsic velocity-independent criterion for superfluid turbulence. *Nature* **2003**, *424*, 1022–1025.
35. Vogelsberger, M.; Genel, S.; Springel, V.; Torrey, P.; Sijacki, D.; Xu, D.; Snyder, G.; Bird, S.; Nelson, D.; Hernquist, L. Properties of galaxies reproduced by a hydrodynamic simulation. *Nature* **2014**, *509*, 177–182.
36. Cavagna, A.; Giardina, I.; Ginelli, F.; Mora, T.; Piovani, D.; Tavarone, R.; Walczak, A.M. Dynamical maximum entropy approach to flocking. *Phys. Rev. E* **2014**, *89*, 042707.
37. Barato, A.C.; Seifert, U. Unifying Three Perspectives on Information Processing in Stochastic Thermodynamics. *Phys. Rev. Lett.* **2014**, *112*, 090601.

38. Han, B.; Wang, J. Least dissipation cost as a design principle for robustness and function of cellular networks. *Phys. Rev. E* **2008**, *77*, 031922.
39. England, J.L. Statistical physics of self-replication. *J. Chem. Phys.* **2013**, *139*, 121923.

Hydrodynamic Force Evaluation by Momentum Exchange Method in Lattice Boltzmann Simulations

Binghai Wen, Chaoying Zhang and Haiping Fang

Abstract: As a native scheme to evaluate hydrodynamic force in the lattice Boltzmann method, the momentum exchange method has some excellent features, such as simplicity, accuracy, high efficiency and easy parallelization. Especially, it is independent of boundary geometry, preventing from solving the Navier–Stokes equations on complex boundary geometries in the boundary-integral methods. We review the origination and main developments of the momentum exchange method in lattice Boltzmann simulations. Then several practical techniques to fill newborn fluid nodes are discussed for the simulations of fluid-structure interactions. Finally, some representative applications show the wide applicability of the momentum exchange method, such as movements of rigid particles, interactions of deformation particles, particle suspensions in turbulent flow and multiphase flow, *etc.*

Reprinted from *Entropy*. Cite as: Wen, B.; Zhang, C.; Fang, H. Hydrodynamic Force Evaluation by Momentum Exchange Method in Lattice Boltzmann Simulations. *Entropy* **2015**, *17*, 8240–8266.

1. Introduction

Fluid-structure interaction plays an important role in a variety of physical phenomena and many fields of engineering applications. For the computational fluid dynamics (CFD), hydrodynamic force evaluation is a key junction reflecting the interaction between fluid and structure. Especially, for the simulations of an object moving in fluid, accurate hydrodynamic force evaluation is a prerequisite to exactly depict the behaviors of the object. Over the past two decades, the lattice Boltzmann method (LBM) [1–5] has developed into a promising and alternative numerical approach for the simulations of complex fluid flows [6–9]. Hydrodynamic force evaluation in LBM mainly includes the momentum exchange method [10–13], the stress integration method [14–16], the immersed boundary method [17–19], *etc.* The methods based on the boundary integration have difficulty to solve the Navier–Stokes equations on complex boundary geometries, as well as a challenge to find a suitable computational mesh to compute fluid flow [10]. Owing to the regular lattice, the discrete velocities and the handy density distribution functions in LBM, the momentum exchange method is convenient to implement and is highly efficient for parallel performance. In recent years, some improvements are proposed to make the method more accurate [20–22]. Remarkably, the Galilean invariant improvement [13] promotes the momentum exchange method to become an exact scheme for hydrodynamic force evaluation without any loss of its simplicity and efficiency. Nowadays, it is very easy to implement the momentum exchange method for the simulations of fluid-structure interactions based on LBM.

The review is organized as follows. Section 2 briefly summarizes the lattice Boltzmann method. Section 3 describes the origin, theory and development of the momentum exchange method in detail. Section 4 is focused on the refill of new fluid nodes for the simulations of moving boundaries. Section 5 introduces a few kinds of application of the momentum exchange method, including rigid particle movements, deformable particle interactions, particle suspensions in turbulent flow and in multiphase flow. Finally, Section 6 presents the conclusions.

2. The Lattice Boltzmann Method

With its roots in the cellular automaton concept and kinetic theory, the lattice Boltzmann equation can recover the incompressible Navier–Stokes equations in the nearly incompressible limit [6–8]. Discretized fully in space, time and velocity, the lattice Boltzmann equation (LBE) can be concisely written as

$$f_i(\mathbf{x}+\mathbf{e}_i, t+1) - f_i(\mathbf{x}, t) = \boldsymbol{\Omega}(f_i) \tag{1}$$

where $f_i(\mathbf{x}, t)$ is the particle distribution function at lattice site $\mathbf{x}$ and time t, moving along the direction defined by the discrete speeds $\mathbf{e}_i$ with $i = 0, ..., N$, $\boldsymbol{\Omega}(f_i)$ is the collision operator and the time step takes 1 in the review. The mass density and the momentum density are defined by

$$\rho = \sum f_i, \ \rho\mathbf{u} = \sum \mathbf{e}_i f_i + \tau\mathbf{g} \tag{2}$$

where $\mathbf{g}$ is the acceleration due to force of gravity. One can consider f_i to be a mass component of a lattice node, and $\mathbf{e}_i f_i$ to be the corresponding momentum component.

With the different collision operators, several variations of the LBE can be read as the single-relaxation-time model [2–5], the multiple-relaxation-time model [23,24], the two-relaxation-time model [25], the entropic lattice Boltzmann equation [26,27], *etc.* It should be noted that the momentum exchange method is based on the momentum components and is independent of the given form of collision operator.

Using the two-dimensional model with nine velocities on a square lattice, of which the discrete velocity set is $e = \{(0,0), (1,0), (0,1), (-1,0), (0,-1), (1,1), (-1,1), (-1,-1), (1,-1)\}$, the single-relaxation-time collision operator can be written as [2]

$$\boldsymbol{\Omega}(f_i) = -\frac{1}{\tau}\left[f_i(\mathbf{x}, t) - f_i^{(eq)}(\mathbf{x}, t)\right] \tag{3}$$

while the equilibrium distribution function is

$$f_i^{(eq)} = \rho\omega_i[1 + 3(\mathbf{e}_i \cdot \mathbf{u}) + \frac{9}{2}(\mathbf{e}_i \cdot \mathbf{u})^2 - \frac{3}{2}\mathbf{u}^2] \tag{4}$$

where ω_i is the weighting coefficient $\omega_0 = 4/9$, $\omega_{1,2,3,4} = 1/9$, $\omega_{5,6,7,8} = 1/36$, and $\mathbf{u}$ is the fluid velocity calculated by Equation (2). The viscosity in the macroscopic equations is $\nu = \dfrac{2\tau - 1}{6}$.

With the most general form which is derived from the linearized collision model, the multiple-relaxation-time collision operator can be defined as [28–30]

$$\Omega(f_i) = -\mathrm{M}^{-1} \cdot \mathrm{S} \cdot \left[\mathbf{m} - \mathbf{m}^{(\mathrm{eq})}\right] \tag{5}$$

where $\mathbf{m}$ and $\mathbf{m}^{(\mathrm{eq})}$ represent the velocity moments of the distribution functions and their equilibria, respectively. For the model with two dimensions and nine discrete velocities, i is an integer $0 \leqslant i \leqslant 8$ and the velocity moments are $\mathrm{m} = (\rho, e, \varepsilon, j_x, q_x, j_y, q_y, p_{xx}, p_{xy})^{\mathrm{T}}$. The conserved moments are the density ρ and the flow momentum $\mathbf{j} = (j_x, j_y) = \rho\mathbf{u}$, $\mathbf{u}$ is the local velocity. The equilibria of nonconserved moments depend only on the conserved moments: $e^{(eq)} = -2\rho + \dfrac{3}{\rho}(j_x^2 + j_y^2)$, $\varepsilon^{(eq)} = \rho - \dfrac{3}{\rho}(j_x^2 + j_y^2)$, $q_x^{(eq)} = -j_x$, $q_y^{(eq)} = j_y$, $p_{xx}^{(eq)} = \dfrac{1}{\rho}(j_x^2 - j_y^2)$, $p_{xy}^{(eq)} = \dfrac{1}{\rho}(j_x j_y)$. M is a linear transformation matrix mapping between discrete velocity space and moment space, $\mathrm{m} = \mathrm{M} \cdot \mathrm{f}$ and $\mathrm{f} = \mathrm{M}^{-1} \cdot \mathrm{m}$. S is a diagonal matrix of nonnegative relaxation factors and is given by $\mathrm{S} = \mathrm{diag}(0, s_e, s_\varepsilon, 0, s_q, 0, s_q, s_v, s_v)$. Then the shear viscosity is $\nu = \dfrac{1}{3}\left(\dfrac{1}{s_v} - \dfrac{1}{2}\right)$.

The evolution of the LBE can be decomposed into two elementary steps, namely collision and advection, to reveal the flow phenomena at the mesoscopic scale [6]:

$$\text{collision: } \tilde{f}_i(\mathbf{x}, t) = f_i(\mathbf{x}, t) + \Omega(f_i) \tag{6}$$

$$\text{advection: } f_i(\mathbf{x} + \mathbf{e}_i, t + 1) = \tilde{f}_i(\mathbf{x}, t) \tag{7}$$

where f_i and $\tilde{f}_i$ denote pre-collision and post-collision states of the particle distribution functions, respectively. The collision step, as the dominant part of the computations, is completely local, hence the full discrete equation is natural to parallelize.

3. The Momentum Exchange Method

3.1. The Original Particulate Suspensions by Ladd

Ladd created the original momentum exchange method in the lattice Boltzmann method in order to evaluate hydrodynamic interactions for the numerical simulations of particulate suspensions [10,11]. His pioneer studies promoted the lattice Boltzmann method to develop into a popular tool for the simulations of fluid-solid interaction, which, nowadays, is still one of the most active fields in LBM. Ladd defined the suspension particle by a boundary shell and treated all lattices, both inner and outer of the solid particle, in an identical fluid fashion. The particulate boundary, as shown in Figure 1, is laid approximately and discretely at the middle of every fluid-solid links, each of which crosses the boundary and connects a fluid node with a solid node.

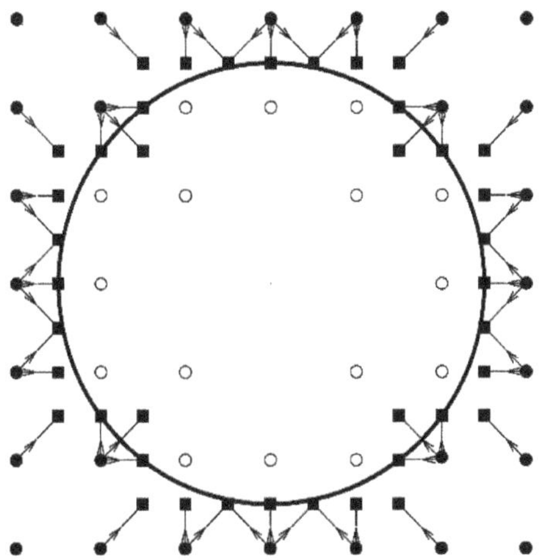

Figure 1. Location of boundary nodes for a curved surface. The velocities along links cutting the boundary surface are indicated by arrows. The locations of the boundary nodes are shown by solid squares, and the fluid nodes by solid circles. The open circles indicate those nodes in the solid adjacent to fluid nodes. (Ladd, 1994 [10]).

Taking into account the movement of the particulate surface, each of the distribution functions on the fluid-solid links is then updated by the following simple rule. Assuming that a moving boundary is intersected at $\mathbf{x}_b$ by a fluid-solid link which connects a solid node $\mathbf{x}_s$ and a fluid node $\mathbf{x}_f$, and the discrete velocity $\mathbf{e}_i$

is from $\mathbf{x}_f$ to $\mathbf{x}_s$, a momentum item computed by the boundary velocity is added to the distribution functions which are bounced back from the particulate boundary

$$f_{\bar{i}}(\mathbf{x}_f,t+1) = \tilde{f}_i(\mathbf{x}_f,t) - \frac{2\omega_i\rho}{c_s^2}(\mathbf{e}_i \cdot \mathbf{u}_b) \tag{8}$$

$$f_i(\mathbf{x}_s,t+1) = \tilde{f}_{\bar{i}}(\mathbf{x}_s,t) - \frac{2\omega_{\bar{i}}\rho}{c_s^2}(\mathbf{e}_{\bar{i}} \cdot \mathbf{u}_b) \tag{9}$$

where c_s is the sound speed and $\mathbf{u}_b$ is the boundary velocity at the intersection. The momentum-exchange occurs during the advection step, and the momentum-exchange value on a fluid-solid link in a time step, namely the force, is written as [10,11]:

$$\mathbf{F}_i(\mathbf{x}_b, t+\frac{1}{2}) = 2\left[\tilde{f}_i(\mathbf{x}_f,t) - \tilde{f}_{\bar{i}}(\mathbf{x}_s,t) - \frac{2\omega_i\rho}{c_s^2}(\mathbf{e}_i \cdot \mathbf{u}_b)\right]\mathbf{e}_i \tag{10}$$

For a circular or spherical rigid particle suspended in fluid, the total hydrodynamic force $\mathbf{F}$ as well as the torque $\mathbf{T}$ exerting on the particle are calculated by

$$\mathbf{F}(t) = \sum \mathbf{F}(\mathbf{x}_b, t) \tag{11}$$

and

$$\mathbf{T}(t) = \sum (\mathbf{x}_b - \mathbf{R}) \times \mathbf{F}(\mathbf{x}_b, t) \tag{12}$$

where $\mathbf{R}$ is the mass center of the solid particle. The summations in Equations (11) and (12) run over all the fluid-solid links. The boundary velocity of point $\mathbf{x}_s$ is computed by the particulate translational velocity $\mathbf{U}$ and the angular velocity $\boldsymbol{\Omega}$ [31]

$$\mathbf{u}_b(t) = \mathbf{U}(t) + \boldsymbol{\Omega}(t) \times (\mathbf{x}_b - \mathbf{R}) \tag{13}$$

The time evolutions of the particle velocity and angle velocity are found by solving Newton's equations of motion,

$$\mathbf{U}(t+1) \equiv \mathbf{U}(t) + \frac{\mathbf{F}(t)}{M} + \frac{(\rho_p - \rho)}{\rho_p}\mathbf{g} \tag{14}$$

and

$$\boldsymbol{\Omega}(t+1) \equiv \boldsymbol{\Omega}(t) + \frac{\mathbf{T}(t)}{I} \tag{15}$$

where I is the moment of inertia, M is the mass of the particle and ρ_p is the particle density. For an uniform circle or sphere with the radius r, the particle mass is computed by $M = \pi r^2 \rho_p$ and $M = \frac{4}{3}\pi r^3 \rho_p$, while the moment of inertia is computed by $I = \frac{1}{2}r^2 M$ and $I = \frac{2}{5}r^2 M$, respectively.

Ladd's method treats both fluid node and particle node as fluid, therefore it is very simple to update the motion of particles. Especially, the method remains the conservations of mass and momentum locally. However, the solid particle is indeed different from the interior fluid. Although the fluid movement in the particle is closely similar to that of a rigid solid [11], the inertial lag of the fluid is obvious at short times, and the contribution of the interior fluid on the force and torque of the particle reduces the stability of the particle velocity update [8]. Nguyen and Ladd [32,33] upgraded the original model by removing the effect of the interior fluid from fluid-particle momentum exchanges and proposed an effective lubrication force for particles in near contact. Başağaoğlu *et al.* [34,35] applied the upgraded model to investigate the lateral migration of a particle in a horizontal channel and a microchannel at different Reynolds numbers.

3.2. The Direct Particle Simulations by Aidun et al.

Aidun *et al.* [12,36] removed the interior fluid from the suspending particle and considered the particle as a real and impermeable one. They arranged the particle boundary approximately at the midpoint of fluid-solid links same as Ladd's method and applied the halfway bounce-back condition to calculate the distribution functions from solid to fluid. A momentum item including the boundary velocity is added into the distribution function which is bounced back from the particle boundary. For a post-collision distribution function $\widetilde{f}_i$, whose discrete velocity $\mathbf{e}_i$ has the direction from a fluid node to a solid one, the momentum component $\mathbf{e}_i\widetilde{f}_i$ moves into the particle and gives the particle a momentum increment. On the opposite direction, $\mathbf{e}_{\bar{i}}\widetilde{f}_{\bar{i}}$ moves out of the particle and gives the particle a momentum decrement. Thus, the momentum change value of the boundary on the fluid-solid link in a time step, namely the force, is written as [12,36,37]:

$$\mathbf{F}_i(\mathbf{x}_b) = \mathbf{e}_i\widetilde{f}_i(\mathbf{x}_f, t) - \mathbf{e}_{\bar{i}}\widetilde{f}_{\bar{i}}(\mathbf{x}_s, t) = \mathbf{e}_i[\widetilde{f}_i(\mathbf{x}_f, t) + \widetilde{f}_{\bar{i}}(\mathbf{x}_s, t)] \tag{16}$$

where $\widetilde{f}_{\bar{i}}(\mathbf{x}_s, t)$ is calculated by the half-way bounce-back boundary condition

$$\widetilde{f}_{\bar{i}}(\mathbf{x}_s, t) = \widetilde{f}_i(\mathbf{x}_f, t) - \frac{2\omega_i \rho}{c_s^2}(\mathbf{e}_i \cdot \mathbf{u}_b) \tag{17}$$

Equation (17) was also derived by the work of Nguyen and Ladd [32], in which fluid occupies the entire region, but fluid inside the particle does not contribute to particle-fluid hydrodynamics. This approach ensures the continuity in the flow field and avoids large artificial pressure gradients, which are caused by the expansion and compression of the fluid near the particle surface. We call the Equation (16) as the conventional momentum exchange (CME) equation.

A common drawback in Ladd's and Aidun's methods is that the boundary geometry, which is located at the middles of fluid-solid links, is zigzag. Mei *et al.* [37] employed the curved boundary conditions [38–40] in the momentum-exchange method, thus, on the grid level, the particulate geometry could be accurately depicted. The distribution functions bounced back from the solid boundary are calculated by curved boundary condition, and the force evaluation is based on the real particulate geometry instead of the previous stepping edges. They also verified that the momentum exchange method, namely Equation (16), is accurate on a stationary boundary for both two-dimensional and three-dimensional flows.

Another improvement by Aidun *et al.* [12] was that the hydrodynamic force evaluation of moving solid particles involved the momenta raised from the lattice type changes of the covered and uncovered nodes. Ding and Aidun [41] further studied lubrication forces between particles in near contact and hydrodynamic interactions between two solid objects in relative motion. Wen *et al.* [20] investigated carefully the effect of these type-changing lattices and applied the curved boundary conditions [38–40,42–45] to simulate moving boundaries.

In the numerical simulations of fluid-structure interactions in LBM, every of the momentum components moving through the boundary will alter the boundary momentum. Figure 2A shows that, in a time step, a boundary shifts from the dotted curve to the real one, the initial fluid node (a_1) is devoured by the solid boundary and changes into a newborn solid lattice node. Every of momentum components on the lattice node moves into the particulate boundary and provides a momentum increment. Thus, the impulse force caused by the type changing of the node (a_1) is written as [12,20]:

$$\mathbf{F}(\mathbf{x}_c) = \sum_i \mathbf{e}_i \tilde{f}_i(\mathbf{x}_c, t) \tag{18}$$

where x_c represents the lattice node altering from fluid to solid and relates to the node (a_1) in Figure 2A.

In the same style, when a boundary shifts, the previous solid lattice node (a_2) is uncovered and changes into a newborn fluid one, as shown in Figure 2B. Every of momentum components on the lattice shifts out of the boundary and provides a momentum decrement. Thus, the impulse force produced by the newborn lattice node (a_2) is written as [12,20]:

$$\mathbf{F}(\mathbf{x}_c) = -\sum_i \mathbf{e}_i \tilde{f}_i(\mathbf{x}_c, t) \tag{19}$$

where x_c represents the lattice node altering from solid to fluid and concerns to the node (a_2) in Figure 2B.

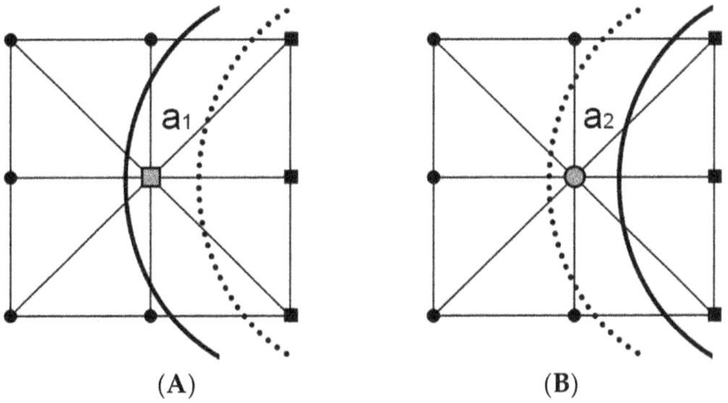

Figure 2. The lattice type is changed if the boundary shifts from the dotted curve to the real one. Squares denote the particle and circles denote the fluid. (**A**) The shaded square a_1 represents a newborn solid lattice node changing from a fluid one. (**B**) The shaded circle a_2 represents a newborn fluid lattice node changing from a solid one. (Wen *et al.*, 2012 [20]).

The impulse forces exerted by the covered/uncovered lattice nodes are added into the conventional momentum exchange equation, therefore the hydrodynamic force on a moving boundary includes two parts: one is calculated on fluid-solid links, and the other is complemented by the impulse forces on type-changing lattices. Therefore, the total hydrodynamic force and torque acting on the particle are now defined as:

$$\mathbf{F} = \sum \mathbf{F}(\mathbf{x}_b) + \sum \mathbf{F}(\mathbf{x}_c) \tag{20}$$

and

$$\mathbf{T} = \sum (\mathbf{x}_b - \mathbf{R}) \times \mathbf{F}(\mathbf{x}_b) + \sum (\mathbf{x}_c - \mathbf{R}) \times \mathbf{F}(\mathbf{x}_c) \tag{21}$$

where the summation of x_b is on all fluid-solid links and the summation of x_c is on all covered/uncovered lattice nodes.

Wen *et al.* [20] verified the accuracy of the method by simulating a series of cylinder sedimentations and the Segré–Silberberg effects [46,47]. However, it is really a discrete event that a lattice node passes through a moving boundary. Therefore, the impulse force leads to a significant force fluctuation and may reduce the simulation stability, so that a time average of velocity is necessary to smooth the velocity profile.

3.3. The Improved Schemes by Caiazzo, Chen, Hu, et al.

In recent years, a few schemes were proposed to improve the accuracy and Galilean invariance of the momentum exchange method. Caiazzo and Junk [48] presented an modified momentum exchange method according to the asymptotic expansion technique,

$$\mathbf{F}_i(\mathbf{x}_b) = \mathbf{e}_i \widetilde{f}_i(\mathbf{x}_f, t) - \mathbf{e}_{\bar{i}} \widetilde{f}_{\bar{i}}(\mathbf{x}_s, t) - 2\omega_i \mathbf{e}_i - \omega_i c_s^{-2}(c_s^{-2} \left|\mathbf{e}_i \cdot \mathbf{u}_b\right|^2 - \mathbf{u}_b^2)\mathbf{e}_i \quad (22)$$

Clausen and Aidun [49] obtained a similar correction to reduce the error of normal stress and investigated the effect on the rheological properties in particle suspensions. Lorrenz *et al.* [50] investigated Galilean invariance and accuracy of the improved method by simulating particle suspensions with Lees-Edwards boundary conditions [51] and a shear flow test.

For a particle suspension model without fluid inside, simulations by Wen *et al.* [20] and Chen *et al.* [22] showed when the impulse force was not included, the numerical results of the conventional exchange equation deviated from both results of finite element method and LBM with the stress integration method, no matter the curved boundary condition or the halfway bounce-back boundary condition was used. Because the conventional momentum exchange equation, namely Equation (16), for stationary boundaries was verified to be accurate [37] and the impulse force was not necessary for LBM with the stress integration method, Chen *et al.* [22] thought that the problem could lie in the calculation of momentum exchange in the moving boundary treatment.

Considering a distribution function as a mass component, the distribution function will gain an additional momentum when it collides with a moving boundary. Due to the constant discrete velocities in LBM, the additional part has to be modified by adjusting the particle distribution function in the bounce back procedure like the last part of Equation (17). The modification leads to a net mass transfer on a fluid-solid link through the physical boundary for the direct simulation of suspending particles without interior fluid. From another angle, the net mass transfer can be seen as a bit of fluid mass which is covered (or uncovered) and is injected, at the time step, back to (or down from) the fluid field [22], as shown in Figure 3. The initial momenta of the net fluid mass must be complemented to CME and a straightforward correct is given by Chen *et al.* [22]:

$$\mathbf{F}_i(\mathbf{x}_b) = \mathbf{e}_i \widetilde{f}_i(\mathbf{x}_f, t) - \mathbf{e}_{\bar{i}} \widetilde{f}_{\bar{i}}(\mathbf{x}_s, t) - \frac{2\omega_i \rho}{c_s^2}(\mathbf{e}_i \cdot \mathbf{u}_b)\mathbf{u}_b \quad (23)$$

Associating with the Aidun's method [12,20], the total impulse produced from Equations (18) and (19) is equal to the total initial momenta of the fluid which is covered or uncovered by the unit length boundary when the boundary shifts from a lattice node to its neighbor.

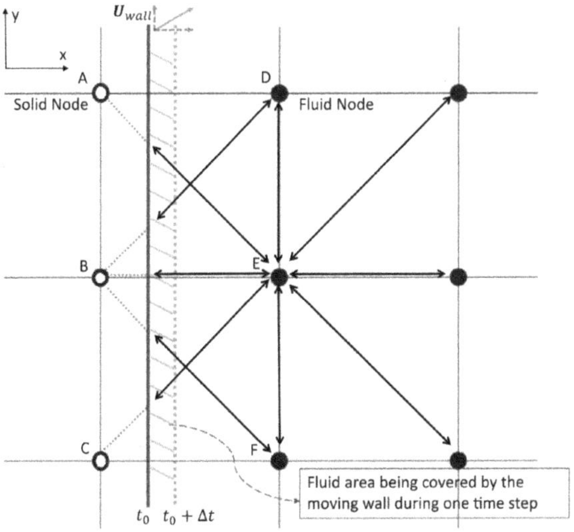

Figure 3. A schematic illustration of a moving boundary on lattice grid. (Chen *et al.*, 2013 [22]).

Based on the finite-volume lattice Boltzmann method, Hu *et al.* [52] proposed a modified momentum exchange method to compute the interactions between fluid and particle. Their aim is to remove the common restriction in the momentum exchange method, in which the boundary points are set at the middle of the grid lines or the intersection of the solid boundaries and the grid lines. The particulate surface is described by some arc (area) elements, and the inside fluid is also used. Considering the control volume, the momentum exchange method is modified by

$$\mathbf{F}_i(\mathbf{x}_b) = 2\mathbf{e}_i \left[\widetilde{f}_{\bar{i}}(\mathbf{x}_s, t) - \frac{\omega_i \rho}{c_s^2}(\mathbf{e}_i \cdot \mathbf{u}_b) \right] V_{\bar{i}} \tag{24}$$

where $V_{\bar{i}}$ is the area (volume) of the local curved edge. By means of numerical integration, the fluid mass which collide with an arc (area) element in the control volume is obtained.

3.4. *The Galilean Invariant Hydrodynamics Force by Wen et al.*

The interfacial momentum transfer can be generalized by a common schematic diagram as shown in Figure 4, in which a moving boundary is located between a fluid node $\mathbf{x}_f$ and a boundary node $\mathbf{x}_s$ and the boundary has a velocity $\mathbf{v}$ at the point of intersection $\mathbf{b}$. In the collision step, the distribution function $\widetilde{f}_{\bar{i}}(\mathbf{x}_b,t)$ has to be calculated by the half-way bounce-back boundary condition [10,12] or the curved boundary conditions [20,22,37,53], in which the forcing terms [54–56] must be included based on the boundary velocity.

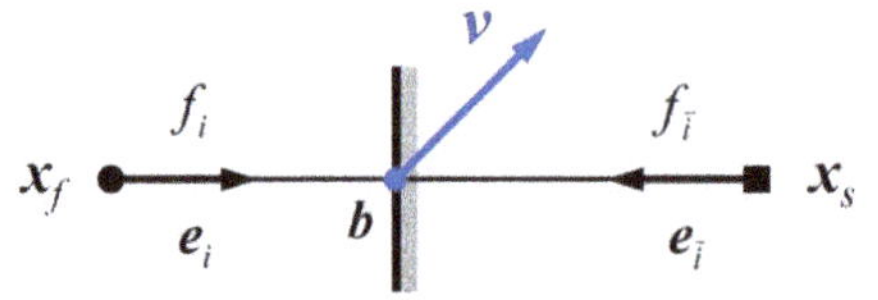

Figure 4. A common schematic diagram to illustrate a moving boundary crossing a fluid-solid link at the point of intersection $\boldsymbol{b}$. $\boldsymbol{x}_f$ and $\boldsymbol{x}_s$ denote the adjacent fluid and solid nodes. The boundary has a velocity $\boldsymbol{v}$ at the point $\boldsymbol{b}$. (Wen *et al.*, 2014 [13]).

Galilean invariance is a fundamental physical property; however, although the dynamics of lattice Boltzmann equation in hydrodynamic restrict meets Galilean invariance [57], this property needs a specific consideration in the treatment of the fluid-structure interactions. According to the theorem of momentum, the momentum transfer through a moving boundary is correlated to the relative velocity and then is independent of the speed of reference frame. Crossing the point of intersection $\mathbf{b}$, the mass component $\widetilde{f}_i(\mathbf{x}_f,t)$ has the velocity $(\mathbf{e}_i - \mathbf{v})$ relative to the boundary and contributes a momentum increment $(\mathbf{e}_i - \mathbf{v})\widetilde{f}_i(\mathbf{x}_f,t)$ to the boundary. Simultaneously, the mass component $\widetilde{f}_{\bar{i}}(\mathbf{x}_s,t)$ has the relative velocity $(\mathbf{e}_{\bar{i}} - \mathbf{v})$ and decreases a momentum $(\mathbf{e}_{\bar{i}} - \mathbf{v})\widetilde{f}_{\bar{i}}(\mathbf{x}_s,t)$ from the boundary. Thus, the Galilean invariant momentum exchange method (GME) can be defined by

$$\mathbf{F}_i(\mathbf{x}_b) = (\mathbf{e}_i - \mathbf{v})\widetilde{f}_i(\mathbf{x}_f,t) - (\mathbf{e}_{\bar{i}} - \mathbf{v})\widetilde{f}_{\bar{i}}(\mathbf{x}_s,t) \tag{25}$$

It is clear that GME turns into CME when the boundary is motionless. GME evaluates the hydrodynamic force in the fluid-structure interaction and works on the motion state of a moving boundary, but has not any direct influence on distribution functions. The similar relative velocity was mentioned in the study of Krithivasan *et al.* [58].

A simple theoretical analysis is employed to compare GME and CME. Suppose the system in Figure 4 is relative static and is physically related to a reference frame with arbitrary uniform velocity $-\mathbf{v}$. It is equivalent to an equilibrium system in which the fluid and boundary have the same uniform velocity $\mathbf{v}$. Substituting Equation (4) into Equation (16), the hydrodynamic force on a fluid-solid link can be evaluated:

$$\begin{aligned}\mathbf{F}_i &= \mathbf{e}_i f_i^{(eq)}(\mathbf{x}_f,t) - \mathbf{e}_{\bar{i}} f_{\bar{i}}^{(eq)}(\mathbf{x}_s,t)\\ &= \mathbf{e}_i \rho\omega_i[1 + 3(\mathbf{e}_i\cdot\mathbf{v}) + \frac{9}{2}(\mathbf{e}_i\cdot\mathbf{v})^2 - \frac{3}{2}\mathbf{v}^2] - \mathbf{e}_{\bar{i}}\rho\omega_{\bar{i}}[1 + 3(\mathbf{e}_{\bar{i}}\cdot\mathbf{v}) + \frac{9}{2}(\mathbf{e}_{\bar{i}}\cdot\mathbf{v})^2 - \frac{3}{2}\mathbf{v}^2]\\ &= 2\rho\omega_i\mathbf{e}_i + 3\rho\omega_i[3(\mathbf{e}_i\cdot\mathbf{v})^2 - \mathbf{v}^2]\mathbf{e}_i.\end{aligned} \tag{26}$$

Because of the term $3\rho\omega_i[3(\mathbf{e}_i \cdot \mathbf{v})^2 - \mathbf{v}^2]\mathbf{e}_i$, the hydrodynamic force is abnormally connected to the speed of the reference frame, and thus the conventional equation presents an inherent flaw of Galilean invariance.

As the discrete velocity $\mathbf{e}_i$ is constant, Galilean invariance cannot be satisfied on a single fluid-solid link, just like a single $\mathbf{e}_i$ cannot express the fluid velocity of a lattice node. However, since the discrete velocity set is symmetrical, the Galilean invariant force evaluation can be achieved locally on the lattice. Without loss of generality, the boundary is assumed to intersect with the discrete velocities, $\mathbf{e}_1$, $\mathbf{e}_5$ and $\mathbf{e}_8$. Substituting Equation (4) into Equation (25) and summating the three directions, the local hydrodynamic force can be analyzed

$$\begin{aligned}\mathbf{F} &= \mathbf{F}_1 + \mathbf{F}_5 + \mathbf{F}_8 \\ &= \sum_{i=1,5,8} [(\mathbf{e}_i - \mathbf{v}) f_i^{(eq)}(\mathbf{x}_f,t) - (\mathbf{e}_{\bar{i}} - \mathbf{v}) f_{\bar{i}}^{(eq)}(\mathbf{x}_s,t)] \\ &= \sum_{i=1,5,8} \{2\rho\omega_i \mathbf{e}_i + 3\rho\omega_i[3(\mathbf{e}_i \cdot \mathbf{v})^2 - \mathbf{v}^2]\mathbf{e}_i - 6\rho\omega_i(\mathbf{e}_i \cdot \mathbf{v})\mathbf{v}\} \\ &= \sum_{i=1,5,8} 2\rho\omega_i \mathbf{e}_i.\end{aligned} \tag{27}$$

The local hydrodynamic force remains constant regardless of the reference speed, therefore GME is proven to be completely Galilean invariant in the equilibrium system.

Another simple simulation can quantitatively show the difference between GME and CME. A vertical thin plate is placed in the relatively static fluid without boundary. GME and CME are used to compute the one-sided pressure of the plate and the equilibrium system is connected to a reference frame. The percentage of the computational errors by GME and CME are shown in Figure 5. It is clear that CME violates Galilean invariance whereas GME fully satisfies in the equilibrium system. The case is independent of the relaxation time and the plate's length.

GME is further examined in a dynamic fluid field in which a cylinder sedimentates in a vertical channel. The width of the channel is 0.4 cm and the cylinder diameter is 0.1 cm. The fluid and particle densities are 1 $\mathrm{g/cm^3}$ and 1.03 $\mathrm{g/cm^3}$, and the kinematic viscosity is 0.01 $\mathrm{cm^2/s}$. The cylinder is located at 0.076 cm to the left channel wall, and then it moves under the gravity acceleration $|G| = 980\ \mathrm{cm^2/s}$. The width of the channel is 120 lattice units and the length is 10 times the width. The simulations apply the second-order interpolation boundary condition [56] on the single-relaxation-time model with the relaxation time $\tau = 0.6$. The particle mass is 8.0896×10^{-3} g and the inertia moment is $1.0112 \times 10^{-5}\ \mathrm{g{\cdot}cm^2}$. The Reynolds numbers is defined by $Re = du_p/v$, where d is the channel width, u_p is the final velocity of the particle and v is the kinematic viscosity. The highly consistent results are obtained by using the multireflection boundary condition [59] on the multiple-relaxation-time model with the diagonal relaxation matrix **S** = diag(0, 1.64, 1.54, 0, 1.9, 0, 1.9,1/τ,1/τ) [23,30].

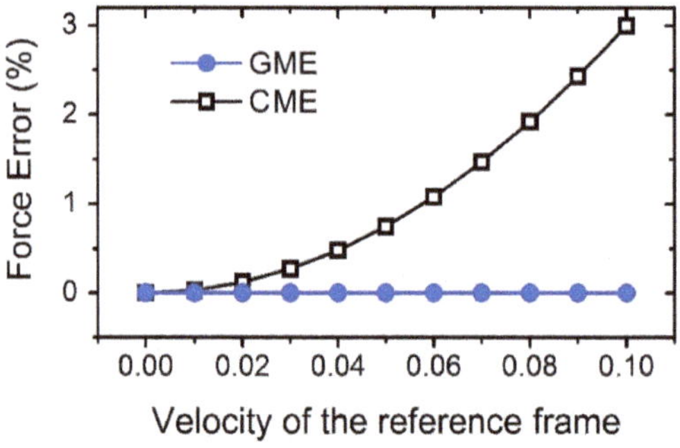

Figure 5. Relative errors in the one-sided pressure on a vertical thin plate in the relatively stationary fluid without boundaries. This equilibrium system is connected to various velocities of the reference frame. Since it properly considers the boundary velocity, Galilean invariant momentum exchange method (GME) is the Galilean invariant and thus has a very high computational accuracy. (Wen *et al.*, 2014 [13]).

Figure 6a,b draws the compare with the simulating results from the momentum exchange methods by Aidun *et al.* (ALD) [12] and the lattice-type-dependent momentum exchange method (LME) [20], together with the benchmarks calculated by the arbitrary Lagrangian–Eulerian technique (ALE) [60,61]. The hydrodynamic forces computed by GME extremely agree with the benchmarks, while the results by ALD and LME have large fluctuations. Please note that all of the data from GME are raw, whereas the data from ALD and LME have been smoothed by using the adjacent-averaging method per 30 points for the horizontal forces and per 100 points for the vertical forces. As so much improvement in the force evaluation is achieved, the force fluctuation of GME is very small and the time average becomes unnecessary.

Figure 7a,b draws the accuracy of the velocity and angle velocity computed by GME, ALD and LME. All velocities from GME are very smooth and in excellent agreement with the ALE benchmarks, whereas the results from ALD and LME clearly fluctuate with some deviations. The GME results are so accurate and steady that the time average of the velocities is totally unnecessary.

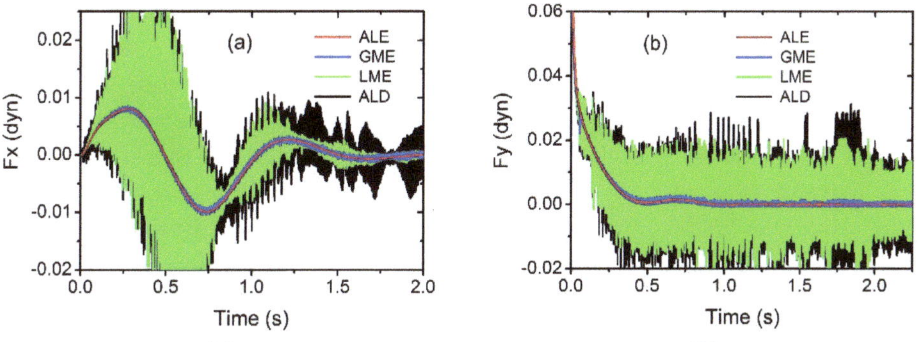

Figure 6. (**a**) Time-dependent horizontal forces and (**b**) time-dependent vertical forces evaluated by GME, lattice-type-dependent momentum exchange method (LME), and Aidun's method (ALD), compared with the ALE benchmark. The density of the cylinder is 1.03 g/cm^3. The GME data is raw, whereas the ALD and LME data have been smoothed by the adjacent-averaging method. (Wen *et al.*, 2014 [13]).

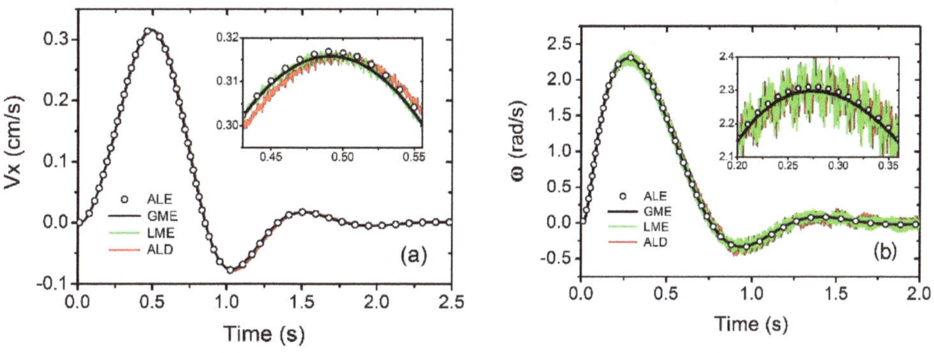

Figure 7. Time-dependent (**a**) horizontal velocities and (**b**) angular velocities evaluated by GME, LME, and ALD, compared with the ALE benchmark. (Wen *et al.*, 2014 [13]).

The deviations of forces and velocities are quantitatively analyzed by a relative L_2-norm error, which is defined by

$$E = \frac{\{\int [f(t) - F(t)]^2 dt\}^{\frac{1}{2}}}{\{\int [F(t)]^2 dt\}^{\frac{1}{2}}} \quad (28)$$

To investigate the effects of lattice scale, the lattice width of the channel increases gradually from 50 to 200 lattice units, while the length remains 10 times the width. For Equation (28), *f(t)* is a LBM result and *F(t)* is an ALE result. Figure 8a illustrates that the relative errors of the GME results rapidly decrease with the increase of the lattice scale. However, the relative errors of the ALD and LME results always remain very high and are more than one order larger than those from GME. To investigate the influences of Reynolds number, we perform a set of simulations in which the particle densities increase from 1.02 to 1.22 g/cm^3, and the corresponding Reynolds number grows gradually from 6.13 to 34.75. As shown in Figure 8b, the relative L_2-norm errors reflect the fluctuation range in particle velocities. Here, *f(t)* is the simulation result and *F(t)* is the smoothed simulation result by the adjacent-averaging method per 20 points. It is clear that the GME results are more accurate and far steadier than the ALD and LME results. Peng *et al.* [62] confirmed the computational accuracy and Galilean invariance of GME by theoretical analyses and numerical simulations.

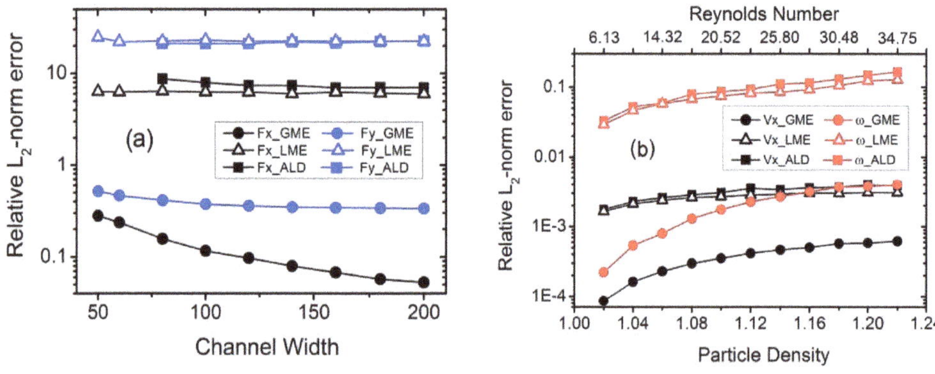

(**a**) (**b**)

Figure 8. (**a**) The relative L_2-norm error of the horizontal forces (Fx, black) and vertical forces (Fy, blue) with increasing lattice scales; (**b**) The relative L_2-norm errors of the horizontal velocity (Vx, black) and the angular velocity (ω, red) with various Reynolds numbers. (Wen *et al.*, 2014 [13]).

4. Refill of New Fluid Nodes

While a suspended particle is moving in fluid some lattices will be covered and uncovered by the particle, and then the type of these lattices will be changed consequently. Ladd's method does not need a further process because all lattice nodes, both inside and outside of a particle, are treated as fluid nodes. When an interior node changes into an exterior node, it is justified if the density of the interior node approximates the right characteristic of the exterior one. This may be only true in the situations that the particulate acceleration is low. If a particle is accelerated, the fluid directly behind the particle typically suffers a lower pressure while the adjacent inner node bears a rather high pressure [50]. This may reduce the stability of the particle velocity update. For the methods without inner fluid node, when a lattice node is changing from a particle node to a fluid one, its properties has to be refilled, such as density, velocity and distribution functions [13,20,22]. Here, we introduce some algorithms used in practice as follows.

Aidun *et al.* [12] presented a simple algorithm to refill the newborn fluid node when they directly simulated particle suspensions. When a boundary node is uncovered due to the movement of the solid particle and becomes a new fluid node, its density is obtained with the following relation:

$$\rho(\mathbf{x},t) = \frac{1}{N}\sum_{N} \rho(\mathbf{x}+\mathbf{e}_i,t) \tag{29}$$

where $\mathbf{x}$ is the newborn fluid lattice node and N indicates the number of fluid nodes adjacent to this lattice node. The equation shows that the fluid density of the newborn node is equal to the average density of its neighboring lattice nodes. The velocity of the solid boundary node at the same time step is used as the macroscopic velocity of the new fluid node,

$$\mathbf{u}(\mathbf{x},t) = \mathbf{U}(t) + \boldsymbol{\Omega}(t) \times (\mathbf{x} - \mathbf{R}(t)) \tag{30}$$

where $\mathbf{U}$ is the particle translational velocity, $\boldsymbol{\Omega}$ is the angular velocity and $\mathbf{R}$ is the particulate mass center. The distribution functions on the newly uncovered node are set as the equilibrium distribution functions calculated by the density and velocity above.

Lallemand and Luo [56] used a second-order normal extrapolation to calculate the missed distribution functions. Along the direction of a chosen discrete velocity $\mathbf{e}_i$, the extrapolation considers the boundary normal direction to maximize the quantity $\hat{\mathbf{n}} \cdot \mathbf{e}_i$, where $\hat{\mathbf{n}}$ is the normal vector out of the wall. For instance, the unknown distribution functions $\{f_i(\mathbf{x})\}$ at node $\mathbf{x}$ as drawn in Figure 9 can be calculated by the extrapolation formula as follow:

$$f_i(\mathbf{x}) = 3f_i(\mathbf{x}') - 3f_i(\mathbf{x}'') + f_i(\mathbf{x}''') \tag{31}$$

where $\mathbf{x}'''$ is the next lattice node along the direction $\mathbf{x}'$ to $\mathbf{x}''$.

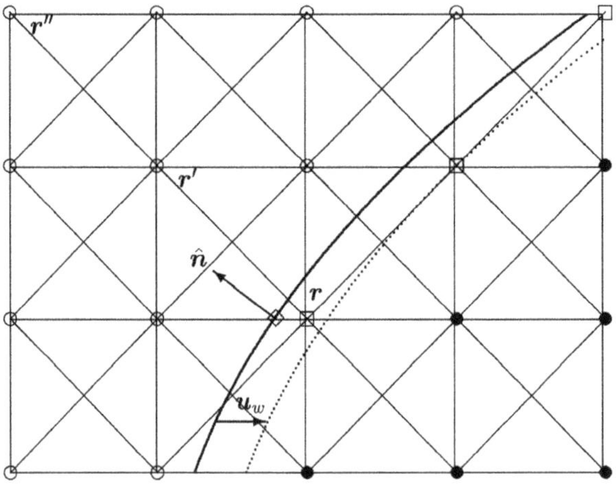

Figure 9. A boundary is moving with velocity u_w. The circles (○) and disks (●) denote fluid and boundary nodes, respectively. The squares (□) denote the nodes turning from boundary nodes to fluid ones at a time step. The solid and dotted curves are the positions of wall boundary at time t and t + 1, respectively. (Lallemand and Luo, 2003 [56]).

Caiazzo [63] suggested a refill scheme by reconstructing the equilibrium and non-equilibrium parts separately according to an asymptotic analysis prediction. The initialization of the populations includes an approximation of the non-equilibrium part which, in practice, can be copied simply from a neighbor of the new fluid node. The equilibrium part is computed basing on the density and velocity which can be gotten by a simple first order extrapolation.

$$f_i(\mathbf{x}, t) = f_i^{eq}(\mathbf{x}, t, \rho, \mathbf{u}) + f_i^{neq}(\mathbf{x} + \mathbf{e}_j, t) \tag{32}$$

where $\mathbf{e}_j$ indicates the extrapolation direction according to the boundary velocity at a point of interface close to the new fluid node. Lorenz *et al.* [50,51] employed a similar approach, in which the equilibrium part was calculated basing on the local pressure.

Krithivasan *et al.* [58] recently developed a diffuse bounce-back boundary condition to simulate moving boundaries instead of imposing the no-slip conditions. This scheme ensures positive-definite populations and retains the simplicity of bounce-back technique. Meanwhile it suggests a refill algorithm to model distributions at the fluid nodes uncovered due to solid movement by quasi-equilibrium distributions. The scheme is demonstrated to reduce force fluctuations and diminish the requirements of interpolation or extrapolation. However, a diffuse boundary condition will introduce some degree of boundary slip [64–66], which would damage simulating accuracy and Galilean invariance.

Fang *et al.* [13,67] used the fluid nodes around the newborn fluid nodes to extrapolate the distribution functions of the newborn fluid nodes. If the extrapolating participants are more than one, the newborn distribution functions are assigned as their average. Three extrapolation algorithms, namely neighbor-node average (A1), linear extrapolation (A2) and second order extrapolation (A3), are as follows [13]

$$\text{A1}: f_i(\mathbf{x}) = \frac{1}{N}\sum_N f_i(\mathbf{x}') \tag{33}$$

$$\text{A2}: f_i(\mathbf{x}) = \frac{1}{N}\sum_N 2f_i(\mathbf{x}') - f_i(\mathbf{x}'') \tag{34}$$

$$\text{A3}: f_i(\mathbf{x}) = \frac{1}{N}\sum_N 3f_i(\mathbf{x}') - 3f_i(\mathbf{x}'') + f_i(\mathbf{x}''') \tag{35}$$

where N is the number of extrapolation participants. The density and velocity of the newborn fluid node at the present time step are computed by the resulting distribution functions,

$$\rho(\mathbf{x}, t) = \sum_i f_i(\mathbf{x}) \tag{36}$$

$$\mathbf{u}(\mathbf{x}, t) = \frac{1}{\rho(\mathbf{x}, t)}\sum_i \mathbf{e}_i f_i(\mathbf{x}) \tag{37}$$

They investigated the three algorithms in the simulations of cylinder sedimentation as shown in Figure 10. It is evident that the algorithm with second order extrapolation can remarkably reduce the fluctuations.

Peng *et al.* [62] proposed a refill scheme by velocity-constrained normal extrapolation based on the multiple-relaxation-time LBM. After the missing distribution functions of a newborn fluid node are completed by the normal extrapolation refill scheme, all moments at the newborn node are computed by multiplying the transfer matrix M,

$$\mathbf{m}(\mathbf{x},t) = \mathrm{M} \cdot \hat{\mathbf{f}}(\mathbf{x},t) \tag{38}$$

where $\hat{\mathbf{f}}$ indicates the temporary distribution functions. Then the momentum moments are constrained to use the velocity of the nearest boundary,

$$\mathbf{J} = \rho_0 \mathbf{u}_b \tag{39}$$

This makes a new moment vector $\mathbf{m}^*(\mathbf{x},t)$. Finally, transfer $\mathbf{m}^*(\mathbf{x},t)$ back to the distribution functions as

$$\mathbf{f}(\mathbf{x},t) = \mathrm{M}^{-1} \cdot \mathbf{m}^*(\mathbf{x},t) \tag{40}$$

They found that this constraint could significantly reduce the fluctuations in the hydrodynamic forces.

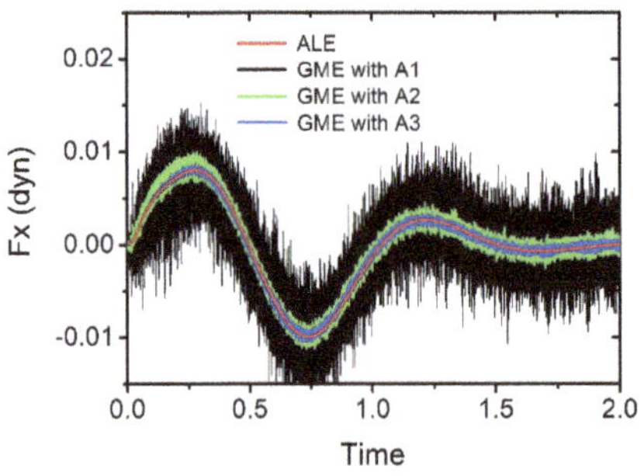

Figure 10. The horizontal forces in cylinder sedimentation. GME is coupled with the different algorithms to fill newborn fluid nodes, neighbor-node average (A1), linear extrapolation (A2) and second-order extrapolation (A3). (Wen *et al.*, 2014 [13]).

Obviously, the scheme is not unique to compute the unknown values, such as density, velocities and distribution functions, on the lattice nodes which move from non-fluid to fluid region. In the particle sedimentation in low Reynolds number, these schemes will produce similar macroscopic results. However, at the micro-scale, they lead to different fluctuating ranges in hydrodynamic forces, which inevitably influence the accuracy and stability of simulations.

5. Applications

As a method to evaluate hydrodynamic force, the momentum exchange method possesses the advantages of simplicity, efficiency, accuracy and robustness. Remarkably, it is independent of the boundary geometry, preventing from solving the Navier–Stokes equations in complex boundary geometries in the boundary-integral methods [10]. Since the pioneer works of Ladd [10,11] and Aidun [12,36], the lattice Boltzmann method has developed into a popular tool for simulations of particle suspensions. In this section, we review some representative applications, such as movements of rigid particles, interactions of deformation particles, particle suspensions in turbulent flow and multiphase flow, *etc*. Particle Brownian motion due to thermal fluctuations and particle surface charges are not covered.

5.1. Rigid Particle Movements

The sedimentation of a single rigid particle is a common case in the simulations of particle suspensions [68–71] and is often used as a benchmark [13,20,22]. Qi [72] simulated sedimentations of spherical and non-spherical particles in finite-Reynolds-number flows and observed phenomena of drafting, kissing and tumbling motion of two particles in a smooth-walled channel. Recently, Wen *et al.* [13] used circle sedimentations to verify Galilean invariance of the momentum exchange method. The configuration is the same to that in Section 3.4. The simulation system is related to several uniform frames of reference in order to investigate Galilean invariance of the hydrodynamic force [13]. Explicitly, the additional uniform velocities **V** = 0, 0.01, 0.02 are initially assigned to the fluid, the particle and the channel.

Figure 11 presents the time-dependent trajectories, angular velocities, horizontal velocities and vertical velocities relative to the channel, comparing with the results by CME and ALE [60,61]. Obviously, regardless of the reference speeds, the GME results are always in excellent agreement with the ALE benchmarks. This supports that GME is highly accurate and Galilean invariant in dynamic fluid. Oppositely, even if the reference frame is motionless, the CME results show noticeable deviations from the benchmark. And the differences become larger and larger when the reference speed increases. These suggest that CME is non-Galilean invariant and is not suitable for moving boundaries.

The lateral migration of a particle suspended in a Poiseuille flow is a classic case. A suspending biconcave particle in a tube flow is studied by using the multiple-relaxation-time lattice Boltzmann method together with GME. The biconcave shape of a red blood cell (RBC) follows the descriptive equation proposed by Fung *et al.* in 1980s [73] and the characteristic radius is 15 lattice units.

The biconcave particles are located at 0.04 and 0.02 cm away from the low channel wall and at the center of the channel in the horizontal direction. Figure 12 draws the particle trajectories at Reynolds numbers 12 and 3, respectively. The biconcave particle shows lateral migrating movement and equilibrium state, which are close to the classic Segré–Silberberg effect [46,47]. Because of the interaction of the parabolic velocity distribution of Poiseuille flow and the biconcave shape, the particulate movement exhibits regular waves and nonuniform rotation. Remarkably, they observe two lateral equilibrium positions corresponding to the particulate releasing positions [74]. The biconcave particle is in successive postures in a rotating period and Figure 13 illustrates a set of velocity contours to draw the dynamic flow field.

GME can be easily implemented in 3D simulations. The Segré–Silberberg effect [46,47] is repeated by simulating a neutrally buoyant rigid sphere migrating laterally in a tube Poiseuille flow. Figure 14 draws two trajectories of the spheres, which are released at the dimensionless radial positions of $r^*/R = 0.66$ and 0.21, where r^* is the radial distance from the tube centerline. The numerical results are highly consistent with the experiments by Karnis *et al.* [47]. This verifies that GME is competent to 3D simulations of particle suspensions.

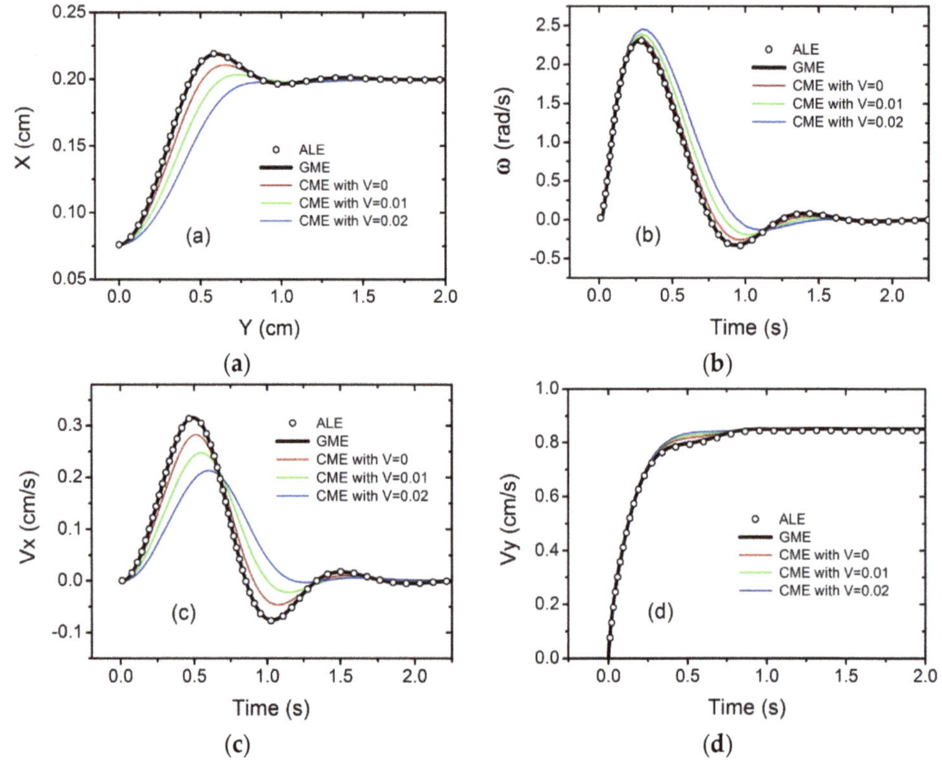

Figure 11. Time-dependent (**a**) trajectories; (**b**) angular velocities; (**c**) horizontal velocities; and (**d**) vertical velocities relative to the channel. The density of the cylinder is 1.03 g/cm^3 and the terminal Reynolds number is 8.33. The dynamic simulation system is connected to three velocities of the reference frame, *i.e.*, **V** = 0, 0.01, and 0.02. (Wen *et al.*, 2014 [13]).

Chen *et al.* [22] simulated an elliptical particle sedimentating in a vertical channel in order to verified their improved momentum exchange method. The major axis is 0.05 cm and the minor axis is 0.025 cm. The kinematic viscosity of the fluid is 1×10^{-6} m^2/s. The channel width is 0.4 cm and is 104 lattice units in the numerical simulation. The density ratio of the solid particle and fluid is 1.1. Figure 15 shows an elliptical sedimentation with a moderate Reynolds number 6.6. It is clear that the results of the three improved schemes agree very well with the benchmarks of the finite element method.

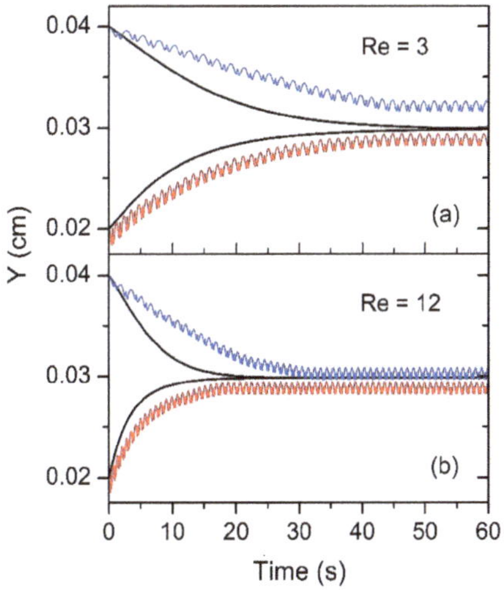

Figure 12. The trajectories of a suspending biconcave particle migrating in a Poiseuille flow. The Reynolds numbers are (**a**) Re = 3 and (**b**) Re = 12. The biconcave particles are located at 0.04 and 0.02 cm away from the low channel wall. The classic Segré–Silberberg effect with a circular particle is represented by the black trajectories. (Wen *et al.*, 2013 [74]).

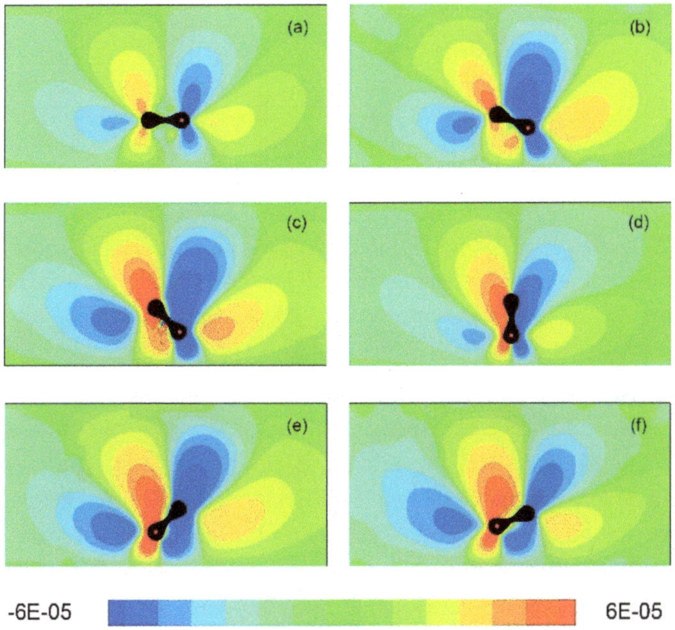

Figure 13. The velocity contours of the fluid around the biconcave particles. (**a**)–(**f**) represent six different postures in a half rotation period. (Wen *et al.*, 2013 [74]).

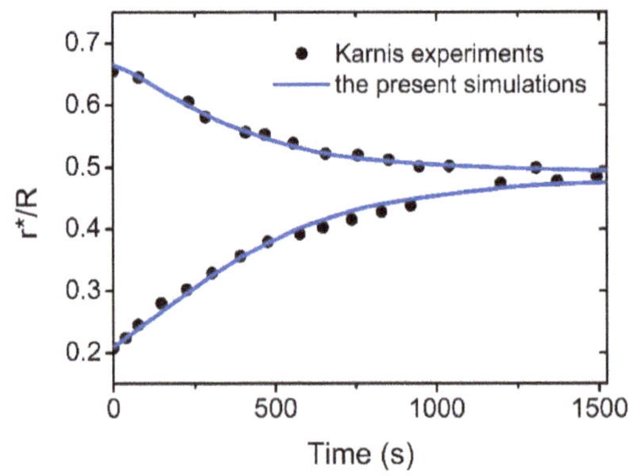

Figure 14. Three-dimensional simulations of the Segré–Silberberg effect by the lattice Boltzmann equation with GME. (Wen *et al.*, 2014 [13]).

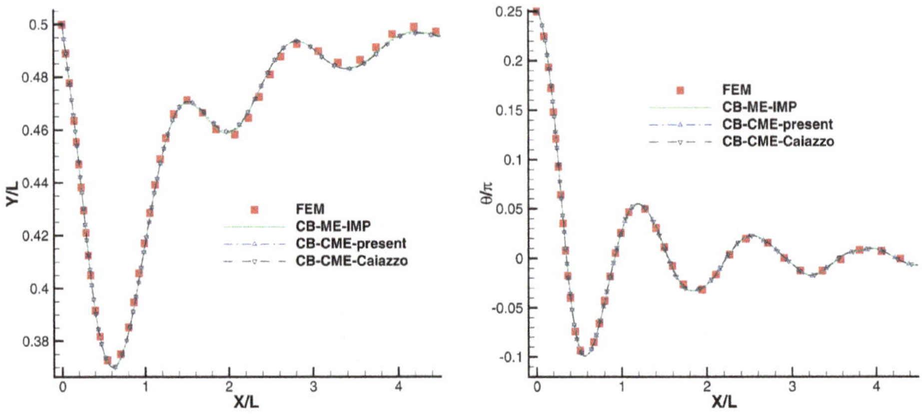

Figure 15. The trajectories and orientations of the particle with Reynolds number 6.6 obtained by the finite element method [75] and LBM with the curved boundary condition. The results of the three improved schemes [12,20,22,48] are in good agreement with the benchmark. (Chen *et al.*, 2013 [22]).

To determinate the critical parameters in platelet margination, Reasor *et al.* [76] investigated the margination dependences on hematocrit, platelet shape, and viscosity ratio of plasma to cytoplasm. The hydrodynamic force is evaluated by the momentum exchange method. Their results emphasize that an increase in hematocrit increases the rate of margination. The viscosity ratio between the interior cytoplasm and suspending fluid can considerably alter the rate of margination. Spherical particles tend to migrate more quickly than disks. The effect of platelet aspect ratio is demonstrated in Figure 16. The spherical-shaped particles have a diameter of 1.73 μm. The platelet-shaped particles are oblate spheroids and rigid, with a major diameter of 2.3 μm and a thickness of 1.0 μm. The disk-shaped particles with a major diameter of 3.26 μm and a thickness of 0.523 μm. It can be seen that both the mean distance to the wall and the standard deviation are reduced with decreasing aspect ratio. The peak concentrations at the wall are larger for the spherical particles and the distributions evolve more rapidly than the disk-shaped particles. The simulations also show the lateral migration of particles with various shapes in tube flow.

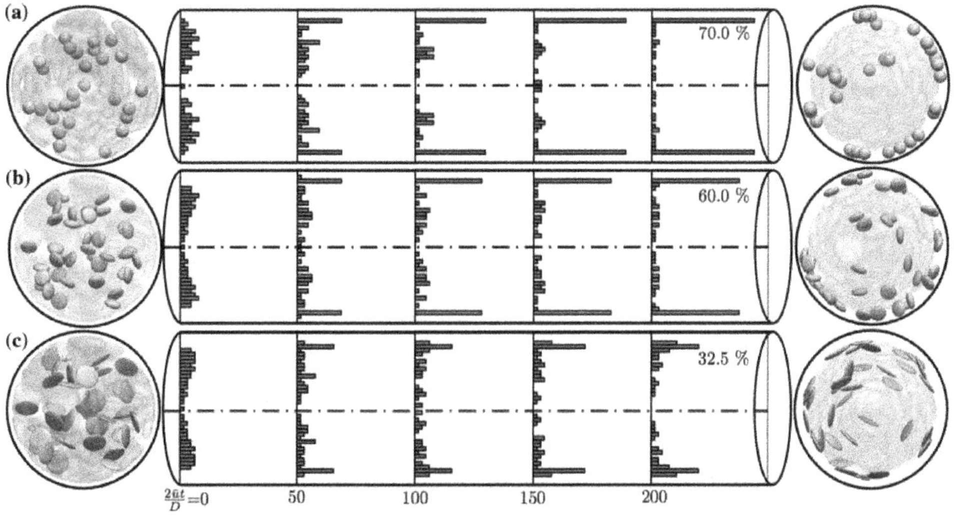

Figure 16. Temporal evolution of platelet distributions for the concentration 0.2 for (**a**) spheres; (**b**) platelets; and (**c**) disk shaped particles. The initial and ending end views of the tube are also given to display the distribution of platelets before and after margination. (Reasor *et al.*, 2013 [76]).

5.2. Deformation Particle Interactions

A growing interest in the simulations of deformation suspensions has promoted the development of coupled methods, in which LBM is combined with an appropriate boundary model to capture the particle deformations. Aidun *et al.* implemented a spectrin-link red blood cell membrane method coupled with the lattice Boltzmann method [7,76–79]. In the method, the particle is represented by a triangular shell mesh, on which the local hydrodynamic force is calculated by momentum exchange method.

To capture the deformation process, Reasor *et al.* [77,78] took a spherical capsule to create a baseline RBC mesh whose surface includes a network of 613 points. Then, 59% of its initial volume was deflated while the surface area keeps constant. The RBC radius was 12 lattice units which approximately equates to 4 μm. Figure 17 illustrates the continuous deformation from a sphere to a biconcave RBC. The equilibrium biconcave shape was reached after 25,000 time steps and is an artifact of minimizing the Helmholtz free energy which involves contributions due to bending.

The spectrin-link method is further compared with experiments [80] and other computational approaches [81] by an optical tweezer experiment performed by Mills *et al.* [80]. The RBC is initially static and is immersed in fluid which simulates blood plasma and hemoglobin. The axial and transverse diameters of an RBC are stretched by optical tweezers with silica beads attached at both ends. As shown in Figure 18, the simulating results are in good agreement with previous researches. MacMeccan *et al.* [77] simulated multiply deformable particle suspensions by coupling the lattice Boltzmann method with finite element analysis. They studied more than 200 fluid-filled and initially spherical capsules in unbounded shear flow. The capsules occupied 40% volume fraction and had identical properties with RBC membranes. Aidun and Clausen [7] further investigated the deformation and interaction of more than 2000 deformable particle suspensions. Melchionna *et al.* [82,83] simulated cardiovascular blood flow, aiming to cardiovascular diagnosis for commodity clinical applications.

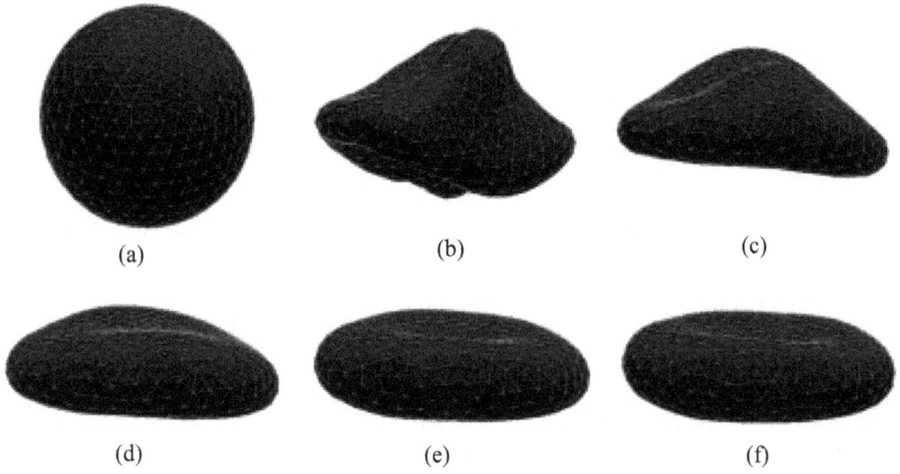

Figure 17. A spherical particle is deflated and become biconcave particle of red blood cell. (**a**) t = 0; (**b**) t = 5000; (**c**) t = 10000; (**d**) t = 15000; (**e**) t = 20000; (**f**) t = 25000. (Reasor *et al.*, 2012 [78]).

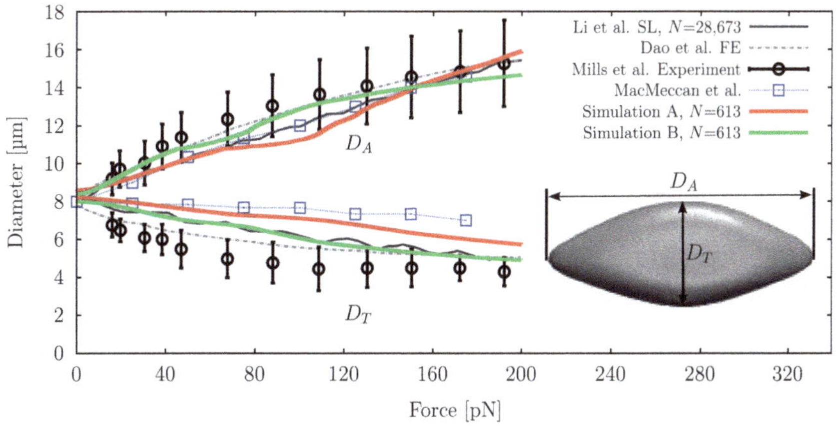

Figure 18. The axial and transverse diameters, D_A and D_T, in μm of the RBC plotted against the applied force. The simulating results are compared to the high-resolution spectrin-level modeling [81], the high-resolution the finite element method [84], the experiments [80], and using the LB–FE implementation [77]. (Reasor *et al.*, 2012 [78]).

5.3. Particle Suspensions in Turbulent Flow

Turbulent flows laden with gas bubbles, small droplets or solid particles are relevant to a wide variety of engineering applications and natural processes, such as plankton dynamics, dust storms, pollutant transport [85]. Using the lattice Boltmzann method coupled with the momentum exchange method, Wang *et al.* [85–87] made a comparative analysis to single-phase turbulence and particle-laden turbulence and studied a decaying isotropic turbulence laden with finite-size particles of Kolmogorov to Taylor microscale sizes. Zhang *et al.* [88,89] investigated the differential settling of cohesive sediment and the non-equilibrium flocculation of cohesive sediments in homogeneous turbulent flows.

For the particle-laden turbulence simulations, Gao *et al.* [85] randomly released particles into the fluid domain after the single-phase flow field developed till a converged velocity-derivative skewness. They implemented a second-order interpolation boundary condition [56] and MPI parallel acceleration based on multiple-relaxation-time LBM. A short-range repulsive force is indroduced to prevent particles from voerlap. Figure 19 draws the vorticity contours of the two evolution times, 1.27 *Te* and 2.12 *Te*, where *Te* is the eddy turnover time. The compuational domain is 256^3 and contains 2304 rigid particles with uniform diameter 11 lattice units. The density ratio of particle and fluid is 2.56. A top layer of fluid has been removed in order to show the scatter of a portion of the particles. It is clear that the magnitude of vorticity diminishes with the time evolution.

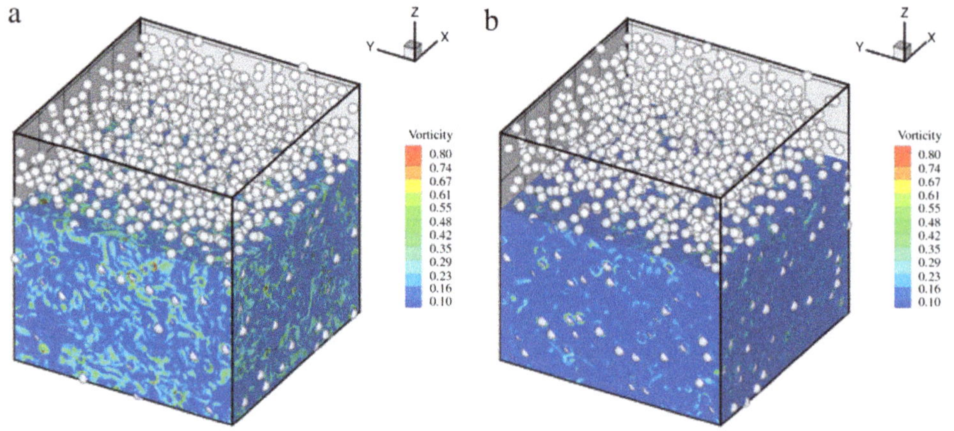

Figure 19. Vorticity contour at (**a**) 1.27 *Te* and (**b**) 2.12 *Te*. A layer of fluid is removed to show the particles in the top portion. (Gao *et al.*, 2013 [85]).

The two-dimensional visualizations of the vorticity magnitude and the particle scatter are provided by horizontally cutting the fluid through the slice near the center of the domain [86,87]. The compuational domain is 256^3 and contains 6400 rigid particles with uniform diameter 8 lattice units. The density ratio of particle and fluid is 5. The Reynolds number of particles are about 10. The turbulent flow is driven by the well-known stochastic forcing scheme of Eswaran and Pope [90]. The particles are observed often associating with high voritcity values (the red spots), as shown in Figure 20. This suggests that, in turbulent flow, motions of finite size particles can produce small-scale flow structures near their surfaces.

5.4. *Particle Suspensions in Multiphase Flow*

Particles or colloids suspended in multiphase flow are commonly encountered in scientific researches and engineering applications such as particle self-assembly, emulsion stabilized by particles and microbe transport in air-water flow. Joshi and Sun [91] combined the Shan-Chen multiphase flow model [9,92–95] with particle suspension to study the capillary interactions between two suspended particles, which were on a liquid-vapor interface and suffered different external forces. The pseudopotential model is applied to calculate the nonideal force in multiphase flow and the momentum exchange method is employed to evaluate the fluid-particle interactions, coupling with the lubrication and Hookean force between pairs of particles. The computational domain uses a 300^2 lattice with periodic boundaries. 36 suspended particles of radius 4.8 lattice units are arrayed in the multiphase flow. The liquid-vapor density ratio is about 30 and the relaxation time is 1. Figure 21 exhibits particle movements during a phase transition process. The adhesive force in the multiphase system is -0.04 and is corresponding to a 77° equilibrium contact angle. Therefore, the particles are hydrophilic. With the development of the two phases, particles are inclined to stay at the gas-liquid interface and then obstruct the phases from growing.

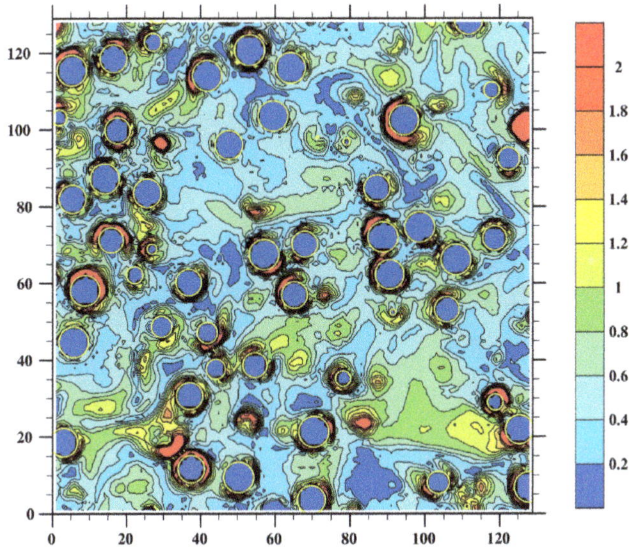

Figure 20. Horizontal snapshots of vorticity contour and particle location in particle-laden turbulent channel flow. The presence of particles can be observed often associating with high vorticity values (represented by the colors towards the red end). This indicates relatively larger dissipation near particle surfaces. (Wang *et al.*, 2014 [86]).

The combined model is further extended to a three-dimensional droplet simulations, considering drop evaporation and particle deposition [96,97]. Adjusting the substrate surface wettability, namely the surface energy, the substrate patterning can control the particle deposition. The computational domain is $250 \times 250 \times 100$ in lattice unit for investigating the surface wetting effect on drop dynamics. In the pattern, the central band is 70 lattice units in width and is hydrophilic, while the side bands are relatively hydrophobic. The equilibrium contact angle of the central hydrophilic band is 30°, and those of the side bands are 60°, 90°, and 120°, respectively. The liquid drop includes 90 particles (10% by volume). Due to an initial offset, the drop impacts the substrate outside the central band. Figure 22 shows that repelling from the bands with relatively low energy, the initially offset drop with suspending particles gradually and automatically moves into the hydrophilic central band. The moving velocities of the drops are affected both by the effective viscosity of the drop and the relative wetting strengths of the bands. Similarly, Liang *et al.* [98] simulated the self-assemblies of colloidal particles on the substrate and investigated the lateral capillary forces and many-body effects.

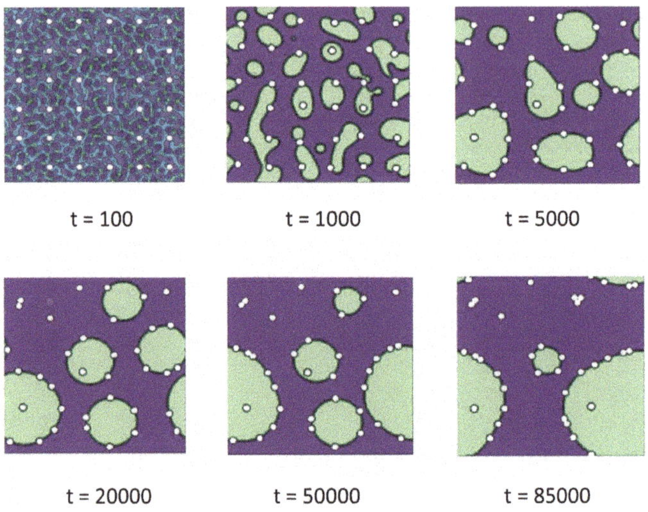

Figure 21. Spinodal decomposition with suspended particles simulated by LBM. The particles are initially arranged in a uniform array. The particles tend to inhibit coarsening of the interface and accumulate at the liquid-vapor interface, as the green liquid domains begin to coalesce and form. (Joshi *et al.*, 2009 [91]).

Particle stabilized emulsions are ubiquitous in the food and cosmetics industry. Jansen and Hurting [99] simulated interactions of multiple spheroidal particles in a multiphase flow lattice Boltzmann method. They demonstrated that the transition from a Bijel to a Pickering emulsion is dependent of the particle concentration, the contact angle, and the ratio of the solvents. Günther *et al.* [100] investigated anisotropic particles at liquid interfaces by simulating emulsions stabilized by particles with complex shapes. The computational domain is a cubic volume with a side length 256 lattice units and periodic boundary conditions are applied. The particles are ellipsoids with major axis 12 and minor axis 6 lattice units and occupy a volume concentration of 0.2. The particle surface has equilibrium contact angles of 90°. As shown in Figure 23 in which the ratio of two phase fluids is 5:2, the ellipsoid particles assemble to some clusters surrounding a fluid phase in an other phase and become Pickering emulsions. They also simulated the Bijel in which the ratio of two phase fluids is 1:1.

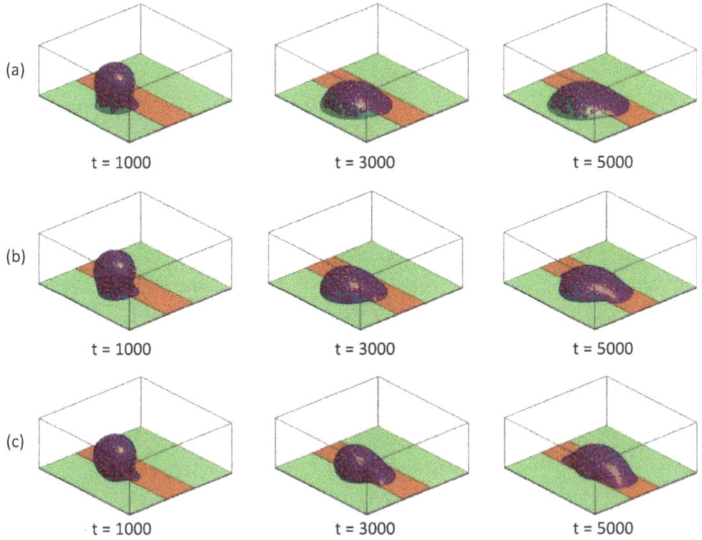

Figure 22. Effect of surface wetting strength on drop dynamics for a liquid drop containing suspending particles. The hydrophilic bands (red) keep 30° contact angle, while the hydrophobic bands are (**a**) 60°, (**b**) 90°, and (**c**) 120° contact angle. The faster driving velocity can be clearly seen in the case (**c**). (Joshi *et al.*, 2010 [96]).

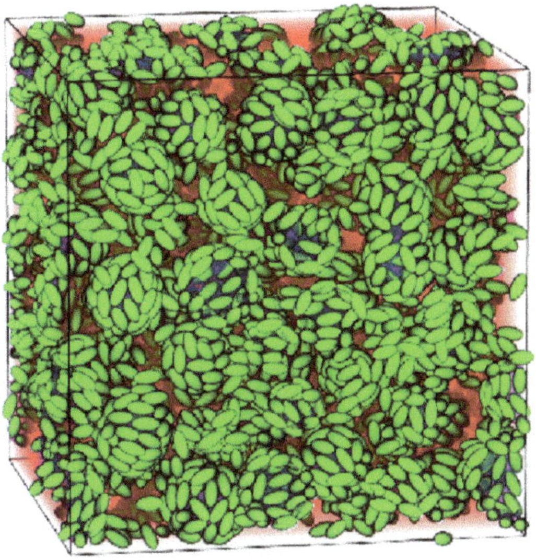

Figure 23. A snapshot of Pickering emulsions stabilized by ellipsoidal particles with an aspect ratio 2 and a volume concentration 0.2. (Günther *et al.*, 2013 [100]).

6. Conclusions

The momentum exchange method is a native scheme in the lattice Boltzmann method. It directly uses discrete velocity and distribution function to evaluate hydrodynamic force. This is totally different from the stress integration method and the immersed boundary method, which have to compute the stress tensor before force evaluation. Therefore, the momentum exchange method has a gift to obtain the simplest and most accurate hydrodynamic force in LBM. Since hydrodynamic force is evaluated based on the momentum transfer on each fluid-solid link, the momentum exchange method only needs local data and is independent of boundary geometry. This endues it with excellent computational efficiency and parallel performance.

Thus far, the momentum exchange method has promoted LBM to become a popular tool for numerical simulations of fluid-structure interactions. Its computational accuracy and Galilean invariance on stationary and moving boundaries in moderate Reynolds numbers have been verified [13,37]. In practice, all kinds of simulations of fluid-structure interactions can benefit from the efficient method. An open issue is to further investigate its computation accuracy in more complex circumstances, such as high Reynolds number flow even turbulence, slip boundary, multiphase flow and so on.

Acknowledgments: This work was supported by the National Natural Science Foundation of China (Grant Nos. 11290164, 11162002, 11362003, 11462003), CAS-Shanghai Science Research Center (Grant No. CAS-SSRC-YJ-2015-01), Guangxi Natural Science Foundation (Grant No. 2014GXNSFAA118018), Guangxi Science and Technology Foundation of College and University (Grant No. KY2015ZD017), Guangxi "Bagui Scholar" Teams for Innovation and Research Project and Guangxi Collaborative Innovation Center of Multi-source Information Integration and Intelligent Processing.

Author Contributions: All authors contributed equally to this work. All authors have read and approved the final manuscript.

Conflicts of Interest: The authors declare no conflict of interest.

References

1. Higuera, F.J.; Succi, S.; Benzi, R. Lattice gas dynamics with enhanced collisions. *Europhys. Lett.* **1989**, *9*, 345–349.
2. Qian, Y.H.; d'Humières, D.; Lallemand, P. Lattice BGK models for Navier–Stokes equation. *Europhys. Lett.* **1992**, *17*, 479–484.
3. Chen, S.Y.; Chen, H.D.; Martinez, D.; Matthaeus, W. Lattice boltzmann model for simulation of magnetohydrodynamics. *Phys. Rev. Lett.* **1991**, *67*, 3776–3779.
4. Chen, H.D.; Chen, S.Y.; Matthaeus, W.H. Recovery of the Navier–Stokes equations using a lattice-gas Boltzmann method. *Phys. Rev. A* **1992**, *45*, R5339–R5342.
5. Benzi, R.; Succi, S.; Vergassola, M. The lattice Boltzmann equation: Theory and applications. *Phys. Rep.* **1992**, *222*, 145–197.

6. Chen, S.; Doolen, G.D. Lattice Boltzmann method for fluid flows. *Annu. Rev. Fluid Mech.* **1998**, *30*, 329–364.
7. Aidun, C.K.; Clausen, J.R. Lattice-Boltzmann method for complex flows. *Annu. Rev. Fluid Mech.* **2010**, *42*, 439–472.
8. Dünweg, B.; Ladd, A.J.C. Lattice Boltzmann Simulations of Soft Matter Systems. In *Advanced Computer Simulation Approaches for Soft Matter Sciences III*; Springer: Berlin/Heidelberg, Germany, 2009; pp. 89–166.
9. Chen, L.; Kang, Q.; Mu, Y.; He, Y.-L.; Tao, W.-Q. A critical review of the pseudopotential multiphase lattice Boltzmann model: Methods and applications. *Int. J. Heat Mass Transf.* **2014**, *76*, 210–236.
10. Ladd, A.J.C. Numerical simulations of particulate suspensions via a discretized Boltzmann equation. Part 1. Theoretical foundation. *J. Fluid Mech.* **1994**, *271*, 285–309.
11. Ladd, A.J.C. Numerical simulations of particulate suspensions via a discretized Boltzmann equation. Part 2. Numerical results. *J. Fluid Mech.* **1994**, *271*, 311–339.
12. Aidun, C.K.; Lu, Y.N.; Ding, E.-J. Direct analysis of particulate suspensions with inertia using the discrete Boltzmann equation. *J. Fluid Mech.* **1998**, *373*, 287–311.
13. Wen, B.; Zhang, C.; Tu, Y.; Wang, C.; Fang, H. Galilean invariant fluid–solid interfacial dynamics in lattice Boltzmann simulations. *J. Comput. Phys.* **2014**, *266*, 161–170.
14. He, X.; Doolen, G. Lattice Boltzmann method on curvilinear coordinates system: Flow around a circular cylinder. *J. Comput. Phys.* **1997**, *134*, 306–315.
15. Inamuro, T.; Maeba, K.; Ogino, F. Flow between parallel walls containing the lines of neutrally buoyant circular cylinders. *Int. J. Multiph. Flow* **2000**, *26*, 1981–2004.
16. Li, H.; Lu, X.; Fang, H.; Qian, Y. Force evaluations in lattice Boltzmann simulations with moving boundaries in two dimensions. *Phys. Rev. E* **2004**, *70*, 026701.
17. Peskin, C.S. Numerical analysis of blood flow in the heart. *J. Comput. Phys.* **1977**, *25*, 220–252.
18. Feng, Z.-G.; Michaelides, E.E. The immersed boundary-lattice Boltzmann method for solving fluid–particles interaction problems. *J. Comput. Phys.* **2004**, *195*, 602–628.
19. Zhou, Q.; Fan, L.-S. A second-order accurate immersed boundary-lattice Boltzmann method for particle-laden flows. *J. Comput. Phys.* **2014**, *268*, 269–301.
20. Wen, B.; Li, H.; Zhang, C.; Fang, H. Lattice-type-dependent momentum-exchange method for moving boundaries. *Phys. Rev. E* **2012**, *85*, 016704.
21. Caiazzo, A.; Junk, M.; Rheinländer, M. Comparison of analysis techniques for the lattice Boltzmann method. *Comput. Math. Appl.* **2009**, *58*, 883–897.
22. Chen, Y.; Cai, Q.; Xia, Z.; Wang, M.; Chen, S. Momentum-exchange method in lattice Boltzmann simulations of particle-fluid interactions. *Phys. Rev. E* **2013**, *88*, 013303.
23. Lallemand, P.; Luo, L.-S. Theory of the lattice Boltzmann method: Dispersion, dissipation, isotropy, Galilean invariance, and stability. *Phys. Rev. E* **2000**, *61*, 6546–6562.
24. D'Humières, D.; Ginzburg, I.; Krafczyk, M.; Lallemand, P.; Luo, L.-S. Multiple-relaxation-time lattice Boltzmann models in three dimensions. *Philos. Trans. R. Soc. Lond. A* **2002**, *360*, 437–451.

25. Ginzburg, I.; Verhaeghe, F.; d'Humières, D. Two-relaxation-time lattice Boltzmann scheme: About parametrization, velocity, pressure and mixed boundary conditions. *Commun. Comput. Phys.* **2008**, *3*, 427–478.
26. Karlin, I.V.; Ferrante, A.; Öttinger, H.C. Perfect entropy functions of the lattice Boltzmann method. *Europhys. Lett.* **1999**, *47*, 182–188.
27. Succi, S.; Karlin, I.V.; Chen, H. *Colloquium*: Role of the *H* theorem in lattice Boltzmann hydrodynamic simulations. *Rev. Mod. Phys.* **2002**, *74*, 1203–1220.
28. D'Humières, D. Generalized lattice-Boltzmann equations. *Rarefied Gas Dyn.* **1994**, *159*, 450–458.
29. Luo, L.-S. Theory of the lattice Boltzmann method: Lattice Boltzmann models for nonideal gases. *Phys. Rev. E* **2000**, *62*, 4982–4996.
30. Luo, L.-S.; Liao, W.; Chen, X.; Peng, Y.; Zhang, W. Numerics of the lattice Boltzmann method: Effects of collision models on the lattice Boltzmann simulations. *Phys. Rev. E* **2011**, *83*, 056710.
31. Ladd, A.J.C.; Verberg, R. Lattice-Boltzmann simulations of particle-fluid suspensions. *J. Stat. Phys.* **2001**, *104*, 1191–1251.
32. Nguyen, N.-Q.; Ladd, A.J.C. Lubrication corrections for lattice-Boltzmann simulations of particle suspensions. *Phys. Rev. E* **2002**, *66*, 046708.
33. Nguyen, N.-Q.; Ladd, A.J.C. Sedimentation of hard-sphere suspensions at low Reynolds number. *J. Fluid Mech.* **2005**, *525*, 73–104.
34. Başağaoğlu, H.; Meakin, P.; Succi, S.; Redden, G.R.; Ginn, T.R. Two-dimensional lattice Boltzmann simulation of colloid migration in rough-walled narrow flow channels. *Phys. Rev. E* **2008**, *77*, 031405.
35. Başağaoğlu, H.; Allwein, S.; Succi, S.; Dixon, H.; Carrola, J.T., Jr.; Stothoff, S. Two-and three-dimensional lattice Boltzmann simulations of particle migration in microchannels. *Microfluid. Nanofluid.* **2013**, *15*, 785–796.
36. Aidun, C.K.; Lu, Y. Lattice Boltzmann simulation of solid particles suspended in fluid. *J. Stat. Phys.* **1995**, *81*, 49–61.
37. Mei, R.; Yu, D.; Shyy, W.; Luo, L.-S. Force evaluation in the lattice Boltzmann method involving curved geometry. *Phys. Rev. E* **2002**, *65*, 041203.
38. Filippova, O.; Hänel, D. Grid refinement for lattice-BGK models. *J. Comput. Phys.* **1998**, *147*, 219–228.
39. Mei, R.; Luo, L.-S.; Shyy, W. An accurate curved boundary treatment in the lattice Boltzmann method. *J. Comput. Phys.* **1999**, *155*, 307–330.
40. Mei, R.; Shyy, W.; Yu, D.; Luo, L.-S. Lattice Boltzmann method for 3-D flows with curved boundary. *J. Comput. Phys.* **2000**, *161*, 680–699.
41. Ding, E.-J.; Aidun, C.K. Extension of the lattice-Boltzmann method for direct simulation of suspended particles near contact. *J. Stat. Phys.* **2003**, *112*, 685–708.
42. Bouzidi, M.; Firdaouss, M.; Lallemand, P. Momentum transfer of a Boltzmann-lattice fluid with boundaries. *Phys. Fluids* **2001**, *13*, 3452–3459.
43. Guo, Z.L.; Zheng, C.G.; Shi, B.C. An extrapolation method for boundary conditions in lattice Boltzmann method. *Phys. Fluids* **2002**, *14*, 2007–2010.

44. Kao, P.-H.; Yang, R.-J. An investigation into curved and moving boundary treatments in the lattice Boltzmann method. *J. Comput. Phys.* **2008**, *227*, 5671–5690.
45. Bao, J.; Yuan, P.; Schaefer, L. A mass conserving boundary condition for the lattice Boltzmann equation method. *J. Comput. Phys.* **2008**, *227*, 8472–8487.
46. Segré, G.; Silberberg, A. Radial particle displacements in poiseuille flow of suspensions. *Nature* **1961**, *189*, 209–210.
47. Karnis, A.; Goldsmith, H.L.; Mason, S.G. The flow of suspensions through tubes: V. inertial effects. *Can. J. Chem. Eng.* **1966**, *44*, 181–193.
48. Caiazzo, A.; Junk, M. Boundary forces in lattice Boltzmann: Analysis of momentum exchange algorithm. *Comput. Math. Appl.* **2008**, *55*, 1415–1423.
49. Clausen, J.R.; Aidun, C.K. Galilean invariance in the lattice-Boltzmann method and its effect on the calculation of rheological properties in suspensions. *Int. J. Multiph. Flow* **2009**, *35*, 307–311.
50. Lorenz, E.; Caiazzo, A.; Hoekstra, A.G. Corrected momentum exchange method for lattice Boltzmann simulations of suspension flow. *Phys. Rev. E* **2009**, *79*, 036705.
51. Lorenz, E.; Hoekstra, A.G.; Caiazzo, A. Lees-edwards boundary conditions for lattice Boltzmann suspension simulations. *Phys. Rev. E* **2009**, *79*, 036706.
52. Hu, Y.; Li, D.; Shu, S.; Niu, X. Modified momentum exchange method for fluid-particle interactions in the lattice Boltzmann method. *Phys. Rev. E* **2015**, *91*, 033301.
53. Huang, H.; Yang, X.; Krafczyk, M.; Lu, X.-Y. Rotation of spheroidal particles in couette flows. *J. Fluid Mech.* **2012**, *692*, 369–394.
54. Luo, L.-S. Unified theory of lattice Boltzmann models for nonideal gases. *Phys. Rev. Lett.* **1998**, *81*, 1618–1621.
55. Guo, Z.; Zheng, C.; Shi, B. Discrete lattice effects on the forcing term in the lattice Boltzmann method. *Phys. Rev. E* **2002**, *65*, 046308.
56. Lallemand, P.; Luo, L.-S. Lattice Boltzmann method for moving boundaries. *J. Comput. Phys.* **2003**, *184*, 406–421.
57. Qian, Y.-H.; Zhou, Y. Complete Galilean-invariant lattice BGK models for the Navier–Stokes equation. *Europhys. Lett.* **1998**, *42*, 359–364.
58. Krithivasan, S.; Wahal, S.; Ansumali, S. Diffused bounce-back condition and refill algorithm for the lattice Boltzmann method. *Phys. Rev. E* **2014**, *89*, 033313.
59. Ginzburg, I.; d'Humières, D. Multireflection boundary conditions for lattice Boltzmann models. *Phys. Rev. E* **2003**, *68*, 066614.
60. Hu, H.H.; Joseph, D.D.; Crochet, M.J. Direct simulation of fluid particle motions. *Theor. Comput. Fluid Dyn.* **1992**, *3*, 285–306.
61. Hu, H.H.; Patankar, N.A.; Zhu, M.Y. Direct Numerical Simulations of Fluid–Solid Systems Using the Arbitrary Lagrangian–Eulerian Technique. *J. Comput. Phys.* **2001**, *169*, 427–462.
62. Peng, C.; Teng, Y.; Hwang, B.; Guo, Z.; Wang, L.-P. Implementation issues and benchmarking of lattice Boltzmann method for moving rigid particle simulations in a viscous flow. *Comput. Math. Appl.* **2015**. in press.
63. Caiazzo, A. Analysis of lattice Boltzmann nodes initialisation in moving boundary problems. *Prog. Comput. Fluid Dyn.* **2008**, *8*, 3–10.

64. Ansumali, S.; Karlin, I.V. Kinetic boundary conditions in the lattice Boltzmann method. *Phys. Rev. E* **2002**, *66*, 026311.
65. Tang, G.H.; Tao, W.Q.; He, Y.L. Lattice Boltzmann method for gaseous microflows using kinetic theory boundary conditions. *Phys. Fluids* **2005**, *17*, 058101.
66. Guo, Z.L.; Shi, B.; Zhao, T.S.; Zheng, C. Discrete effects on boundary conditions for the lattice Boltzmann equation in simulating microscale gas flows. *Phys. Rev. E* **2007**, *76*, 056704.
67. Fang, H.; Wang, Z.; Lin, Z.; Liu, M. Lattice Boltzmann method for simulating the viscous flow in large distensible blood vessels. *Phys. Rev. E* **2002**, *65*, 051925.
68. Wan, R.-Z.; Fang, H.-P.; Lin, Z.; Chen, S. Lattice Boltzmann simulation of a single charged particle in a Newtonian fluid. *Phys. Rev. E* **2003**, *68*, 011401.
69. Li, H.; Fang, H.; Lin, Z.; Xu, S.X.; Chen, S. Lattice Boltzmann simulation on particle suspensions in a two-dimensional symmetric stenotic artery. *Phys. Rev. E* **2004**, *69*, 031919.
70. Zhang, C.-Y.; Shi, J.; Tan, H.-L.; Liu, M.-R.; Kong, L.-J. Sedimentation of a single charged elliptic cylinder in a Newtonian fluid by lattice Boltzmann method. *Chin. Phys. Lett.* **2004**, *21*, 1108–1110.
71. Zhang, C.-Y.; Tan, H.-L.; Liu, M.-R.; Kong, L.-J.; Shi, J. Lattice Boltzmann simulation of sedimentation of a single charged elastic dumbbell in a Newtonian fluid. *Chin. Phys. Lett.* **2005**, *22*, 896–899.
72. Qi, D. Lattice-Boltzmann simulations of particles in non-zero-Reynolds-number flows. *J. Fluid Mech.* **1999**, *385*, 41–62.
73. Fung, Y.C. *Biomechanics*; Springer: New York, NY, USA, 1981.
74. Wen, B.-H.; Chen, Y.-Y.; Zhang, R.-L.; Zhang, C.-Y.; Fang, H.-P. Lateral migration and nonuniform rotation of biconcave particle suspended in Poiseuille flow. *Chin. Phys. Lett.* **2013**, *30*, 064701.
75. Xia, Z.; Connington, K.W.; Rapaka, S.; Yue, P.; Feng, J.J.; Chen, S. Flow patterns in the sedimentation of an elliptical particle. *J. Fluid Mech.* **2009**, *625*, 249–272.
76. Reasor, D.A.; Mehrabadi, M.; Ku, D.N.; Aidun, C.K. Determination of critical parameters in platelet margination. *Ann. Biomed. Eng.* **2013**, *41*, 238–249.
77. MacMeccan, R.M.; Clausen, J.R.; Neitzel, G.P.; Aidun, C.K. Simulating deformable particle suspensions using a coupled lattice-Boltzmann and finite-element method. *J. Fluid Mech.* **2009**, *618*, 13–39.
78. Reasor, D.A., Jr.; Clausen, J.R.; Aidun, C.K. Coupling the lattice-Boltzmann and spectrin-link methods for the direct numerical simulation of cellular blood flow. *Int. J. Numer. Meth. Fluids* **2012**, *68*, 767–781.
79. Wu, J.; Aidun, C.K. Simulating 3D deformable particle suspensions using lattice Boltzmann method with discrete external boundary force. *Int. J. Numer. Meth. Fluids* **2010**, *62*, 765–783.
80. Mills, J.P.; Qie, L.; Dao, M.; Lim, C.T.; Suresh, S. Nonlinear elastic and viscoelastic deformation of the human red blood cell with optical tweezers. *Mech. Chem. Biosys.* **2004**, *1*, 169–180.

81. Li, J.; Dao, M.; Lim, C.; Suresh, S. Spectrin-level modeling of the cytoskeleton and optical tweezers stretching of the erythrocyte. *Biophys. J.* **2005**, *88*, 3707–3719.
82. Melchionna, S.; Bernaschi, M.; Succi, S.; Kaxiras, E.; Rybicki, F.J.; Mitsouras, D.; Coskun, A.U.; Feldman, C.L. Hydrokinetic approach to large-scale cardiovascular blood flow. *Comput. Phys. Commun.* **2010**, *181*, 462–472.
83. Bernaschi, M.; Melchionna, S.; Succi, S.; Fyta, M.; Kaxiras, E.; Sircar, J.K. MUPHY: A parallel MUlti PHYsics/scale code for high performance bio-fluidic simulations. *Comput. Phys. Commun.* **2009**, *180*, 1495–1502.
84. Dao, M.; Lim, C.T.; Suresh, S. Mechanics of the human red blood cell deformed by optical tweezers. *J. Mech. Phys. Solids* **2003**, *51*, 2259–2280.
85. Gao, H.; Li, H.; Wang, L.-P. Lattice Boltzmann simulation of turbulent flow laden with finite-size particles. *Comput. Math. Appl.* **2013**, *65*, 194–210.
86. Wang, L.-P.; Ayala, O.; Gao, H.; Andersen, C.; Mathews, K.L. Study of forced turbulence and its modulation by finite-size solid particles using the lattice Boltzmann approach. *Comput. Math. Appl.* **2014**, *67*, 363–380.
87. Wang, L.-P.; Peng, C.; Guo, Z.; Yu, Z. Lattice Boltzmann simulation of particle-laden turbulent channel flow. *Comput. Fluids* **2016**, *124*, 226–236.
88. Zhang, J.-F.; Zhang, Q.-H. Lattice Boltzmann simulation of the flocculation process of cohesive sediment due to differential settling. *Cont. Shelf Res.* **2011**, *31*, S94–S105.
89. Zhang, J.; Zhang, Q.; Qiao, G. A lattice Boltzmann model for the non-equilibrium flocculation of cohesive sediments in turbulent flow. *Comput. Math. Appl.* **2014**, *67*, 381–392.
90. Eswaran, V.; Pope, S.B. An examination of forcing in direct numerical simulations of turbulence. *Comput. Fluids* **1988**, *16*, 257–278.
91. Joshi, A.S.; Sun, Y. Multiphase lattice Boltzmann method for particle suspensions. *Phys. Rev. E* **2009**, *79*, 066703.
92. Shan, X.; Chen, H. Lattice Boltzmann model for simulating flows with multiple phases and components. *Phys. Rev. E* **1993**, *47*, 1815–1819.
93. Shan, X.; Chen, H. Simulation of nonideal gases and liquid-gas phase transitions by the lattice Boltzmann equation. *Phys. Rev. E* **1994**, *49*, 2941–2948.
94. Shan, X. Analysis and reduction of the spurious current in a class of multiphase lattice Boltzmann models. *Phys. Rev. E* **2006**, *73*, 047701.
95. Shan, X. Pressure tensor calculation in a class of nonideal gas lattice Boltzmann models. *Phys. Rev. E* **2008**, *77*, 066702.
96. Joshi, A.S.; Sun, Y. Wetting dynamics and particle deposition for an evaporating colloidal drop: A lattice Boltzmann study. *Phys. Rev. E* **2010**, *82*, 041401.
97. Joshi, A.S.; Sun, Y. Numerical simulation of colloidal drop deposition dynamics on patterned substrates for printable electronics fabrication. *J. Disp. Tech.* **2010**, *6*, 579–585.
98. Liang, G.; Zeng, Z.; Chen, Y.; Onishi, J.; Ohashi, H.; Chen, S. Simulation of self-assemblies of colloidal particles on the substrate using a lattice Boltzmann pseudo-solid model. *J. Comput. Phys.* **2013**, *248*, 323–338.

99. Jansen, F.; Harting, J. From bijels to pickering emulsions: A lattice Boltzmann study. *Phys. Rev. E* **2011**, *83*, 046707.
100. Günther, F.; Janoschek, F.; Frijters, S.; Harting, J. Lattice Boltzmann simulations of anisotropic particles at liquid interfaces. *Comput. Fluids* **2013**, *80*, 184–189.

A Lattice Gas Automata Model for the Coupled Heat Transfer and Chemical Reaction of Gas Flow Around and Through a Porous Circular Cylinder

Hongsheng Chen, Zhong Zheng, Zhiwei Chen and Xiaotao T. Bi

Abstract: Coupled heat transfer and chemical reaction of fluid flow in complex boundaries are explored by introducing two additional properties, *i.e.*, particle type and energy state into the Lattice gas automata (LGA) Frisch–Hasslacher–Pomeau (FHP-II) model. A mix-redistribute of energy and type of particles is also applied on top of collision rules to ensure randomness while maintaining the conservation of mass, momentum and energy. Simulations of heat transfer and heterogeneous reaction of gas flow passing a circular porous cylinder in a channel are presented. The effects of porosity of cylinder, gas inlet velocity, and reaction probability on the reaction process are further analyzed with respect to the characteristics of solid morphology, product concentration, and temperature profile. Numerical results indicate that the reaction rate increases with increasing reaction probability as well as gas inlet velocity. Cylinders with a higher value of porosity and more homogeneous structure also react with gas particles faster. These results agree well with the basic theories of gas–solid reactions, indicating the present model provides a method for describing gas–solid reactions in complex boundaries at mesoscopic level.

Reprinted from *Entropy*. Cite as: Chen, H.; Zheng, Z.; Chen, Z.; Bi, X.T. A Lattice Gas Automata Model for the Coupled Heat Transfer and Chemical Reaction of Gas Flow Around and Through a Porous Circular Cylinder. *Entropy* **2016**, *18*, 2.

1. Introduction

The simulation of heat transfer and chemical reaction of fluid flow in porous media is of considerable importance in many practical applications such as combustion chambers, heat exchangers, food processing, catalytic reactors, refrigeration, air cooling and thermal energy storage devices. Simulation results can be found for porous catalyst particles, packed catalyst beds, and arranged pipes. Among these studies, methods based on conventional partial differential equations such as volume-averaging theory [1] and Darcy models [2–4], or discrete methods, e.g., lattice Boltzmann methods (LBM) [5–8] have been the major approaches. However, to the best of our knowledge, the porous media reported in the literature are relatively simple and usually consist of regular arranged solid cylinders. Few studies have been carried out to investigate the characteristics of heat transfer coupling by chemical reaction in porous media. Additionally, the conventional methods based on nonlinear partial differential equations (PDEs) also suffer difficulties such as truncation error and high sensitivity to boundary conditions, making it difficult to describe the detailed structure of porous media or simulate complex processes in porous media.

Lattice gas automata (LGA) is a mesoscopic simulation method from the viewpoint that fluids consist of a large number of particles that "live" on regular lattices with interactions conserving mass and momentum [9]. It is a "bottom-up" and "equation-free" method capturing both macroscopic and mesoscopic characteristics of complex/multi-scale systems, quite distinctive from molecular dynamics (MD), kinetic theory of gases and other methods based on the discretization of partial differential equations. Although LGA suffers drawbacks like statistical noise, the lake of Galilean invariance and velocity-dependent pressure, LGA preserves the particle nature and numerical stability compared with lattice Boltzmann methods [10]. More detailed microscopic interaction among particles or between particles and walls can be obtained when using LGA. Thus, various investigations on flow past obstacles or reaction using lattice gas automata have been carried out, such as flow over cylinders or plat plate [11–13], reaction and diffusion systems [14–16], first order reactions [17], motivation phenomena of atom and molecule [18,19], kinetically and thermodynamically controlled reactions [20–22], and Lindemann theory [23,24]. However, investigations on flow, heat transfer and chemical reaction around a porous cylinder using LGA were rarely reported.

Therefore, in this paper, we intend to explore the application of a LGA method to the heat transfer and chemical reaction of fluid flow around and through a porous circular cylinder in a channel. An algorithm based on the Frisch–Hasslacher–Pomeau (FHP-II) LGA model was developed to deal with the coupled heat transfer and chemical reaction. Quartet structure generation set (QSGS) was used to construct the porous circular cylinder. The influences of porous structure, *i.e.*, porosity, pore size and homogeneity, as well as that of reaction probability and flow velocity were further discussed.

2. Simulation Method

2.1. Lattice Gas Automata Model for Heat Transfer and Chemical Reaction

In FHP-II, particles interact with each other according to a number of pre-defined collision and propagation rules detailed by Frish *et al.* [25], as shown in Figure 1. Based on these simple rules, LGA is capable of displaying complex fluid flow behavior, and consequently, it can be used as a simulation tool for describing physical phenomena. The mass and momentum conservation can be written as below, respectively,

$$\sum_i n_i(t+1, \mathbf{r}+\mathbf{c}_i) = \sum_i n_i(t, \mathbf{r}) \tag{1}$$

$$\sum_i \mathbf{c}_i n_i(t+1, \mathbf{r}+\mathbf{c}_i) = \sum_i \mathbf{c}_i n_i(t, \mathbf{r}) \tag{2}$$

where $n_i(t, \mathbf{r})$ is the occupation state of the cell in i-th direction at time t and place $\mathbf{r}$ if the cell is empty, its value is 0, otherwise, its value is 1. $\mathbf{c}_i$ is the lattice velocity in i-th direction, and

$$\mathbf{c}_i = \begin{cases} \left(\cos\frac{\pi}{3}i, \sin\frac{\pi}{3}i\right), & i = 1, \ldots, 6 \\ 0 & , \; i = 0 \end{cases}$$

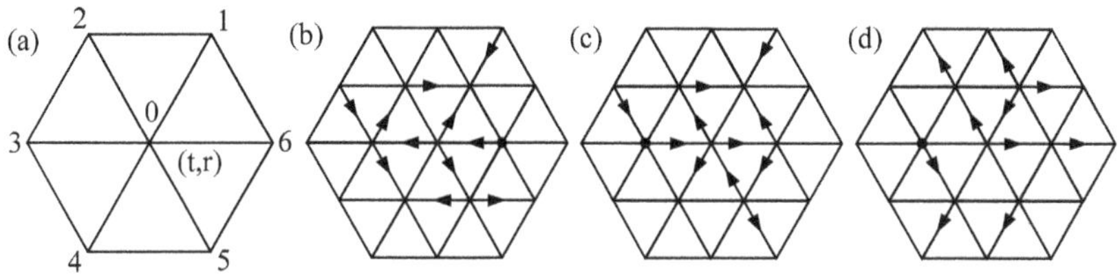

Figure 1. Evolution process of FHP-II, where arrows denote moving particles and points represent static particles: (**a**) node at $(t, \mathbf{r})$; (**b**) initialization; (**c**) after collision; and (**d**) after propagation.

In the conventional LGA model, particles in the system are of same mass (assumed as 1) and velocity scale (assumed as $|\mathbf{c}_i| = 1$), in other words, they are indistinguishable. In order to describe problems involving heat transfer and chemical reactions, the particles should be distinguishable on temperature and substance type. In current model, every particle is at either of the two energy states [26] e_i, *i.e.*, 0 or 1, which represent low (minimum) and high (maximum) temperatures, respectively. Particularly, if the cell is empty, e_i is fixed as 0. Particles are also marked by finite kinds of substance types s_i, and in this paper, s_i is equal to 0 or 1, representing reactant and product, respectively. Besides mass and momentum, extra conservations are also taken into account with respect to energy state and substance type, given as

$$\sum_i n_i^{\alpha,\beta}(t+1, \mathbf{r}) = \sum_i n_i^{\alpha,\beta}(t, \mathbf{r}) \tag{3}$$

where α and β are the substance type and energy state of the cell in i-th direction, respectively. To ensure the conservation of the number of particles with different energy states and substance types at each node during the collision step of FHP-II model, the substance type and energy state need to follow Equations (4) and (5), respectively, known as component conservations

$$\sum_i s_i^{\alpha}(t+1,\mathbf{r}) = \sum_i s_i^{\alpha}(t,\mathbf{r}) \quad , \alpha = 1 \; or \; 0 \tag{4}$$

$$\sum_i e_i^{\beta}(t+1,\mathbf{r}) = \sum_i e_i^{\beta}(t,\mathbf{r}) \quad , \beta = 1 \; or \; 0 \tag{5}$$

Afterwards, the energy states and substance types of each node are mixed and re-distributed. In fact, the energy states and substance types will be arbitrarily attached to the particles at the node after collision. The overall conservations of mass and momentum will still follow Equations (1) and (2). However, for the propagation process, the particle will move with energy state and substance type, which is described as Equation (6). The evolution process of this model is illustrated in Figure 2.

$$n_i^{s_i(t+1,\mathbf{r}+\mathbf{c}_i),e_i(t+1,\mathbf{r}+\mathbf{c}_i)}(t+1,\mathbf{r}+\mathbf{c}_i) = n_i^{s_i(t,\mathbf{r}),e_i(t,\mathbf{r})}(t,\mathbf{r}) \tag{6}$$

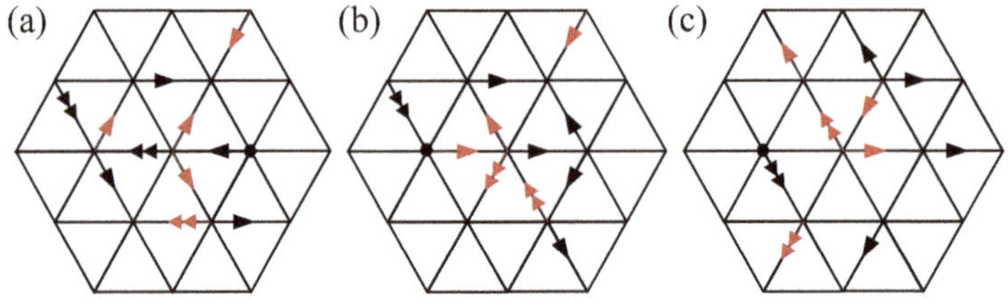

Figure 2. Evolution process of present model, where red represents particles with high energy state, black represents particles with low energy state, and double arrows mean product particles and single arrows denote reactant particles: (**a**) initialization; (**b**) after collision; and (**c**) after propagation.

In FHP-II, the density and momentum are defined as

$$\rho(t,\mathbf{r}) = \sum_{i=0}^{6} n_i(t,\mathbf{r}) \tag{7}$$

$$\rho\mathbf{u} = \sum_{i=0}^{6} \mathbf{c}_i n_i(t,\mathbf{r}) \tag{8}$$

The temperature and product concentration at a node are described as the proportion of particles with high energy state ($\beta = 1$) and the proportion of product particles ($\alpha = 1$), respectively, as the following dimensionless forms:

$$T = \frac{\sum_{i} n_i^{\alpha,1}(t,\mathbf{r})}{\sum_{i,\beta} n_i^{\alpha,\beta}(t,\mathbf{r})} \tag{9}$$

$$C = \frac{\sum_{i} n_i^{1,\beta}(t,\mathbf{r})}{\sum_{i,\alpha} n_i^{\alpha,\beta}(t,\mathbf{r})} \tag{10}$$

2.2. Chemical Reaction Scheme

The scheme of chemical reaction is based on the algorithm proposed by Bresolin and Oliveira [27] to simulate unimolecular and bimolecular reactions. For first-order reactions considering unimolecular collision, the rate constant, k, is described as

$$k = \int_0^{\infty} v(E)P(E)dE \tag{11}$$

where $v(E)$ is the frequency of collisions with energy E above the minimum energy E^*, $P(E)$ is the energy distribution of the molecules, which is given by the Maxwell–Boltzmann distribution, and for a molecule with n classic energy states, the fraction of molecules with energy states E_1,E_2, ... ,E_n can be written as

$$P(E_1, E_2, \ldots E_n) = \frac{e^{\frac{-(E_1+E_2+\ldots E_n)}{RT}}}{RT} = \frac{e^{\frac{-E}{RT}}}{RT} \tag{12}$$

Integrating Equation (12) over all energy values yields:

$$P(E) = \left(\frac{E}{RT}\right)^{n-1} \frac{e^{\frac{-E}{RT}}}{(n-1)!RT} \tag{13}$$

According to Rice–Ramsperger–Kassel (RRK) model [28–30], the frequency of collisions $v(E)$ is suggested to be the formula as follows:

$$\begin{cases} \overline{v(E)} = 0 & if\ E < E^* \\ \overline{v(E)} = C(1-\frac{E^*}{E})^{n-1} & if\ E \geqslant E^* \end{cases} \tag{14}$$

where C is a constant. For the simulation of chemical reaction in LGA, the collisions between molecules can be interpreted as taking place among particles at a node.

In this paper, each reactant particle propagates with an associated probability r deciding it reacts and converts to a product particle or not, generally described as the probability of effective collision. This probability is determined by a probability distribution function and a threshold for reaction. Herein, for simplification, the standard Gaussian distribution was used as the probability distribution function instead of Maxwell–Boltzmann distribution, a threshold K^* was used instead of E^* to decide the critical energy state of a reactant particle. During the collision step of LGA, a random number K following the probability distribution function is generated and compared to K^*; if $K > K^*$, the reaction is consider to be able to take place, otherwise, no reaction occurs.

Moreover, the frequency of collisions in LGA is supposed to be 1.0 per iteration when $K > K^*$, otherwise its value is 0, similar to Equation (14). Thus, the specific reaction constant equals to the integration of energy distribution function above K^* according to Equation (11). Therefore, reaction probability r is equal to the proportion of the particles with energy above K^*, as well as the specific reaction constant. As shown in Figure 3, the curve represents the probability distribution function (PDF), and K^* is a threshold for reaction, the shaded area (beyond K^*) represents the frequency of collisions of hot populations, *i.e.*, reaction probability r. For a given K^*, reaction probability is obtained, and vice versa. In this work, K^* was determined by a given reaction probability from the inverse of normal distribution [31]. The applications of the chemical reaction scheme will be further discussed.

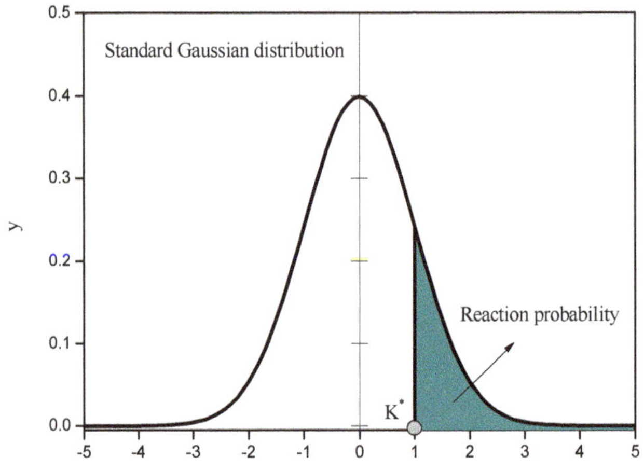

Figure 3. Standard Gaussian distribution function for determining the probability of chemical reaction.

3. Results and Discussion

3.1. Validation of the Chemical Reaction Scheme

The chemical reaction scheme was applied to irreversible first-order reaction $A \rightarrow B$ and reversible first-order reaction $A \rightleftharpoons B$, such as nuclear decay and transformation of isomers. The reactions take place in a square enclosure with area 200×200 (lattice units). Every site was initially filled with six A-type particles and bounce-back type boundary condition was employed to all walls, where the momentum of particle is directly reversed while the substance state keeps unchanged during the collision process. Additionally, no heat transfer was taken into consideration in these two cases.

Molecules collide and react at random. Nevertheless, the time evolution of macroscopic amounts or concentrations of molecules is usually quite reproducible due to the reproducibility of experimental conditions. Such laws are called deterministic. More detailed introduction can be found in reference [32]. For reaction $A \rightarrow B$, the deterministic half-life period can be obtained by $t_{1/2} = \ln 2/R$, where r represents the probability of one A-type particle changes to one B-type particle at each time, while the half-life period means the iterations needed for reducing the number of A-type particle by half during the simulation process. The reaction probability, r, from A to B was set to be 0.03, 0.04 and 0.05, and K^* was determined as 1.88, 1.75 and 1.65, respectively. During the collision process, a normal distributed random number was generated to compare with K^* to decide to react or not. As shown in Figure 4, it can be observed that the concentration of particle A, obtained by Equation (10), decreases with increasing reaction time, and the reaction rate, *i.e.*, the slope of curve in Figure 4a, increases as reaction probability increases. Figure 4b shows reasonable agreement has been achieved between deterministic method and present simulation. Additionally, for a system with 100×100, and reduction probability $r = 0.001$, the deterministic half-life period is 693.1 iterations, Seybold *et al.* [17] reported 688.3 ± 9.7 iterations, and 691 iterations using present model, which also indicates the feasibility of present model in reaction systems.

While $A \underset{k_2}{\overset{k_1}{\rightleftharpoons}} B$ is an equilibrium system, from the law of mass action, the deterministic equilibrium coefficient is defined as $K_{eq} = k_1/k_2 = R(A,B)/R(B,A)$, and the equilibrium coefficient can be obtained by the ratio of the final concentration of B and A as $K_{eq} = [B]/[A]$ in a stochastic system, like simulations using lattice gas automata. The reaction probability from A to B was set as 0.05, 0.06, and 0.07, corresponding to the reaction probability from B to A 0.04, 0.03, and 0.02, and for reaction probability of 0.02 and 0.07, K^* is 2.05 and 1.48, respectively. Similar results are obtained compared to $A \rightarrow B$, as shown in Figure 5, however, Figure 5a presents a platform as time advances, meaning an equilibrium state has been reached. Figure 5b indicates that good agreement has been achieved between the stochastic method and deterministic method. The chemical reaction scheme will be further used for the simulations of gas–solid reaction described latter.

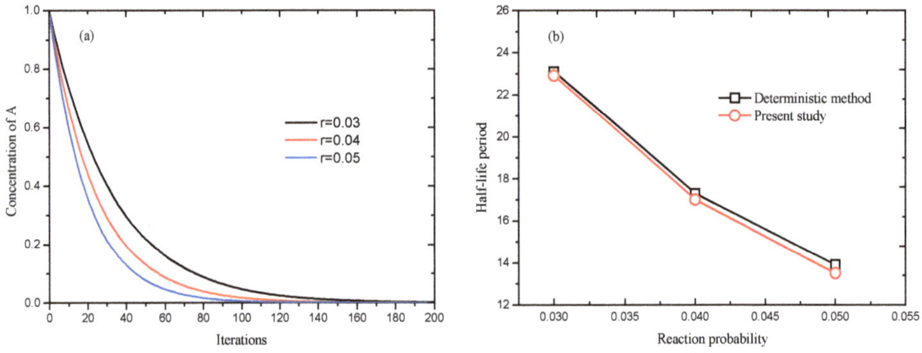

(a) (b)

Figure 4. Simulation of $A \rightarrow B$, (**a**) processes at different reaction probabilities; and (**b**) comparison of results obtained by deterministic method and present model.

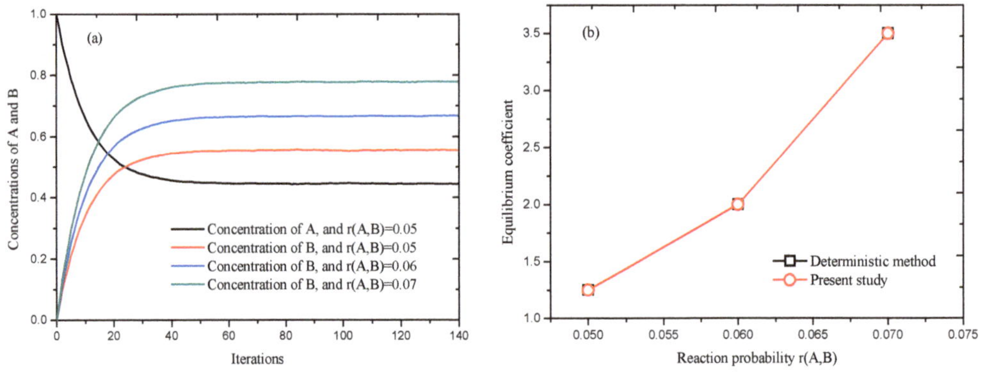

(a) (b)

Figure 5. Simulation of $A \rightleftharpoons B$, (**a**) processes at different reaction probabilities; and (**b**) comparison of results obtained by deterministic method and present model.

3.2. Heat Transfer and Reaction across a Porous Circular Cylinder

The characteristics of the flow and heat transfer over a solid circular cylinder in a square enclosure and a rectangular channel have been investigated by us using this model previously, showing reasonable feasibility and reliability of the application of this model to the problems of flow and heat transfer, details can be found in reference [33]. Herein, simulations of flow, heat transfer and reaction around and through a porous circular cylinder in a channel were carried out. The schematic diagram of the simulated system is shown as Figure 6, where a porous circular cylinder with diameter D = 1/4W is placed at the coordinates (x = 1/3L, y = 1/2W) of a channel with width W = 400 and length L = 1200 (lattice unit). Reactant gas entering from the inlet at a velocity **u**, flows around and through and react with the porous cylinder. The effects of porous structure, reaction probability and gas velocity at inlet on the characteristics of the system will be further discussed in detail.

The porous media investigated were generated by a comprehensive approach termed as quartet structure generation set (QSGS) [34,35], which has been demonstrated capable of generating morphological features close to many real porous media [34]. Following the steps illustrated by Wang *et al.* [23], a porous two-dimensional cylinder can be generated with a set of three construction parameters, including (i) growing phase (fluid) distribution probability, C_d, which decides initial number of fluid seeds in the system; (ii) directional growth probability of fluid, D_i, which is considered the same for all directions in this work; and (iii) fluid volume fraction P.

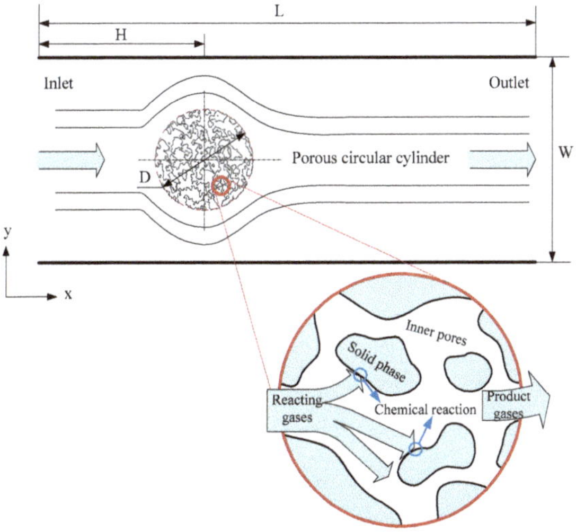

Figure 6. Schematic diagram of the processes happened in the circular cylinder with porous media structure.

Figure 7 presents the inner morphology of the porous circular cylinders of diameter D = 100 (lattice unit) resulted from different combinations of construction parameters, where shaded area is solid phase and the rest to be pores. As can be seen in Figure 7a–d, the solid phase disappears homogenously with increasing porosity, leading to larger pore sizes; and for Figure 7e–h, more agglomeration of solid phase, *i.e.*, less surface area and bigger pore size, appeared for larger directional growth probability; on the contrary, for Figure 7i–l, more homogenous structure as well as smaller pore size of porous media can be obtained as the distribution probability increases, as a result, more react-able surface area is generated. Porous cylinders of different porosity, pore size, and surface area could be obtained by adjusting the three parameters (C_d, D_i, P), for the purpose of comparison.

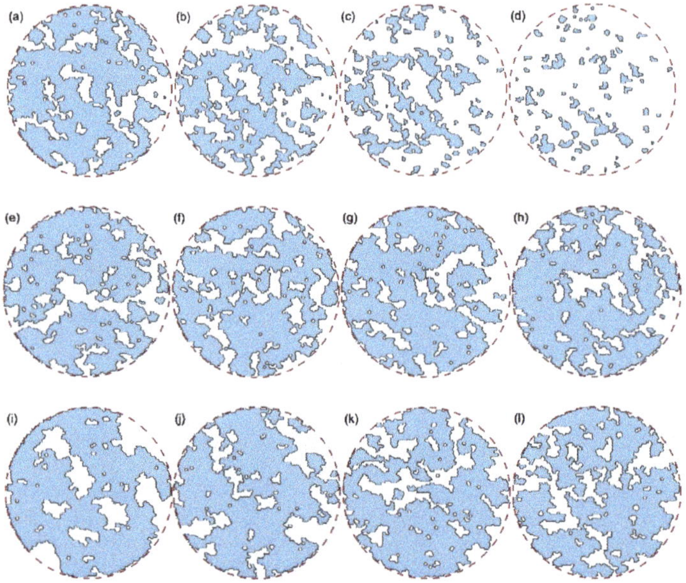

Figure 7. Porous cylinder generated with different construction parameter. (**a**–**d**) Effect of porosity on the morphology of cylinder, where $C_d = 0.02$, $D_i = 0.2$ and (**a**) $P = 0.2$, (**b**) $P = 0.4$, (**c**) $P = 0.6$, and (**d**) $P = 0.8$. (**e**–**h**) Effect of directional growth probability on the morphology of cylinder, where $C_d = 0.02$, $P = 0.2$ and (**e**) $D_i = 0.05$, (**f**) $D_i = 0.15$, (**g**) $D_i = 0.25$, and (**h**) $D_i = 0.35$. (**i**–**l**) Effect of distribution probability on the morphology of cylinder, where $D_i = 0.2$, $P = 0.2$ and (**i**) $C_d = 0.005$, (**j**) $C_d = 0.015$, (**k**) $C_d = 0.025$, and (**l**) $C_d = 0.035$.

Table 1 lists the simulation cases with different parameter sets, with Cases 1 to 4 selected from Figure 7 to investigate the influence of the parameters of QSGS algorithm on the reaction process, Cases 1, 5 and 6 to study the effect of reaction probability, and Cases 1, 7 and 8 are used to investigate the inlet velocity.

Table 1. Parameter sets for simulation cases.

Case No.	Parameters of QSGS			r	u
	C_d	D_i	P		
1	0.02	0.2	0.2	0.1587	1.0
2	0.02	0.2	0.4	0.1587	1.0
3	0.035	0.2	0.2	0.1587	1.0
4	0.02	0.35	0.2	0.1587	1.0
5	0.02	0.2	0.2	0.0668	1.0
6	0.02	0.2	0.2	0.3085	1.0
7	0.02	0.2	0.2	0.1587	0.5
8	0.02	0.2	0.2	0.1587	0.8

A conceptual first order heterogeneous reaction between the reaction gas and the porous cylinder illustrated in Equation (10) is considered in this research, with A being the inlet reactant gas, B the porous solid material, and C as the product gas.

$$A(g) + B(s) \rightarrow C(g) \tag{15}$$

This reaction happens at the solid surface and is simply interpreted as one reactant gas particle reacting with one solid site and generating one product gas particle at mesoscopic level, where the (microscopic level) inner structures or properties of the gas particle or solid site are ignored. The conversion of solid phase B can be defined by the ratio of the number of reacted solid sites to the initial number of solid sites, and a formula as follows is defined

$$X = 1 - \frac{N_t}{N_0} \tag{16}$$

where N_t is the number of solid sites at time t, and N_0 is the initial number of solid sites. Thus, the reaction rate can be further obtained from the time derivative of Equation (11) as dX/dt.

Instead of heat generated during the reaction process, a simplified heat transfer case is considered, where solid sites are set as hot heat source with constant temperature (*i.e.*, 1.0). Gaseous particles will be enhanced to high energy state at impact with solid surface. The gas particles initially entering the channel are of constant low temperature state as 0.

Bounce-back and adiabatic boundary conditions are applied to the channel walls except the outlet which is a free boundary where all particles are released. Solid sites are set as bounce-back boundaries, where gaseous particles will also decide to reaction with according to the chemical reaction scheme.

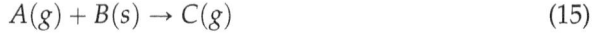

The simulation carried out in such an order that reaction is not considered in the first 2000 iterations to test the coupling of flow and heat transfer only, also to ensure that the reaction takes place at a stable flow field. Reaction is introduced thereafter to investigate its effect to the flow and heat transfer.

The statistical result of temperature, component concentration, and solid phase conversion are obtained from the simulation by space average. It is noted that the statistics level has impacts on the detail of macroscopic picture. This effect is, however, not discussed in this work. Instead, to ensure consistency, the space averaged results in every 2×2 grids are presented for all cases.

3.2.1. Effect of Inner Porous Structure

As discussed in the previous part, the inner porous structure will be influenced by the parameters of QSGS. In this section, the influence of inner structure on the behavior of flow, heat transfer and chemical reaction around and through the cylinder will be investigated. The inlet velocity is constant as 1.0 with a site density $\rho = 1.0$, and based on FHP-II model, $Re = 158.2$ using D as the characteristics parameter.

Figure 8 shows the temperature contours around and inside the porous cylinder before chemical reaction takes place at t = 2000 iterations. A large number of fluid particles are activated to high energy state when they flow over and in the porous circular cylinder, forming a high temperature zone at surrounding and inside the porous structure. As time elapses, the high temperature zone extends to the neighboring field gradually, and no clear eddies have developed behind the cylinder. The field of gas velocity also appeared to have impacts on the profile of temperature.

The heterogeneous reaction is started after 2000 iterations. A product zone is then formed as fluid particles flow over and through the porous cylinder and react with the solid sites according to the reaction scheme. Figure 9 shows the product concentration around the solid cylinder at different times for Case 4. The solid sites disappear gradually as time goes on. It can be observed that the product emerges mainly at some hot points, and distributes homogeneously around the cylinder at the beginning, and then diffuses off from the reaction interface to the surrounding. Figure 10 shows the corresponding temperature contour around the cylinder for Figure 9. The structure of cylinder changes with time, thus the heat transfer characteristics changes as a result.

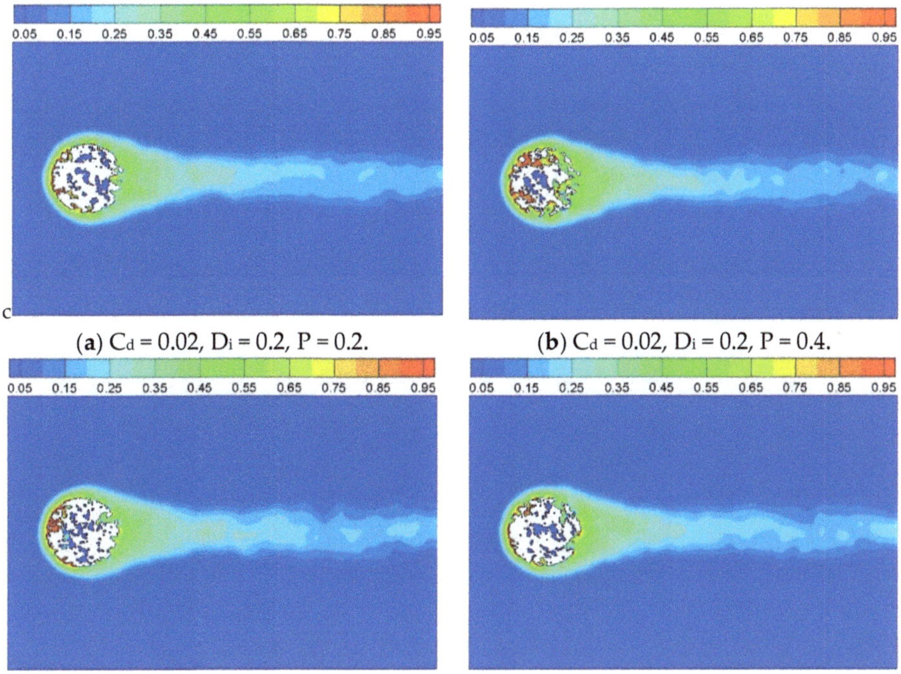

(**a**) $C_d = 0.02$, $D_i = 0.2$, $P = 0.2$.

(**b**) $C_d = 0.02$, $D_i = 0.2$, $P = 0.4$.

(**c**) $C_d = 0.035$, $D_i = 0.2$, $P = 0.2$.

(**d**) $C_d = 0.02$, $D_i = 0.35$, $P = 0.2$.

Figure 8. Temperature contour before reaction with different inner structures, $t = 2000$ iterations.

In order to compare the reaction characteristics of computational cases with porous inner structure, contours of product concentration are shown as Figure 11, for $t = 2500$. It can be observed that the solid cylinder has disappeared completely at $t = 2500$ for Case 2 which has higher porosity than the other three cases. Figure 12 shows the conversion of porous cylinders with different inner structure, as a function of time. It is also notable that solid conversion of higher porosity (Case 2) is faster. This is considered due to the less amount of solid phase, as well as the larger pore size, which facilitates the diffusion of reactant and product. The other three cases with same porosity of 0.2 proceed similarly, but it can still be noted that Case 3 with larger distribution probability progress faster than the other two cases for the former part of time, about 2500 iterations. This can be attributed to the higher homogeneity resulting from larger distribution probability, leading to a larger surface area (reacting sites) to mass ratio. However, this improvement on reaction speed is limited by the relatively smaller pore size slowing down the gas diffusion. Knowing this, it is understandable that Case 4, with larger solid agglomeration and bigger pore size due to higher value of directional growth probability, has the exact opposite performance compare to Case 2.

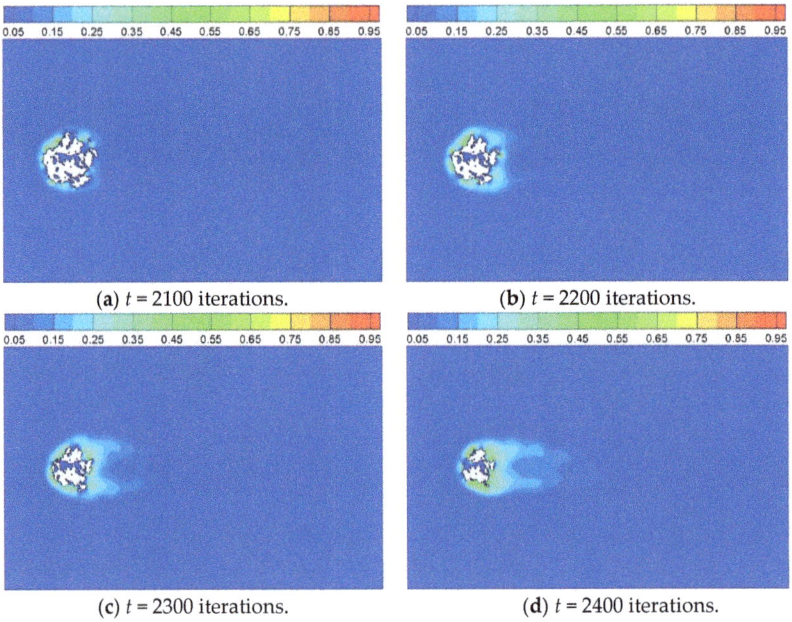

(**a**) $t = 2100$ iterations. (**b**) $t = 2200$ iterations.

(**c**) $t = 2300$ iterations. (**d**) $t = 2400$ iterations.

Figure 9. Product concentration at different times for Case 4, where $C_d = 0.02$, $D_i = 0.35$, and $P = 0.2$.

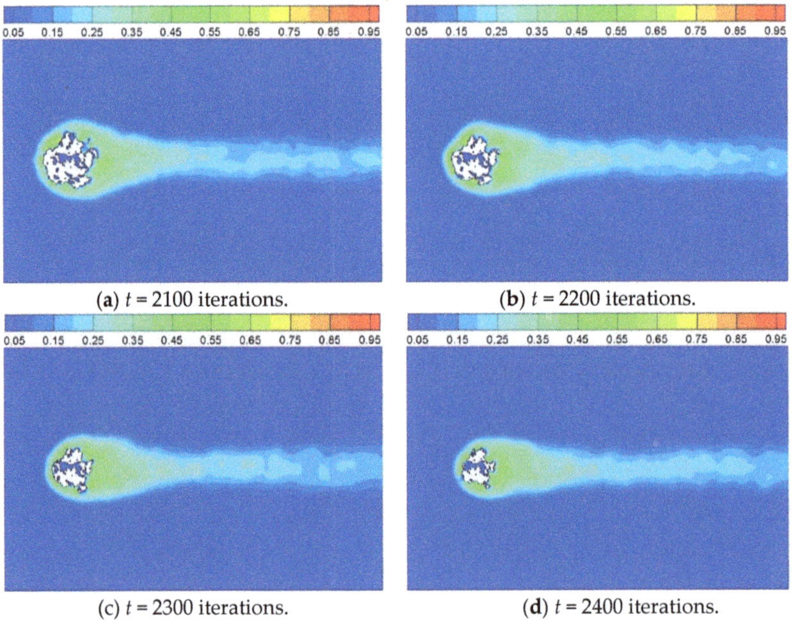

(**a**) $t = 2100$ iterations. (**b**) $t = 2200$ iterations.

(**c**) $t = 2300$ iterations. (**d**) $t = 2400$ iterations.

Figure 10. Temperature contour at different times for Case 4, where $C_d = 0.02$, $D_i = 0.35$, and $P = 0.2$.

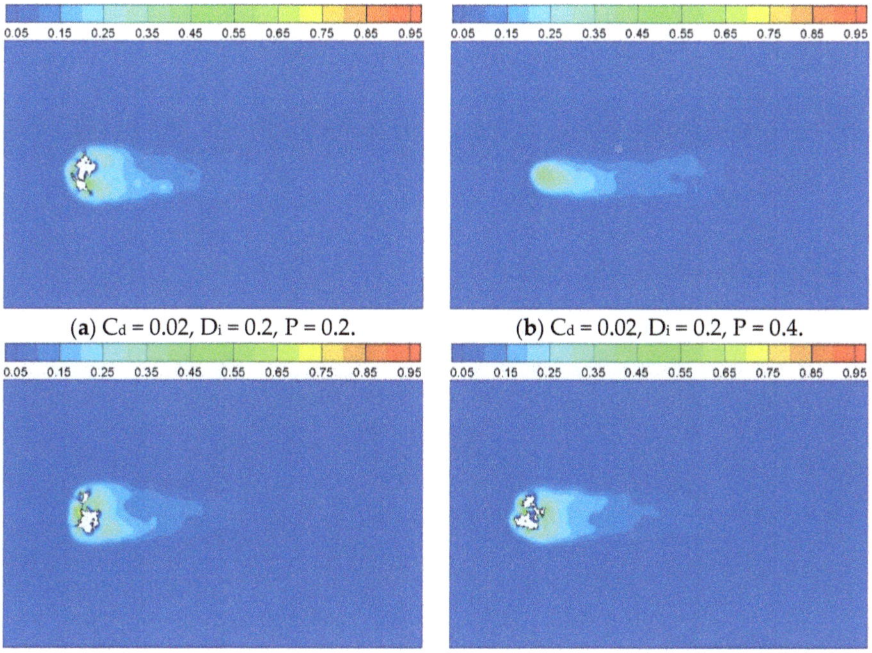

(**a**) $C_d = 0.02$, $D_i = 0.2$, $P = 0.2$. (**b**) $C_d = 0.02$, $D_i = 0.2$, $P = 0.4$.

(**c**) $C_d = 0.035$, $D_i = 0.2$, $P = 0.2$. (**d**) $C_d = 0.02$, $D_i = 0.35$, $P = 0.2$.

Figure 11. Product concentration with different inner structures, t = 2500.

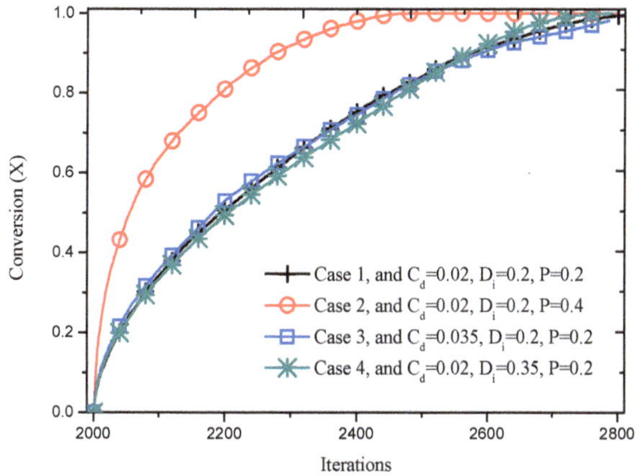

Figure 12. Dependence of conversion on reacting time for cylinders with different inner structure.

3.2.2. Effect of Reaction Probability

In order to obtain the influence of reaction probability on the process evolution, the value of K^* was set to be 0.5 and 1.5, and the reaction probability was determined as 0.3085 and 0.0668 according to the normal distribution function. The reaction probability decides the reaction velocity, as Equation (9). The product concentration and temperature contour around the cylinder at t = 2500 are shown as Figure 13. It can be obviously noted that the reaction processes faster with a higher value of probability, which also can be seen from Figure 14. The conversion increases as reacting time advances and reaction probability increases. For instance, the complete reaction time increases from 696 iterations to 1140 iterations when the reaction probability decreases from 0.3085 to 0.0668.

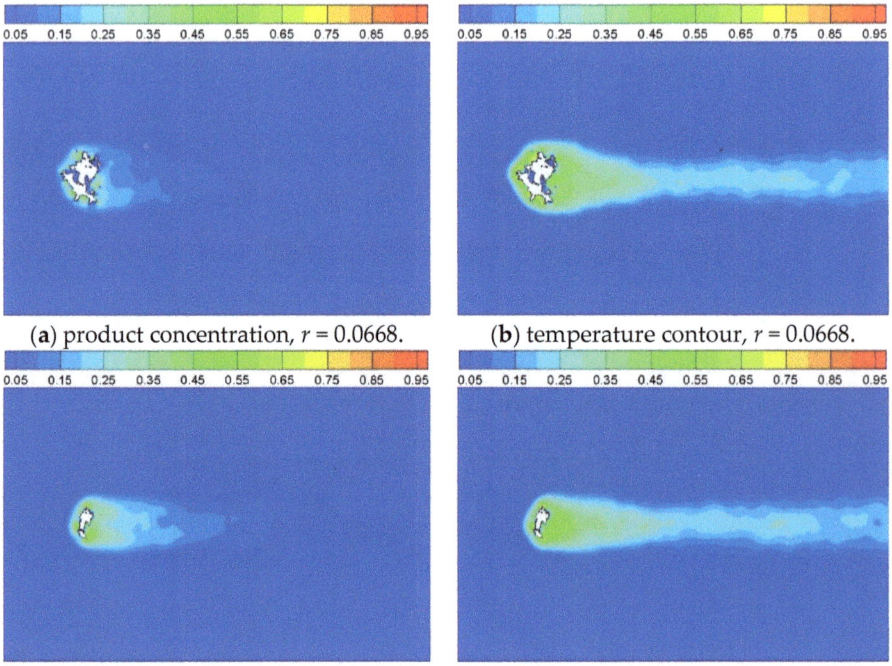

(**a**) product concentration, r = 0.0668. (**b**) temperature contour, r = 0.0668.

(**c**) product concentration, r = 0.3085. (**d**) temperature contour, r = 0.3085.

Figure 13. Product concentration and temperature contour of Case 5 and Case 6, where reaction probability is 0.0668 and 0.3085, respectively, and t = 2500.

3.2.3. Effect of Inlet Gas Velocity

The equilibrium mean occupation numbers are calculated by Fermi–Dirac distribution, as follows

$$n_i^{eq} = \frac{1}{1 + \exp(h + \mathbf{q} \cdot \mathbf{c}_i)} \quad (17)$$

where h is a real number and $\mathbf{q}$ is a D-dimensional vector. The two parameters are termed as Lagrange multipliers. For simplification, the Lagrange multipliers of the equilibrium distributions for lattice gas automata can be obtain by the algebra formula [9]

$$n_i(\mathbf{u}) = d + 2d\mathbf{c}_i \cdot \mathbf{u} + 2d\frac{1-2d}{1-d}c_{i\alpha}^2 u_\alpha^2 - d\frac{1-2d}{1-d}\mathbf{u}^2 \quad (18)$$

where d is equal to $\rho/7$ for FHP-II, and $\mathbf{u}$ is the node velocity. Thus, the variable n_i at the inlet nodes can be initialized by Equation (18) with a given particle density d and a node velocity $\mathbf{u}$.

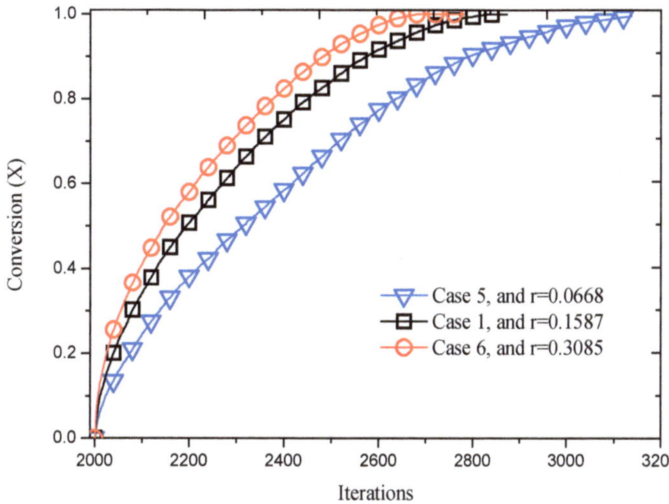

Figure 14. Reaction processes with different reaction probabilities.

For this section, the effect of velocity at inlet on the reaction process is discussed, and the parameters are listed in Table 1, as Cases 1, 7 and 8. Particle density ρ is fixed as 1.0 for all numerical computation cases, thus d is equal to $1/7$. The product and temperature contour of cases with different inlet velocities are shown as Figure 15, it can be seen that Case 8 reacts faster than Case 7 and slower than Case 1 compared with Figure 11a, indicating that the reaction velocity increases as the velocity at inlet increases. This information can also be obtained from the dependence of conversion on the reacting time, as shown in Figure 16, which shows that the extent of conversion increases with increasing reacting time, as well as inlet velocity. This agrees well with the theories of surface reaction in gas–solid systems, such as unreacted shrinking core model [36] and pore model [37], which indicate that the increase of gas velocity promotes the collision between gas particles as well as between gas particles and solid sites, facilitating the diffusion and external dispersion of reactant and product.

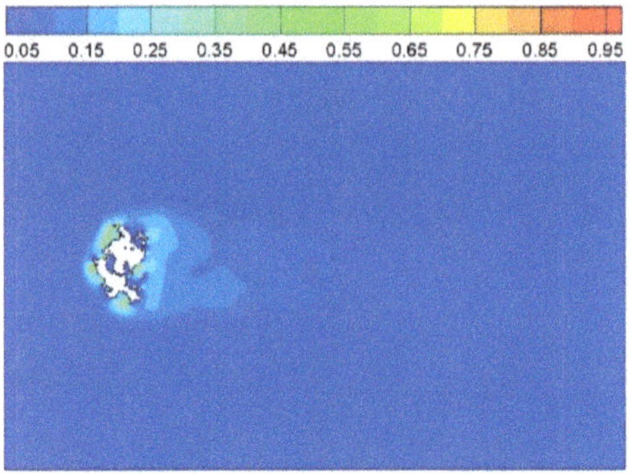

(**a**) product concentration, **u** = 0.5.

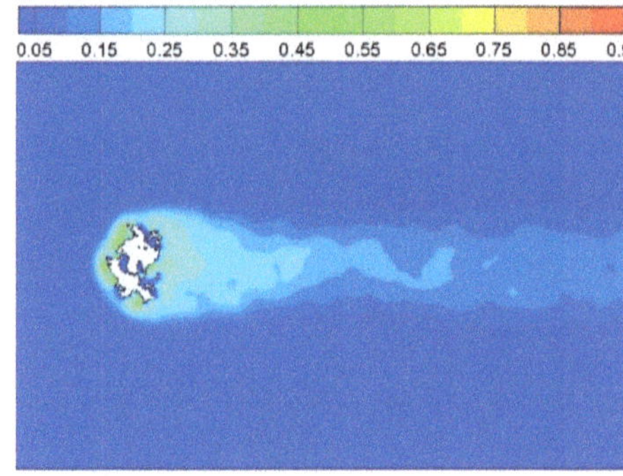

(**b**) temperature contour, **u** = 0.5.

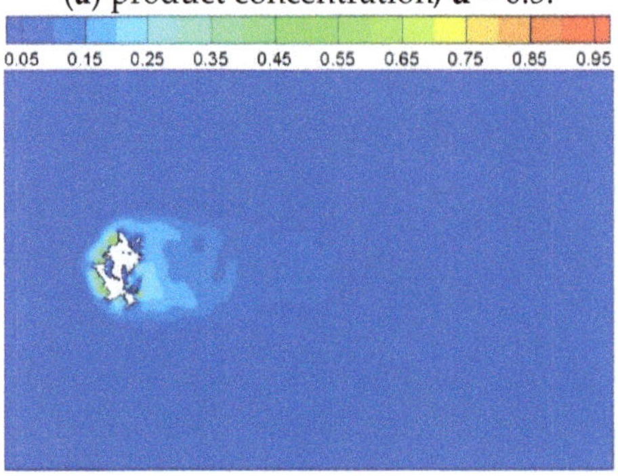

(**c**) product concentration, **u** = 0.8.

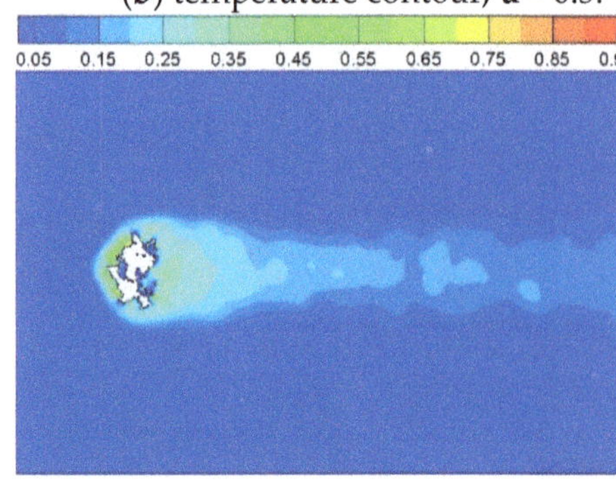

(**d**) temperature contour, **u** = 0.8.

Figure 15. Product concentration and temperature contour of Case 7 and Case 8, where inlet gas velocity is 0.5 and 0.8, respectively, and t = 2500 iterations.

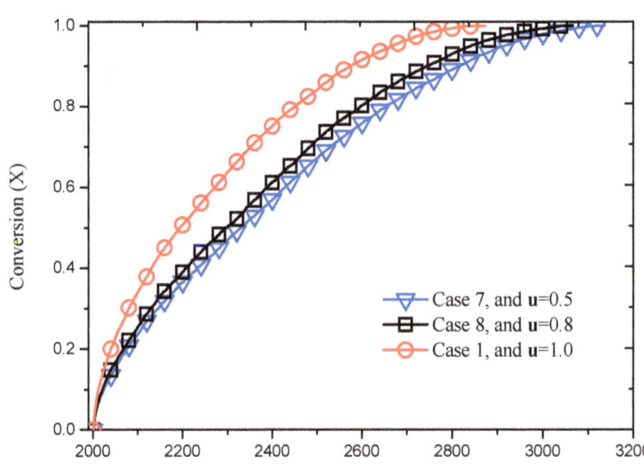

Figure 16. Reaction processes at different inlet velocities.

4. Conclusions

A two-dimensional lattice gas automata model is developed to simulate the flow, heat transfer and chemical reaction around and through porous cylinders constructed by QSGS algorithm. In this model, two additional particle properties, *i.e.*, particle type and energy state, are introduced to account for involved chemical species and fluid temperature, respectively. Heat transfer on the interface of gas–solid is then simulated by the change of gas particle energy state at impact. A chemical reaction scheme based on collision theory is developed, where chemical reaction is also interpreted as the change of particle type according to the probability of reaction.

By controlling the construction parameters of the QSGS method, porous cylinders of different pore sizes, solid agglomeration, porosity and surface to mass ratio are generated for the investigation. Their effects, together with that of reaction probability and inlet fluid velocity on the profiles of temperature, solid conversion rate, and reaction product concentration, are discussed. Numerical results indicate that cylinders with a higher porosity, larger pore size, and more surface area to mass ratio react with gas particles faster. Moreover, the reaction velocity increases with increasing reaction probability as well as gas velocity at inlet. These results agree well with the basic theories of the gas dispersion, pore diffusion, and solid surface reaction. The proposed LGA model is therefore believed to provide a prospective modeling strategy for describing gas–solid chemical reaction occurring at complex boundaries from the viewpoint of mesoscopic level.

Acknowledgments: This work is made possible by the financial support from the National Natural Science Foundation of China (Grant No. 51074201 and No. 51274264), and the National Instrumentation Grant Program (No. 2011YQ120039).

Author Contributions: Zhong Zheng and Hongsheng Chen conceived and designed the model and simulations; Zhiwei Chen and Hongsheng Chen analyzed the data; Xiaotao T. Bi and Zhiwei Chen made editing and improvement corrections of manuscript; Zhiwei Chen and Hongsheng Chen wrote the paper. All authors have read and approved the final manuscript.

Conflicts of Interest: The authors declare no conflict of interest.

References

1. Yang, C.; Thovert, J.-F.; Debenest, G. Upscaling of mass and thermal transports in porous media with heterogeneous combustion reactions. *Int. J. Heat Mass Transf.* **2015**, *84*, 862–875.
2. Mahmud, S.; Fraser, R.A. Free convection and irreversibility analysis inside a circular porous enclosure. *Entropy* **2003**, *5*, 358–365.
3. Liu, H.; Patil, P.R.; Narusawa, U. On Darcy-Brinkman equation: Viscous flow between two parallel plates packed with regular square arrays of cylinders. *Entropy* **2007**, *9*, 118–131.

4. Makinde, O.D.; Eegunjobi, A.S. Entropy generation in a couple stress fluid flow through a vertical channel filled with saturated porous media. *Entropy* **2013**, *15*, 4589–4606.
5. Li, M.; Wu, Y.; Zhao, Z. Effect of endothermic reaction mechanisms on the coupled heat and mass transfers in a porous packed bed with Soret and Dufour effects. *Int. J. Heat Mass Transf.* **2013**, *67*, 164–172.
6. Li, X.; Cai, J.; Xin, F.; Huai, X.; Guo, J. Lattice Boltzmann simulation of endothermal catalytic reaction in catalyst porous media. *Appl. Therm. Eng.* **2013**, *50*, 1194–1200.
7. Machado, R. Numerical simulations of surface reaction in porous media with lattice Boltzmann. *Chem. Eng. Sci.* **2012**, *69*, 628–643.
8. Xin, F.; Li, X.-F.; Xu, M.; Huai, X.-L.; Cai, J.; Guo, Z.-X. Simulation of gas exothermic chemical reaction in porous media reactor with lattice Boltzmann method. *J. Therm. Sci.* **2013**, *22*, 42–47.
9. Wolf-Gladrow, D.A. *Lattice-Gas Cellular Automata and Lattice Boltzmann Models*; Springer: Berlin/Heidelberg, Germany, 2000.
10. McNamara, G.R.; Garcia, A.L.; Alder, B.J. Stabilization of thermal lattice Boltzmann models. *J. Statist. Phys.* **1995**, *81*, 395–408.
11. McCarthy, J.F. Flow through arrays of cylinders: Lattice gas cellular automata simulations. *Phys. Fluids* **1994**, *6*, 435–437.
12. Vogeler, A.; Wolf-Gladrow, D.A. Pair interaction lattice gas simulations: Flow past obstacles in two and three dimensions. *J. Statist. Phys.* **1993**, *71*, 163–190.
13. Eissler, W.; Drtina, P.; Frohn, A. Cellular automata simulation of flow around chains of cylinders. *Int. J. Numer. Methods Eng.* **1992**, *34*, 773–791.
14. Dab, D.; Lawniczak, A.; Boon, J.-P.; Kapral, R. Cellular-automaton model for reactive systems. *Phys. Rev. Lett.* **1990**, *64*, 2462–2465.
15. Boon, J.P.; Dab, D.; Kapral, R.; Lawniczak, A. Lattice gas automata for reactive systems. *Phys. Rep.* **1996**, *273*, 55–147.
16. Weimar, J.R. Cellular automata for reaction-diffusion systems. *Parallel Comput.* **1997**, *23*, 1699–1715.
17. Seybold, P.G.; Kier, L.B.; Cheng, C.-K. Simulation of first-order chemical kinetics using cellular automata. *J. Chem. Inf. Comput. Sci.* **1997**, *37*, 386–391.
18. Seybold, P.G.; Kier, L.B.; Cheng, C.-K. Stochastic cellular automata models of molecular excited-state dynamics. *J. Phys. Chem. A* **1998**, *102*, 886–891.
19. Seybold, P.G.; Kier, L.B.; Cheng, C.-K. Aurora borealis: Stochastic cellular automata simulations of the excited-state dynamics of oxygen atoms. *Int. J. Quantum Chem.* **1999**, *75*, 751–756.
20. Neuforth, A.; Seybold, P.G.; Kier, L.B.; Cheng, C.-K. Cellular automata models of kinetically and thermodynamically controlled reactions. *Int. J. Chem. Kinet.* **2000**, *32*, 529–534.
21. Roberts, J.D.; Caserio, M.C. *Basic Principles of Organic Chemistry*, 2nd ed.; WA Benjamin: Menlo Park, CA, USA, 1977.
22. Lin, K.-C. Understanding product optimization: Kinetic versus thermodynamic control. *J. Chem. Educ.* **1988**, *65*, 857–860.

23. Hollingsworth, C.A.; Seybold, P.G.; Kier, L.B.; Cheng, C.-K. First-order stochastic cellular automata simulations of the Lindemann mechanism. *Int. J. Chem. Kinet.* **2004**, *36*, 230–237.
24. Lindemann, F.A.; Arrhenius, S.; Langmuir, I.; Dhar, N.R.; Perrin, J.; McC. Lewis, W.C. Discussion on "the radiation theory of chemical action". *Trans. Faraday Soc.* **1922**, *17*, 598–606.
25. Frisch, U.; d'Humières, D.; Hasslacher, B.; Lallemand, P.; Pomeau, Y.; Rivet, J.-P. Lattice gas hydrodynamics in two and three dimensions. *Complex Syst.* **1987**, *1*, 649–707.
26. Zheng, Z.; Gao, X. Lattice gas automata method for modeling fluid flow and heat transfer in metallurgical porous media. *Acta Met. Sin.* **2000**, *36*, 433–437.
27. Bresolin, C.S.; Oliveira, A.A.M. An algorithm based on collision theory for the lattice Boltzmann simulation of isothermal mass diffusion with chemical reaction. *Comput. Phys. Commun.* **2012**, *183*, 2542–2549.
28. Baercor, T.; Mayer, P.M. Statistical Rice–Ramsperger–Kassel–Marcus quasiequilibrium theory calculations in mass spectrometry. *J. Am. Soc. Mass Spectrom.* **1997**, *8*, 103–115.
29. Moon, J.H.; Oh, J.Y.; Kim, M.S. A systematic and efficient method to estimate the vibrational frequencies of linear peptide and protein ions with any amino acid sequence for the calculation of Rice–Ramsperger–Kassel–Marcus rate constant. *J. Am. Soc. Mass Spectrom.* **2006**, *17*, 1749–1757.
30. Moon, J.H.; Sun, M.; Kim, M.S. Efficient and reliable calculation of Rice–Ramsperger–Kassel–Marcus unimolecular reaction rate constants for biopolymers: Modification of Beyer–Swinehart algorithm for degenerate vibrations. *J. Am. Soc. Mass Spectrom.* **2007**, *18*, 1063–1069.
31. Wichura, M.J. Algorithm as 241: The percentage points of the normal distribution. *J. R. Stat. Soc.* **1988**, *37*, 477–484.
32. Lecca, P.; Laurenzi, I.; Jord, F. *Deterministic Versus Stochastic Modelling in Biochemistry and Systems Biology*; Woodhead Publishing: Cambridge, UK, 2013.
33. Chen, H.; Zheng, Z.; Chen, Z.; Bi, X.T. Simulation of flow and heat transfer around a heated stationary circular cylinder by lattice gas automata. *Powder Technol.* **2015**.
34. Wang, M.; Wang, J.; Pan, N.; Chen, S. Mesoscopic predictions of the effective thermal conductivity for microscale random porous media. *Phys. Rev. E* **2007**, *75*, 036702.
35. Chen, L.; Wu, G.; Holby, E.F.; Zelenay, P.; Tao, W.-Q.; Kang, Q. Lattice Boltzmann pore-scale investigation of coupled physical-electrochemical processes in C/Pt and non-precious metal cathode catalyst layers in proton exchange membrane fuel cells. *Electrochim. Acta* **2015**, *158*, 175–186.
36. Homma, S.; Ogata, S.; Koga, J.; Matsumoto, S. Gas–solid reaction model for a shrinking spherical particle with unreacted shrinking core. *Chem. Eng. Sci.* **2005**, *60*, 4971–4980.
37. Petersen, E.E. Reaction of porous solids. *AIChE J.* **1957**, *3*, 443–448.

Three-Dimensional Lattice Boltzmann Simulation of Liquid Water Transport in Porous Layer of PEMFC

Bo Han, Meng Ni and Hua Meng

Abstract: A three-dimensional two-phase lattice Boltzmann model (LBM) is implemented and validated for qualitative study of the fundamental phenomena of liquid water transport in the porous layer of a proton exchange membrane fuel cell (PEMFC). In the present study, the three-dimensional microstructures of a porous layer are numerically reconstructed by a random generation method. The LBM simulations focus on the effects of the porous layer porosity and boundary liquid saturation on liquid water transport in porous materials. Numerical results confirm that liquid water transport is strongly affected by the microstructures in a porous layer, and the transport process prefers the large pores as its main pathway. The preferential transport phenomenon is more profound with a decreased porous layer porosity and/or boundary liquid saturation. In the transport process, the breakup of a liquid water stream can occur under certain conditions, leading to the formation of liquid droplets inside the porous layer. This phenomenon is related to the connecting bridge or neck resistance dictated by the surface tension, and happens more frequently with a smaller porous layer porosity. Results indicate that an optimized design of porous layer porosity and the combination of various pore sizes may improve both the liquid water removal and gaseous reactant transport in the porous layer of a PEMFC.

Reprinted from *Entropy*. Cite as: Han, B.; Ni, M.; Meng, H. Three-Dimensional Lattice Boltzmann Simulation of Liquid Water Transport in Porous Layer of PEMFC. *Entropy* **2016**, *18*, 17.

1. Introduction

The proton exchange membrane fuel cell (PEMFC) is considered one of the most promising energy conversion devices for future transportation applications, due to its high energy efficiency, high power density, and environment-friendly operations. Figure 1a shows a schematic of a single piece of a PEMFC, which is mainly composed of a polymer electrolyte membrane (PEM), bipolar plates with the built-in gas channels, gas diffusion layers (GDLs), and catalyst layers (CLs). The GDLs and CLs are heterogeneous porous media, in which the chemical species diffusion, electrochemical reactions, and migration of protons and electrons occur, as shown in Figure 1b,c. During PEMFC operations, water is produced by the electrochemical reactions in the cathode catalyst layer (CCL). Water is needed to fully hydrate the polymer membrane to increase its proton conductivity and thus improve the PEMFC performance, but excessive water, particularly liquid water in the porous materials (GDLs and CLs), can block hydrogen/oxygen transport to the reaction sites and consequently decrease the cell performance. Therefore, water management in a PEMFC, particularly the liquid water transport and distribution in the porous materials, play a key role in fuel cell operations [1,2].

Over the past decades, many experimental and numerical studies have been conducted to investigate the complex transport phenomena in the porous media of PEMFCs [3–14]. These studies, however, focused mainly on the macro-scale transport processes. To obtain a deeper understanding of the underlying transport mechanisms in the porous materials of a PEMFC, fundamental studies have also been conducted to reveal the microstructures of the porous materials and further examine their effects on the transport processes. Hizir *et al.* [15] used the optical profilometry and SEM to scan the surface morphology of the CL of a PEMFC. Their studies indicated that there are surface cracks in the CL, which could significantly affect the multi-phase liquid water transport behaviors. Based on images from SEM, Thiele *et al.* [16] reconstructed a three-dimensional geometry of the cathode catalyst layer (CCL) of a PEMFC. The pore size distribution in the CCL was revealed in detail. Due to the great complexity of the microstructures in PEMFCs, it is difficult to directly observe liquid water transport inside the porous materials using the present experimental techniques. Therefore, the micro- and meso-scale numerical modeling and simulation approaches are widely used as effective tools to study the detailed transport processes in the porous media. For example, Wang *et al.* [17] and Mukherjee *et al.* [18] developed a direct numerical simulation method to study the pore-scale species transport in the cathode catalyst layer of a PEMEC. The CCL was reconstructed using a stochastic technique.

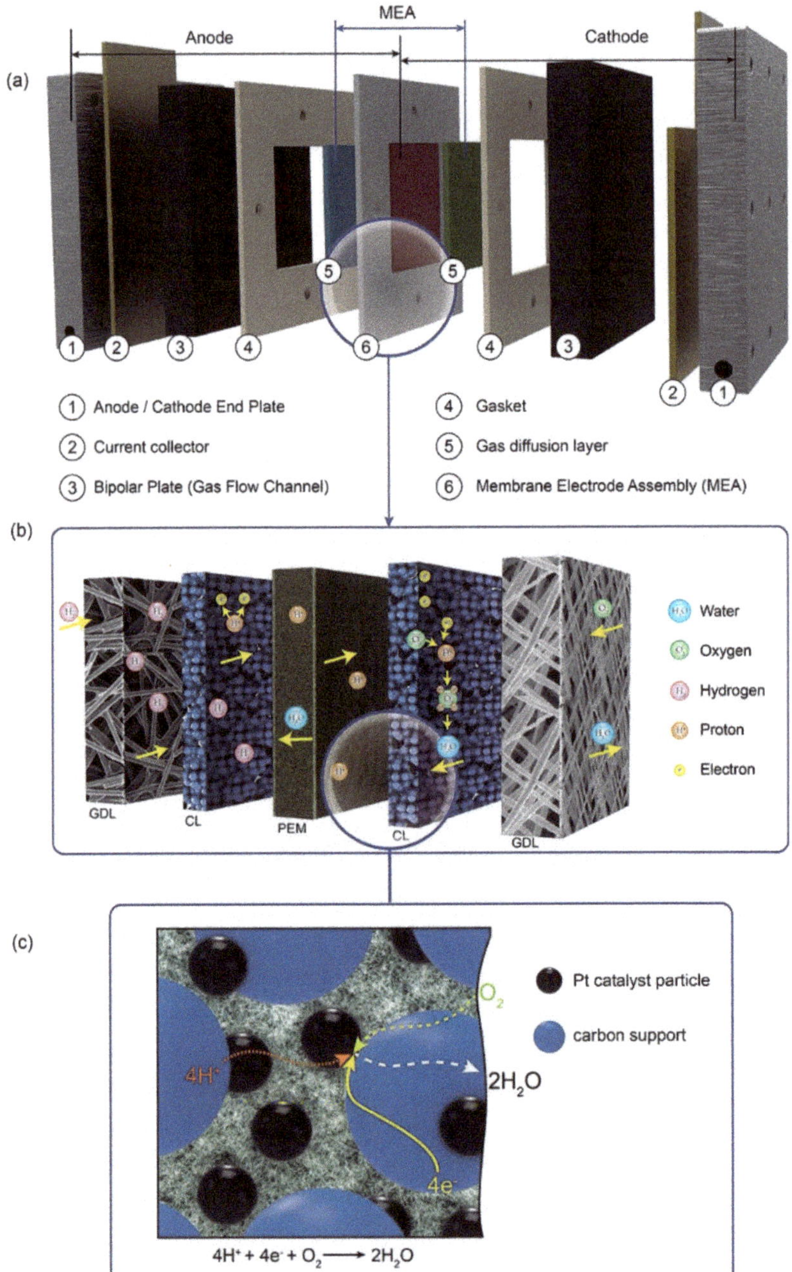

Figure 1. (**a**) Schematic of a single piece of a PEMFC; (**b**) Schematic of the microstructure and mass transport phenomena in porous layers; (**c**) Schematic of transport and reaction mechanism in CL.

Although good progress has been made in studying transport phenomena at the micro scale, it is still a challenging issue in fully understanding the liquid water transport mechanisms inside the porous materials of a PEMFC. Recently, the lattice Boltzmann method (LBM), which is well established for simulating multi-phase and multi-component fluid flows [19–26], has been applied to study liquid water transport in PEMFCs. In LBM, the density distribution functions are directly solved by implementing the collision and streaming procedures at discrete lattices, and thus this method can handle the gas–liquid and solid–liquid interactions automatically and accurately. Sinha *et al.* [27] performed a two-phase LBM simulation to study the effects of wetting property on the liquid water dynamic behaviors in a reconstructed carbon-paper GDL. Park *et al.* [28] conducted an LBM simulation of the liquid droplet transport through a GDL made of woven carbon cloth. Hao *et al.* [29,30] used a free-energy LBM to study the dynamic behaviors of liquid droplets in the gas channel and analyzed the effects of the micro-scale porous structure on the relative permeability of a porous material. Mukherjee *et al.* [31] conducted LBM simulation to examine the two-phase transport process and liquid water flooding phenomena in the porous media of a PEMFC. Han *et al.* [32–34] conducted two-dimensional two-phase LBM simulations to analyze the formation, growth, and interaction of liquid droplets inside the GDL and gas channel. Effects of several important parameters, including the gas flow velocity, surface contact angle, and gas channel shape, on the liquid water transport characteristics were investigated. These early studies prove that LBM simulations are suitable for numerical study of the micro-scale two-phase transport in the complex porous media of a PEMFC.

In this paper, a three-dimensional two-phase lattice Boltzmann method is employed to simulate the liquid water transport dynamics in the porous layer of a PEMFC. A random method is employed to numerically reconstruct the micro-scale structures of a porous medium. The numerical model is validated with two test cases concerning the gas–liquid phase separation and liquid–solid surface interaction. The numerical studies focus on the effects of the porous layer porosity and boundary liquid saturation on liquid water transport behaviors. Results obtained herein can help to improve fundamental understanding of the liquid water transport processes in porous materials of PEMFCs.

2. Lattice Boltzmann Model and Its Validation

2.1. Lattice Boltzmann Model

The present three-dimensional two-phase LBM simulation is based on the single relaxation time Bhatnagar–Gross–Krook (BGK) evolution equation and Shan–Chan model [22,23]. The Boltzmann–BGK equation, which is an approximation of the continuum Boltzmann equation, is discretized in time, space, and velocity domains to obtain a full discrete LBM equation:

$$f_\alpha\left(x + e_\alpha \Delta t,\ t + \Delta t\right) - f_\alpha\left(x,\ t\right) = \frac{1}{\tau}\left(f_\alpha^{eq}\left(x,\ t\right) - f_\alpha\left(x,\ t\right)\right), \tag{1}$$

where f is the density distribution function, x is the space position vector, e_α is the lattice velocity vector at the αth direction, t is the time, and τ is the relaxation time that can be determined using the fluid kinematic viscosity. Generally, there are two important steps in numerical solution of the LBM equation, streaming and collision, which need to be implemented separately in the numerical treatment. In the streaming step, the distribution function representing discrete particles is moved into the neighboring lattices in a fixed velocity scheme. In the collision step, the density distribution function is relaxed toward its equilibrium state at a given relaxation time. In the present simulation, the equilibrium distribution function is defined as:

$$f_\alpha^{eq}\left(x,\ t\right) = \omega_\alpha \rho\left[1 + \frac{e_\alpha \cdot u^{eq}}{c_s^2} + \frac{\left(e_\alpha u^{eq}\right)^2}{2c_s^4} - \frac{\left(u^{eq}\right)^2}{2c_s^2}\right], \tag{2}$$

where ω_α is the weighting factor, ρ is the macroscopic fluid density, u^{eq} is the macroscopic equilibrium velocity, and c_s is the lattice sound velocity. In order to determine these parameters, the D3Q19 scheme that includes 19 discrete velocities in three space dimensions is used, as shown in Figure 2.

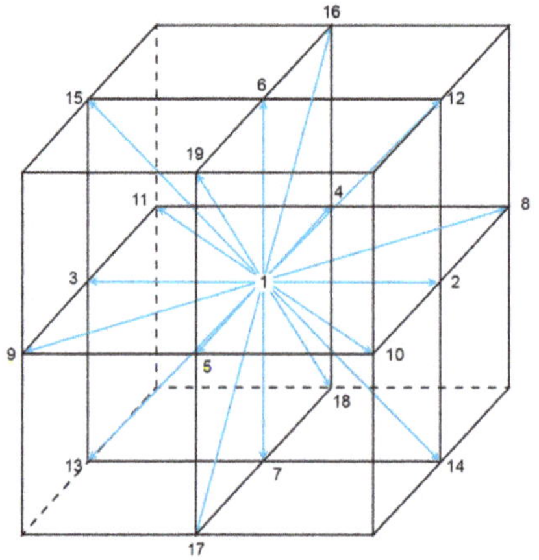

Figure 2. D3Q19 scheme for three-dimensional lattice Boltzmann simulations.

The relevant parameters in Equation (2) are defined in detail in the following:

- The weighting factor:

$$\omega_\alpha = \begin{cases} 1/3, & \alpha = 1 \\ 1/18, & \alpha = 2,3,4,5,6,7 \\ 1/36, & \alpha = 8,9,10,11,12,13,14,15,16,17,18,19 \end{cases} . \tag{3}$$

- The macroscopic fluid density:

$$\rho = \sum_{\alpha=1}^{19} f_\alpha \, . \tag{4}$$

- The lattice velocity vector:

$$e = c\begin{bmatrix} 0 & 1 & -1 & 0 & 0 & 0 & 0 & 1 & -1 & 1 & -1 & 1 & -1 & -1 & 1 & 0 & 0 & 0 & 0 \\ 0 & 0 & 0 & 1 & -1 & 0 & 0 & 1 & -1 & -1 & 1 & 0 & 0 & 0 & 0 & 1 & -1 & 1 & -1 \\ 0 & 0 & 0 & 0 & 0 & 1 & -1 & 0 & 0 & 0 & 0 & 1 & -1 & 1 & -1 & 1 & -1 & -1 & 1 \end{bmatrix}, \tag{5}$$

where c is defined as the ratio of the space distance between lattice nodes to the time step.

In the Shan–Chen two-phase model, extra forces are included in the equilibrium state distribution function to handle the fluid–fluid and fluid–solid interactions. The forces are calculated as:

$$\boldsymbol{F} = \boldsymbol{F}_{int}(\boldsymbol{x}, t) + \boldsymbol{F}_{ads}(\boldsymbol{x}, t) + \boldsymbol{F}_g(\boldsymbol{x}), \tag{6}$$

In Equation (6), the term $\boldsymbol{F}_g(\boldsymbol{x})$ is the gravity force or uniform steady body force, and $\boldsymbol{F}_{int}(\boldsymbol{x}, t)$ is the inter-particle cohesive force that is introduced to account for the interaction between fluid particles. The adsorption force $\boldsymbol{F}_{ads}(\boldsymbol{x}, t)$, which is used to account for the fluid–solid interaction, is expressed as:

$$\boldsymbol{F}_{ads}(\boldsymbol{x}, t) = -G\psi(\rho(\boldsymbol{x}, t)) \sum_\alpha \omega_\alpha \psi(\rho_w) s(\boldsymbol{x} + \boldsymbol{e}_\alpha \Delta t, t)\boldsymbol{e}_\alpha \tag{7}$$

where G is a constant representing the interaction strength between neighboring particles, Ψ is a density-dependent potential function, and ρ_w is a free parameter that can be tuned to obtain different contact angles.

The forces in Equation (6) are implemented into LBM by changing the equilibrium velocity in Equation (2). Therefore, the new equilibrium distribution function for the present two-phase simulation is expressed as:

$$f_\alpha^{eq\prime}(\boldsymbol{x}, t) = \omega_\alpha \rho \left[1 + \frac{\boldsymbol{e}_\alpha \left(\boldsymbol{u} + \frac{\tau \boldsymbol{F}}{\rho} \right)}{c_s^2} + \frac{\left(\boldsymbol{e}_\alpha \left(\boldsymbol{u} + \frac{\tau \boldsymbol{F}}{\rho} \right) \right)^2}{2c_s^4} - \frac{\left(\boldsymbol{u} + \frac{\tau \boldsymbol{F}}{\rho} \right)^2}{2c_s^2} \right], \tag{8}$$

where u is the microscopic velocity, which can be calculated as:

$$u = \frac{1}{\rho} \sum_{\alpha=1}^{19} f_{\alpha} e_{\alpha}. \quad (9)$$

More details concerning the two-phase lattice Boltzmann model can be found in the previous publications [22,23,32].

2.2. Model Validation

The preceding three-dimensional two-phase lattice Boltzmann model has been developed into a computational program in our research group. Before it is used for detailed numerical studies of liquid water transport in the porous layer of a PEMFC, two numerical test cases, concerning the gas–liquid two-phase separation and contact angle effect on liquid droplet behaviors, were conducted for model validations.

The first test concerns the gas–liquid two-phase separation and maintaining of a single liquid droplet after the two-phase separation. A single liquid droplet with a specified radius was initially placed in a gas phase domain, which was divided into $80 \times 80 \times 80$ lattices at the x, y, and z directions, respectively. The boundary condition was set to be fully periodic in the simulations. Without the body force effect, the droplet should reach an equilibrium state with unchanged shape and size. As shown in Figure 3a,b, the figures on the left side illustrate two spherical droplets at a radius of 10 lu and 20 lu, respectively, at the equilibrium state. The two-dimensional views of the liquid droplets in the y-z cross section in the middle of the x direction are shown in the figures on the right side. The red region represents the liquid phase, and the blue one is the gas phase. It can be observed that the interface between gas phase and liquid phase is very sharp and clear, verifying that the present model is capable of handling the gas–liquid two-phase interaction and phase separation.

In addition, according to Laplace's law, the static pressure difference across a droplet interface should be proportional to the fluid surface tension and inversely proportional to the droplet radius. To test Laplace's law, a series of LBM simulations with various droplet radii were performed. As shown in Figure 4, the solid line represents the results directly calculated from Laplace's law, and the dots are obtained from the LBM simulations. Two cases were simulated using different numbers of lattices, including case 1 with $80 \times 80 \times 80$ lattices and case 2 with $60 \times 60 \times 60$ lattices. At a relatively large droplet radius, the simulation results from the two cases are both consistent with Laplace's law. However, at a small droplet radius, results from case 1 show better agreement than those from case 2, because of the finer grid resolution. For instance, when the droplet radius is equal to 10 lu, the pressure difference calculated in case 1 is 1.14, which is very close to the result from Laplace's law, but the pressure difference in case 2 is only 1.04, showing a relatively large error. Therefore, the simulation results indicate that the present two-phase model is sufficiently accurate for numerical studies, using a set of 80 × 80 × 80 lattices.

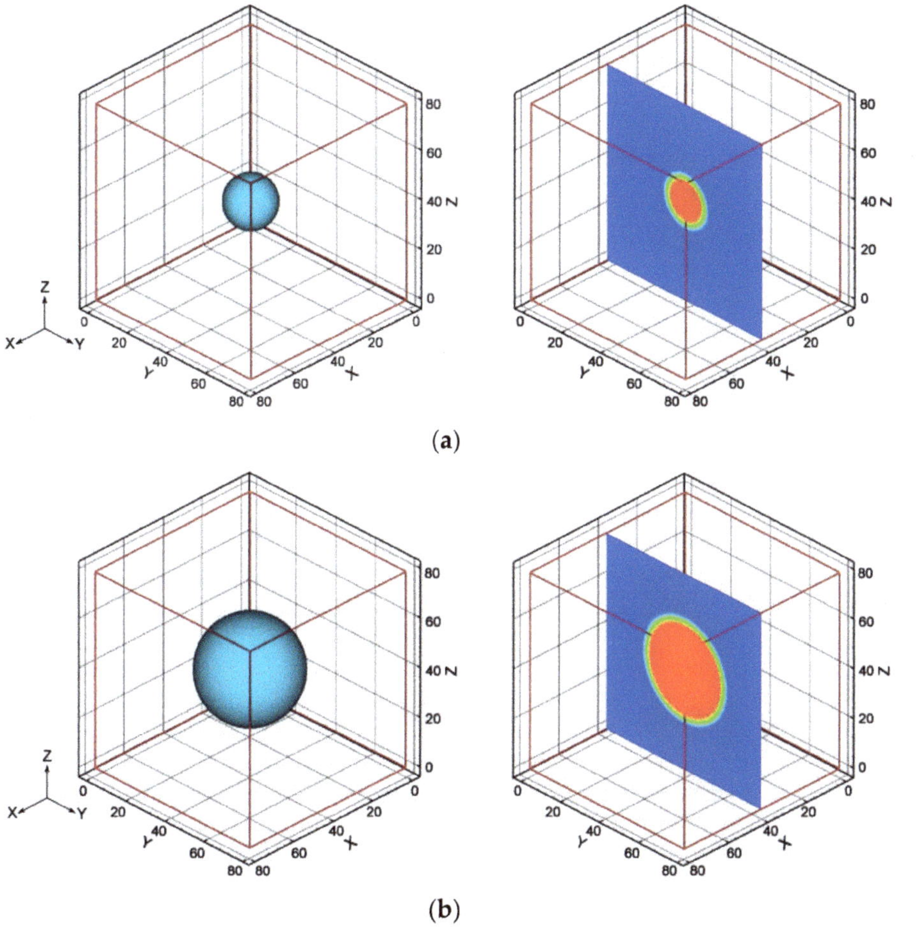

Figure 3. Model validations concerning two-phase separation: (**a**) A liquid droplet with a radius of 10 lu; (**b**) A liquid droplet with a radius of 20 lu.

The second test was carried out to study the effect of the surface contact angle on liquid water droplet behaviors on a solid surface. The computational domain was again divided into a total of 80 × 80 × 80 lattices. The periodic boundary condition was used at the x and y directions, and the wall bounce-back boundary condition was set for the top and bottom boundaries at the z direction. The body force effect was neglected in the simulations. Initially, a spherical liquid droplet was placed on the solid surface at the bottom boundary. The contact angle of the solid surface was adjusted to represent different interaction strengths between the liquid droplet and solid surface. The LBM simulations were performed until the droplet reached the steady state. Figure 5 clearly shows the proper characteristics with interaction between the liquid droplet and solid surface at a contact angle of 90° (Figure 5a) and 180° (Figure 5b), respectively. Results further verify that the present two-phase lattice Boltzmann model can also accurately treat the liquid–solid interaction.

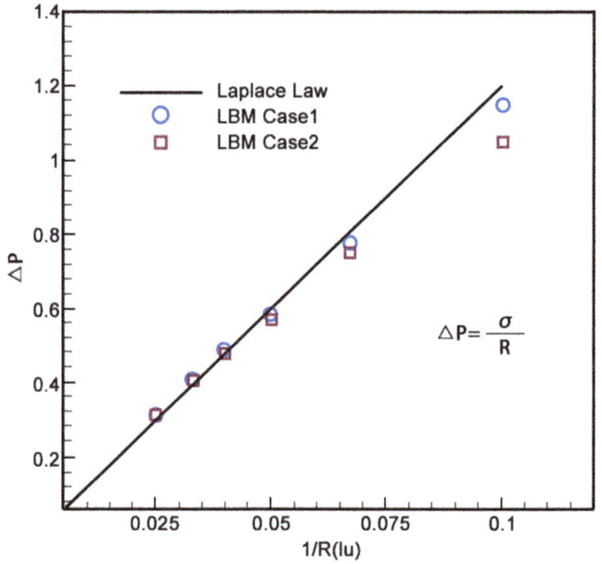

Figure 4. Model validations against Laplace's law.

It should be mentioned that because of the numerical problem in treating a large density ratio in the LBM simulations, the density ratio in the present studies is set at 10, and the dynamic viscosity is also at 10. Therefore, the two-phase flows solved in the above tested cases and following numerical simulations concern the liquid water transport in a dense gaseous phase. However, the previous studies [23,25,27,31–34] and the present test results have clearly shown that the LBM simulations are capable of capturing the fundamental physics in two-phase flows.

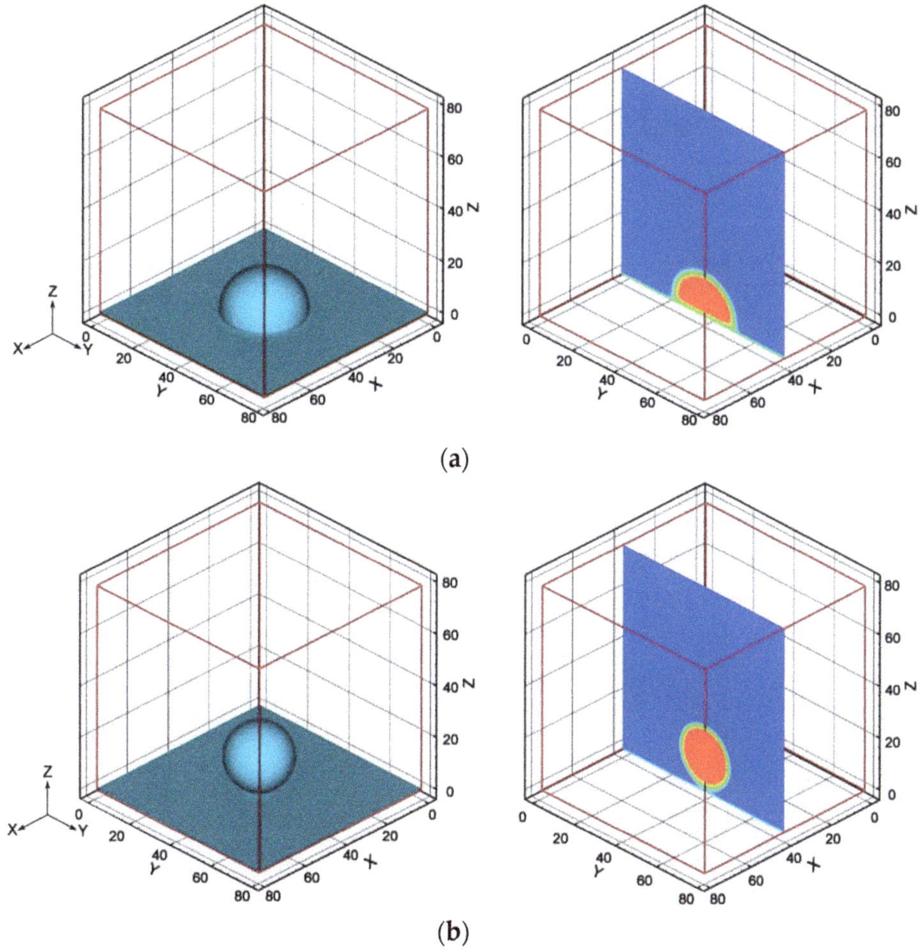

Figure 5. Model validations concerning liquid–solid interactions: (a) At a contact angle of $\theta = 90$; (b) At a contact angle of $\theta = 180$.

3. Results and Discussion

After model validations, numerical studies of liquid water transport in the porous layer of a PEMFC are conducted in this section, focusing mainly on the effects of the porous layer porosity and boundary liquid saturation on the transport behaviors.

3.1. Microstructure Reconstruction

Reconstruction of the microstructure of a porous layer is a prerequisite prior to conducting LBM simulations. In the open literature [35–37], two methods were generally used to reconstruct the microstructures of a porous material. One method is based on the three-dimensional experimental imaging techniques, such as the scanning electron microscopy. Another method is based on the numerical random generation, which has lower cost and higher computational efficiency compared with the experimental method. Due to the complexity of the porous layer structures in PEMFCs, the second method is used in the present study to reproduce the porous structures. The following assumptions are made in the reconstruction procedure:

(a) the porous material is formed by spherical particles with various sizes;
(b) the spherical particles are randomly distributed.

Since the porous material (e.g., GDL) in a PEMFC is generally made of carbon paper with very complex microstructures, the second assumption is reasonable, but the first assumption is only a simplification for the present qualitative study.

Based on these assumptions, an effective numerical model, in which a random function is introduced to account for the pore distribution characteristics, is developed to reproduce the three-dimensional porous layer. In this reconstruction method, the porosity is an important controlling parameter, which is defined as the ratio of the pore volume to the total volume of the porous material.

$$\varepsilon_p = \frac{V_p}{V} = 1 - \frac{V_s\left(r_s\right)}{V}, \tag{10}$$

where V_p represents the pore volume, V is the total volume of the porous medium, and $V_s\left(r_s\right)$ is the solid phase volume that is related to the radius r_s of the spherical particles.

A spherical particle can be generated using a random method to satisfy the following condition:

$$(x - x_0)^2 + (y - y_0)^2 + (z - z_0)^2 \leqslant r_s^2, \tag{11}$$

$$\begin{cases} r_s = r_{is} + ranf \\ x_0 = x_{i0} + ranf \\ y_0 = y_{i0} + ranf \\ z_0 = z_{i0} + ranf \end{cases}, \tag{12}$$

where (x_0, y_0, z_0) and r_s are the center coordinates and radius of a spherical particle, which are allowed to vary with a random function, $ranf$, to account for the complex characteristics of a porous medium in a PEMFC. The solid particles are allowed to contact each other, depending on their positions and radii. To start the reconstruction procedure, the parameters, (x_{i0}, y_{i0}, z_{i0}) and r_{is}, which represent the initial center coordinates and an initial radius of a spherical particle, are specified.

A discriminant function is used to distinguish the fluid and solid lattice nodes. The function is defined as:

$$g(x,y,z) = \begin{cases} 1, & Solid\ lattice \\ 0, & Fluid\ lattice \end{cases}. \tag{13}$$

As shown in Figure 6, two porous materials with different porosities have been reconstructed. The computational domain is set to be 80 × 80 × 80 lattice nodes at the x, y, and z directions, respectively. Since the thickness of a GDL in a PEMFC is around 200 μm, a lattice unit thus represents about 2.5 μm in the present study. Figure 6a shows a reconstructed porous medium with a porosity of 0.4, while Figure 6b shows one with a porosity of 0.7. The difference between the two reconstructed porous materials can be clearly observed, and the reconstruction can capture the random micro-pore structures. It should be noted that the mean pore size in the reconstructed porous material is around 15 μm, which is in the practical range of a PEMFC. In the computational domain, six lattices are located in each dimension of the pore, and it should be sufficient for obtaining reasonably accurate results in the present qualitative study.

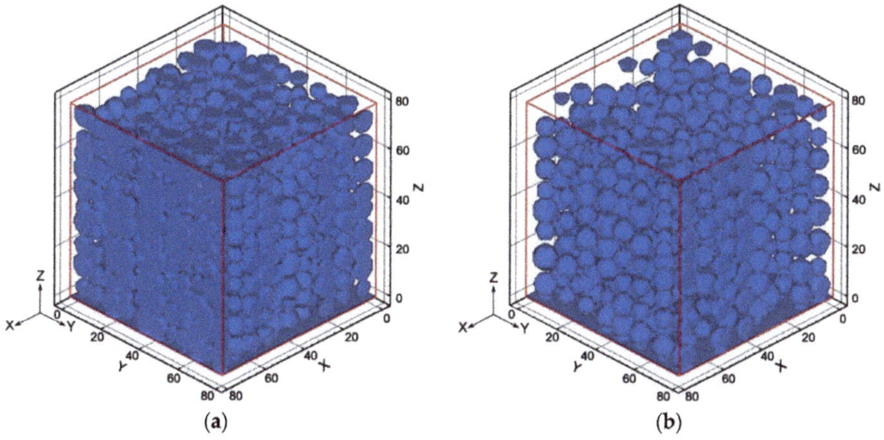

Figure 6. Reconstructed porous layers using a random numerical generation method: (**a**) Porosity: 0.4; (**b**) Porosity: 0.7.

3.2. Liquid Water Transport in Reconstructed Porous Materials

Liquid water transport in the porous layer of a PEMFC is studied in this section, based on the two reconstructed porous layers in Figure 6 and using the three-dimensional two-phase lattice Boltzmann method described and validated in the previous sections. The effects of the porous layer porosity and boundary liquid saturation on liquid water transport behaviors are examined.

In the present LBM simulations, three types of boundary conditions are used, including: (a) the pressure boundary condition that is imposed on the up (inlet) and down (outlet) boundaries in the computational domain; (b) the periodic boundary condition that is set to another four boundaries; (c) the bounce-back boundary condition that is used to simulate the interactions between the fluid particles and solid surfaces inside the reconstructed porous structure. In the present studies, all the solid surfaces in the porous material are assumed to be hydrophobic with a contact angle of 160°. At the inlet, the liquid water distribution is randomly specified, in consistency with the boundary liquid saturation, and a small velocity at $0.5 \times 10^{-5}\ \mathrm{ms}^{-1}$ is defined for the liquid water to start the numerical simulations.

Figure 7 illustrates the transient liquid water transport process in the reconstructed porous layer of a PEMFC, with a porosity of 0.7, as presented in Figure 6b. In this case, the boundary liquid saturation at the top surface, which may represent the chemical reaction sites in a PEMFC, is set to 0.19. In PEMFC operations, liquid water is produced in the CL, penetrates through the porous CL and GDL, and moves into the gas channel (refer to Figure 1b,c). Figure 7a shows the detailed liquid water distribution at a time step of t = 100 ts. It can be observed that at the early stage of the transport process, when the time step is less than 100 ts, the liquid water front moves relatively uniformly from the inlet boundary of the reconstructed porous layer. The figure on the right side in Figure 7a shows the two-dimensional cross-sectional views of liquid water distribution at different x positions. It can be seen that more liquid water exists in the middle section, because in this case, there are larger pores in this region. Liquid water transport in a porous material is mainly controlled by the capillary pressure, and thus prefers the large empty area as the preferential transport pathway [31]. The capillary pressure in the porous layer can be evaluated using the following analytical expression [38]:

$$p_c = \sigma cos\theta \left(\frac{\varepsilon}{K}\right)^{1/2} J\left(s\right), \tag{14}$$

where σ is the surface tension of liquid water, θ is the contact angle of the porous material, ε is the porosity, K is the permeability that is related to the pore sizes, and $J\left(s\right)$ is a function that is related to the liquid saturation s. This formulation has been widely used in the macro-scale numerical studies of liquid water transport in the CL and GDL of a PEMFC [38].

As the time step increases to 2000 ts, liquid water penetrates deeper into the porous layer, as shown in Figure 7b. At this instant, many separate liquid water fronts exist. The preferential transport characteristics of liquid water in the porous layer can be clearly observed. Results on the left side of Figure 7b clearly show that liquid water in some small pores, instead of move forward, retracts and changes its transport pathway to the neighboring large pores. These results clearly elucidate the effect of the microstructures on liquid water transport in a porous layer of a PEMFC. Since the large pores serve as the main pathway for liquid water transport, the small pores are thus available for the gaseous reactant transport. This indicates that a good combination of large and small pores in a porous material can help gaseous species transport and consequently improve the fuel cell performance.

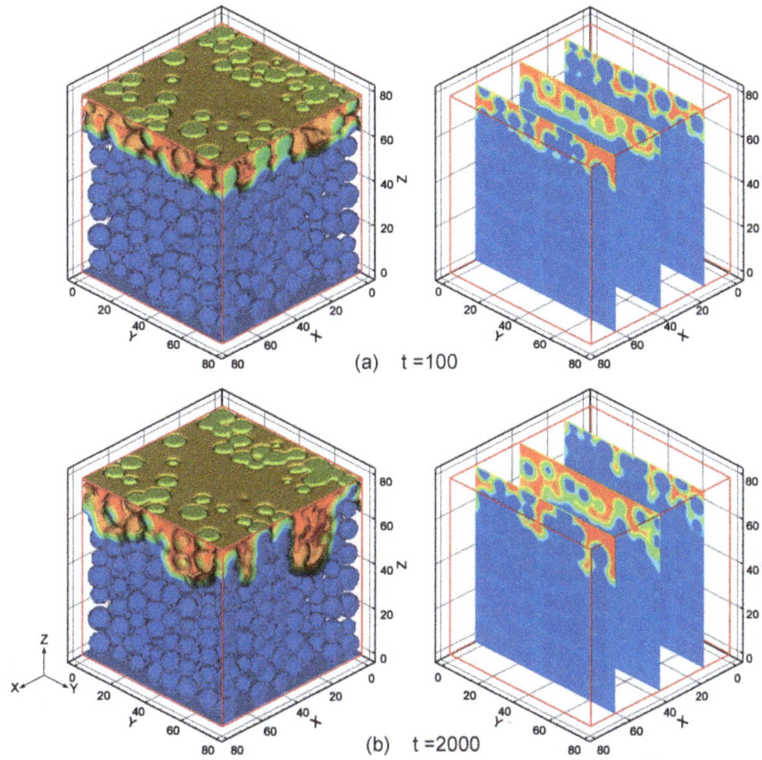

Figure 7. Liquid water evolution and distributions in a reconstructed porous layer with a porosity of 0.7 and a boundary liquid saturation of 0.19. (**a**) At a time step of t = 100 ts; (**b**) At a time step of t = 2000 ts.

Since the amount of liquid water generated by the electrochemical reactions on the cathode side in a PEMFC varies with different cell operation conditions, it is thus important to study the effect of the boundary liquid saturation on liquid water transport and distribution in the porous layer. As shown in Figure 8, the boundary liquid saturation is increased to 0.31, meaning that more liquid water is produced in this case by the electrochemical reactions. According to Equation (14), the variation of liquid water saturation can have a direct effect on the capillary pressure and thus may influence the liquid water transport and distribution.

Figure 8a,b displays the evolution of liquid water in the reconstructed porous medium at two different time steps, with a boundary liquid saturation of 0.31. Liquid water moves more uniformly and faster in this case with a larger amount of liquid water, compared to the case with a boundary liquid saturation at 0.19. At a time step of t = 2000 ts, separate liquid water fronts can be observed, but it appears that the preferential transport characteristics in this case are not as profound as the preceding case, as shown in Figure 7. An interesting physical phenomenon is also clearly observed; during liquid water invasion, it appears that the breakup of a liquid water stream can occur under certain conditions, leading to the formation of a liquid droplet in the porous layer. This phenomenon is related to the connecting bridge or neck resistance dictated by the surface tension. As the feeding pore becomes smaller or the invasion pore becomes larger, the connecting resistance will be weakened, resulting in the detachment of a liquid droplet from the feeding stream.

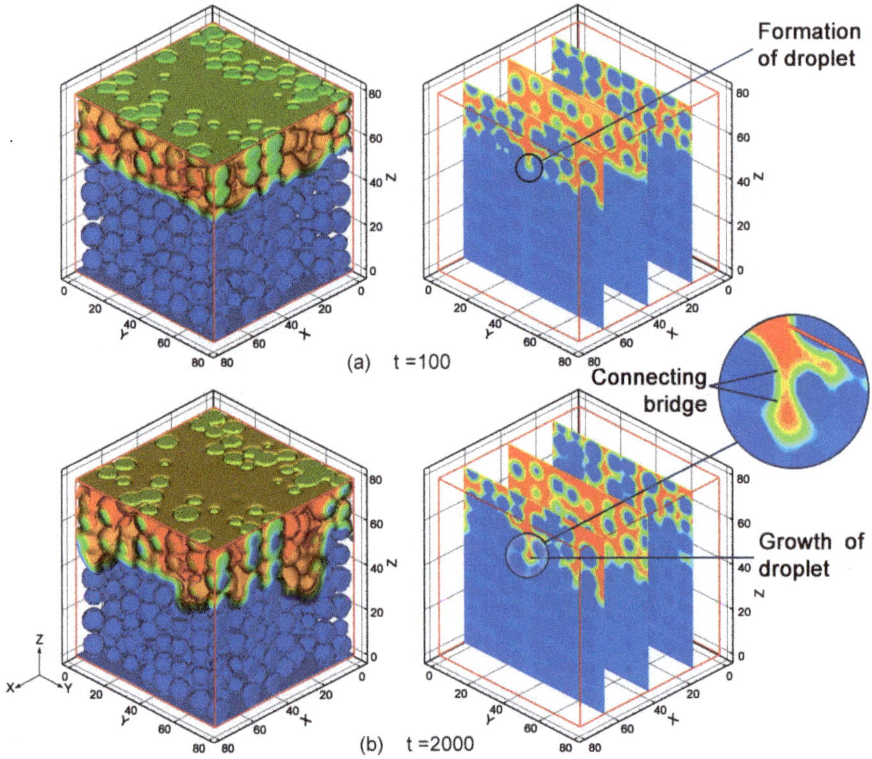

Figure 8. Liquid water evolution and distributions in a reconstructed porous layer with a porosity of 0.7 and a boundary liquid saturation of 0.31. (**a**) At a time step of t = 100 ts; (**b**) At a time step of t = 2000 ts.

The effect of the porous layer porosity on liquid water transport and distribution was studied next. Figure 9 shows results obtained from LBM simulations of liquid water transport in a reconstructed porous medium with a porosity of 0.4. In this case, the boundary liquid saturation is set to be 0.19. It is found that liquid water transport is very sensitive to the porous layer porosity. The preferential liquid water transport characteristics are more profound in this case with a smaller porous layer porosity, compared with the results in Figure 7. According to Equation (14), the variation of the porous layer porosity can also have a direct effect on the capillary pressure and thus can influence the liquid water transport and distribution. It is also observed that the detachment and formation of liquid droplets occur earlier and more frequently in this case, as compared to results calculated with a larger porosity, as shown in Figures 7 and 8.

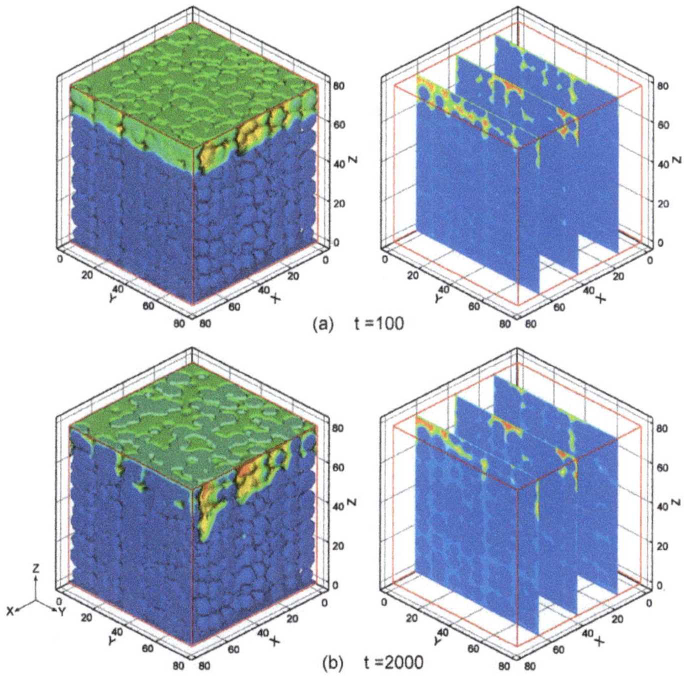

Figure 9. Liquid water evolution and distributions in a reconstructed porous layer with a porosity of 0.4 and a boundary liquid saturation of 0.19. (**a**) At a time step of t = 100 ts; (**b**) At a time step of t = 2000 ts.

Figure 10 shows liquid water evolution in the reconstructed porous layer with a porosity of 0.4, but the boundary liquid saturation is increased to 0.31. Again, liquid water penetrates faster and deeper in this case with more liquid water, similar to the trend shown in Figure 8. However, the preferential liquid water transport characteristics become more profound in this case with a decreased porous layer porosity, compared with the results in Figure 8.

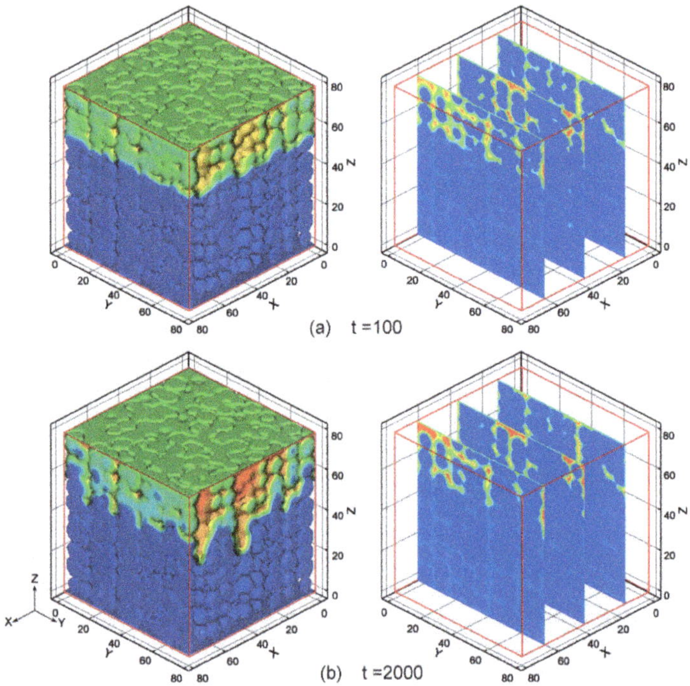

Figure 10. Liquid water evolution and distributions in a reconstructed porous layer with a porosity of 0.4 and a boundary liquid saturation of 0.31. (**a**) At a time step of t = 100 ts; (**b**) At a time step of t = 2000 ts.

4. Conclusions

In this paper, a three-dimensional two-phase lattice Boltzmann model is implemented to qualitatively study liquid water transport behaviors in the porous layer of a PEMFC. The model is validated against two test cases concerning the gas–liquid separation and liquid–solid interaction. Model validations prove that the lattice Boltzmann model is capable of accurately simulating the two-phase transport phenomena. A random numerical generation method is used to reconstruct the microstructures of a porous layer. The solid surfaces in the porous material are assumed to be hydrophobic with a contact angle of 160°. The present LBM simulations focus on the effects of the porous layer porosity and boundary liquid saturation on liquid water transport in porous materials.

Numerical results confirm that liquid water transport is strongly affected by the microstructures in a porous layer, and liquid water prefers the large pores as its main pathway. It is found that during the transport process, liquid water in some small pores may retract and change its pathway to the neighboring large pores. The preferential transport phenomenon is more profound with a decreased porous layer porosity and/or boundary liquid saturation. In the transport process, the breakup of a liquid water stream can occur under certain conditions, leading to the formation of liquid droplets inside the porous layer. This phenomenon is related to the connecting bridge or neck resistance dictated by the surface tension, and happens more frequently with a smaller porous layer porosity.

Results obtained in this paper indicate that an optimized design of porous layer porosity and the combination of various pore sizes may improve both the liquid water removal and gaseous reactant transport in a porous layer. The LBM simulations can help to improve the fundamental understanding of liquid water transport phenomena in the porous materials of PEMFCs.

Acknowledgments: This work was financially supported by the National Natural Science Foundation of China (No.10972197).

Author Contributions: The research was proposed and designed by Meng Ni and Hua Meng, and the numerical study was performed by Bo Han. Bo Han wrote the first draft of this paper, and Hua Meng made revisions. The authors have read and approved the final version.

Conflicts of Interest: The authors declare no conflict of interest.

References

1. Song, G.-H.; Meng, H. Numerical modeling and simulation of PEM fuel cells: Progress and perspective. *Acta Mech. Sin.* **2013**, *29*, 318–334.
2. Mukherjee, P.P.; Kang, Q.; Wang, C.-Y. Pore-scale modeling of two-phase transport in polymer electrolyte fuel cells—progress and perspective. *Energy Environ. Sci.* **2011**, *4*, 346–369.
3. Zhang, F.-Y.; Advani, S.G.; Prasad, A.K.; Boggs, M.E.; Sullivan, S.P.; Beebe, T.P., Jr. Quantitative characterization of catalyst layer degradation in PEM fuel cells by X-ray photoelectron spectroscopy. *Electrochim. Acta* **2009**, *54*, 4025–4030.
4. Sinha, P.K.; Halleck, P.; Wang, C.-Y. Quantification of liquid water saturation in a PEM fuel cell diffusion medium using X-ray microtomograph. *Electrochem. Solid State Lett.* **2006**, *9*, A344–A348.
5. Turhan, A.; Heller, K.; Brenizer, J.S.; Mench, M.M. Quantification of liquid water accumulation and distribution in a polymer electrolyte fuel cell using neutron imaging. *J. Power Sources* **2006**, *160*, 1195–1203.
6. Song, D.; Wang, Q.; Liu, Z.; Navessin, T.; Eikerling, M.; Holdcroft, S. Numerical optimization study of the catalyst layer of PEM fuel cell cathode. *J. Power Sources* **2004**, *126*, 104–111.
7. Lin, G.; He, W.; Nguyen, T.V. Modeling liquid water effects in the gas diffusion and catalyst layers of the cathode of a PEM fuel cell. *J. Electrochem. Soc.* **2004**, *151*, A1999–A2006.

8. Liu, J.; Eikerling, M. Model of cathode catalyst layers for polymer electrolyte fuel cells: The role of porous structure and water accumulation. *Electrochim. Acta* **2008**, *53*, 4435–4446.
9. Jiao, K.; Zhou, B. Effects of electrode wettabilities on liquid water behaviors in PEM fuel cell cathode. *J. Power Sources* **2008**, *175*, 106–119.
10. Ju, H. Analyzing the effects of immobile liquid saturation and spatial wettability variation on liquid water transport in diffusion media of polymer electrolyte fuel cells (PEFCs). *J. Power Sources* **2008**, *185*, 55–62.
11. Wang, Y. Modeling of two-phase transport in the diffusion media of polymer electrolyte fuel cells. *J. Power Sources* **2008**, *185*, 261–271.
12. Das, P.K.; Li, X.; Liu, Z.-S. Analysis of liquid water transport in cathode catalyst layer of PEM fuel cells. *Int. J. Hydrog. Energy* **2010**, *35*, 2403–2416.
13. Meng, H. Multi-dimensional liquid water transport in the cathode of a PEM fuel cell with consideration of the micro-porous layer (MPL). *Int. J. Hydrog. Energy* **2009**, *34*, 5488–5497.
14. Meng, H. Numerical studies of liquid water behaviors in PEM fuel cell cathode considering transport across different porous layers. *Int. J. Hydrog. Energy* **2010**, *35*, 5569–5579.
15. Hizir, F.E.; Ural, S.O.; Kumbur, E.C.; Mench, M.M. Characterization of interfacial morphology in polymer electrolyte fuel cells: Micro-porous layer and catalyst layer surfaces. *J. Power Sources* **2010**, *195*, 3463–3471.
16. Thiele, S.; Zengerle, R.; Ziegler, C. Nano-morphology of a polymer electrolyte fuel cell catalyst layer—Imaging, reconstruction and analysis. *Nano Res.* **2011**, *4*, 849–860.
17. Wang, G.; Mukherjee, P.P.; Wang, C.-Y. Optimization of polymer electrolyte fuel cell cathode catalyst layers via direct numerical simulation modeling. *Electrochim. Acta* **2007**, *52*, 6367–6377.
18. Mukherjee, P.P.; Wang, C.-Y. Stochastic microstructure reconstruction and direct numerical simulation of the PEFC catalyst layer. *J. Electrochem. Soc.* **2006**, *153*, A840–A849.
19. Higuera, F.J.; Succi, S. Simulating the flow around a circular cylinder with a lattice Boltzmann equation. *Europhys. Lett.* **1989**, *8*, 517–521.
20. Succi, S; Foti, E.; Higuera, F. Three-dimensional flows in complex geometries with the lattice Boltzmann method. *Europhys. Lett.* **1989**, *10*, 433–438.
21. Benzi, R.; Succi, S.; Vergassola, M. The lattice Boltzmann equation: Theory and applications. *Phys. Rep.* **1992**, *222*, 145–197.
22. Chen, S.; Doolen, G.D. Lattice Boltzmann method for fluid flows. *Annu. Rev. Fluid Mech.* **1998**, *30*, 329–364.
23. Shan, X.; Chen, H. Lattice Boltzmann model for simulating flows with multiple phases and components. *Phys. Rev. E* **1993**, *47*.
24. Martys, N.S.; Chen, H. Simulation of multicomponent fluids in complex three-dimensional geometries by the lattice Boltzmann method. *Phys. Rev. E* **1996**, *53*.

25. Huang, H.; Li, Z.; Liu, S.; Lu, X.-Y. Shan-and-Chen-type multiphase lattice Boltzmann study of viscous coupling effects for two-phase flow in porous media. *Int. J. Numer. Methods Fluids* **2009**, *61*, 341–354.
26. Chen, L.; Kang, Q.; Tang, Q.; Robinson, B.A.; He, Y.L.; Tao, W.Q. Pore-scale simulation of multicomponent multiphase reactive transport with dissolution and precipittaion. *Int. J. Heat Mass Transf.* **2015**, *85*, 935–949.
27. Sinha, P.K.; Mukherjee, P.P.; Wang, C.-Y. Impact of GDL structure and wettability on water management in polymer electrolyte fuel cells. *J. Mater. Chem.* **2007**, *17*, 3089–3103.
28. Park, J.; Matsubara, M.; Li, X. Application of lattice Boltzmann method to a micro-scale flow simulation in the porous electrode of a PEM fuel cell. *J. Power Sources* **2007**, *173*, 404–414.
29. Hao, L.; Cheng, P. Lattice Boltzmann simulations of liquid droplet dynamic behavior on a hydrophobic surface of a gas flow channel. *J. Power Sources* **2009**, *190*, 435–446.
30. Hao, L.; Cheng, P. Pore-scale simulations on relative permeabilities of porous media by lattice Boltzmann method. *Int. J. Heat Mass Transf.* **2010**, *53*, 1908–1913.
31. Mukherjee, P.P.; Wang, C.-Y.; Kang, Q. Mesoscopic modeling of two-phase behavior and flooding phenomena in polymer electrolyte fuel cells. *Electrochim. Acta* **2009**, *54*, 6861–6875.
32. Han, B.; Yu, J.; Meng, H. Lattice Boltzmann simulations of liquid droplets development and interaction in a gas channel of a proton exchange membrane fuel cell. *J. Power Sources* **2012**, *202*, 175–183.
33. Han, B.; Meng, H. Lattice Boltzmann simulation of liquid water transport in turning regions of serpentine gas channels in proton exchange membrane fuel cells. *J. Power Sources* **2012**, *217*, 268–279.
34. Han, B.; Meng, H. Numerical studies of interfacial phenomena in liquid water transport in polymer electrolyte membrane fuel cells using the lattice Boltzmann method. *Int. J. Hydrog. Energy* **2013**, *38*, 5053–5059.
35. Wang, Y.; Cho, S.; Thiedmann, R.; Schmidt, V.; Lehnert, W.; Feng, X. Stochastic modeling and direct simulation of the diffusion media for polymer electrolyte fuel cells. *Int. J. Heat Mass Transf.* **2010**, *53*, 1128–1138.
36. Thiedmann, R.; Fleischer, F.; Hartnig, C.; Lehnert, W.; Schmidt, V. Stochastic 3D modeling of the GDL structure in PEMFCs based on thin section detection. *J. Electrochem. Soc.* **2008**, *155*, B391–B399.
37. Zhang, Y.; Sun, Q.; Xia, C.; Ni, M. Geometric properties of nanostructured solid oxide fuel cell electrodes. *J. Electrochem. Soc.* **2013**, *160*, F278–F289.
38. Meng, H.; Wang, C.-Y. Model of two-phase flow and flooding dynamics in polymer electrolyte fuel cells. *J. Electrochem. Soc.* **2005**, *152*, A1733–A1741.

Long-Range Electron Transport Donor-Acceptor in Nonlinear Lattices

Alexander P. Chetverikov, Werner Ebeling and Manuel G. Velarde

Abstract: We study here several simple models of the electron transfer (ET) in a one-dimensional nonlinear lattice between a donor and an acceptor and propose a new fast mechanism of electron surfing on soliton-like excitations along the lattice. The nonlinear lattice is modeled as a classical one-dimensional Morse chain and the dynamics of the electrons are considered in the tight-binding approximation. This model is applied to the processes along a covalent bridge connecting donors and acceptors. First, it is shown that the electron forms bound states with the solitonic excitations in the lattice. These so-called solectrons may move with supersonic speed. In a heated system, the electron transfer between a donor and an acceptor is modeled as a diffusion-like process. We study in detail the role of thermal factors on the electron transfer. Then, we develop a simple model based on the classical Smoluchowski–Chandrasekhar picture of diffusion-controlled reactions as stochastic processes with emitters and absorbers. Acceptors are modeled by an absorbing boundary. Finally, we compare the new ET mechanisms described here with known ET data. We conclude that electron surfing on solitons could be a special fast way for ET over quite long distances.

Reprinted from *Entropy*. Cite as: Chetverikov, A.P.; Ebeling, W.; Velardee, M.G. Long-Range Electron Transport Donor-Acceptor in Nonlinear Lattices. *Entropy* **2016**, *18*, 92.

1. Introduction

Taking into account the high interest in the development of molecular electronics, we study here several models for the transfer of electrons along molecular chains modeled as one-dimensional (1d) nonlinear lattices. It is worth noting that, recently, it has been suggested that DNA and possibly other (bio)macromolecules could serve as electronic conductors otherwise said (bio)molecular wires that can conduct charge carriers with virtually no resistance [1–7]. Experiments are now starting to provide the first clues about the mechanisms that underlie charge transport. The field has recently been revived with the advent of measurements on artificial DNA-like molecules. Barton and colleagues [3] measured the fluorescence produced by an excited molecule and found that the fluorescence quenching was due to the charge on the excited donor molecule leaking along the length of the DNA to a nearby acceptor molecule, thus indicating that such DNA is a conducting molecular wire. Their findings, however, are not without criticism as the physical mechanism offering transport has not been clearly identified. The essential aspect here is the study of the temperature influence on transfer processes; for other studies of the thermal dependence of macroscopic transport, see [8–14].

2. One-Dimensional Dynamical Model of Electron Transfer

Long-range electron tunneling through biomolecules (e.g., azurin and DNA) has been studied in substantial detail both experimentally and theoretically over the past twenty years [2,4,15–20]. In previous work, [11–14,21–27] we studied the influence of nonlinear lattice excitations on electron transfer (ET). It was shown that electron trapping by solitons and a new form of electron transport mediated by solitons is possible in an anharmonic 1d lattice. Here, we focus attention on the process that occurs between two units in the chain, the donor D and the acceptor A, with in-between a bridge consisting of n elements denoted by b: D-b-b-b-b-b.........-b-b-b-A. We shall show that solitons may help to transfer electrons along the bridge. We will not consider in detail the processes D-b and b-A, referring to the literature [17,18].

If solitons are excited by an external source, stable bound states of electrons and solitons may be formed, which may move with supersonic or slightly subsonic velocity. In a heated system, solitons are excited thermally. These solitons and the corresponding solectrons have a finite life time and change their direction stochastically. The electrons may surf on thermal solitons from donor to acceptor. Under these conditions, we may consider the ET as a process similar to a diffusion-controlled reaction.

Building upon earlier work [11–14,21–27], our interest is here in developing a dynamical theory which allows the description of ET over long distances, including the influence of thermal excitations. Adopting a 1d nonlinear lattice model to portray the backbone of the biomolecular chain (polypeptide, polynucleotide or other), we investigate the consequences of nonlinear, running, lattice excitations on electron transport. The theoretical model studied here is based the quantum mechanical "tight binding" approximation (TBA) [28] for the electrons which are moving on a lattice with Morse interactions. Note that the use of Morse potentials is just for numerical convenience and is not a must for the results. We study in detail the role of thermal factors on electron transfer and the transition times, and give a thorough discussion of the role of nonlinearity. One of the tasks is to develop a simple model based on the Smoluchowski–Chandrasekhar picture of diffusion-controlled reactions as stochastic processes with emitters and absorbers. We estimate the time-distance relations for the new non-standard mechanisms of ET and the dependence on temperature at moderate temperatures. The lattice interactions allow for phonon-and soliton-longitudinal vibrations with compressions governed by the repulsive part of the potential [11–14,29–32]. Thus, we consider a 1d anharmonic lattice with dynamics described by the following Hamiltonian: $H = H_{lattice} + H_{electron}$. The lattice part includes nearest-neighbor Morse interactions:

$$H_{lattice} = \sum_n \left\{ \frac{p_n^2}{2M} + D\left(1 - \exp\left[-B\left(q_n - q_{n-1}\right)\right]\right)^2 \right\}. \quad (1)$$

Here, M denotes the mass of a lattice unit (all units have equal mass), the coordinates and momenta q_n, p_n; $(n = 1, ..., N)$ describe their respective displacements from equilibrium positions $(n\sigma)$ and momenta, B characterizes the stiffness of the spring-like constant in the Morse potential, D is the depth of the potential well, and σ defines equilibrium lattice spacing. These compressions were shown to be responsible for electron trapping by the lattice excitations thus leading to the formation of dynamic bound states (solectrons) of the electron with the soliton (the same phenomenon is valid also for the solitonic peaks of a periodic cnoidal wave moving through the lattice). Furthermore, we consider the whole system in a "thermal bath" characterized by a Gaussian white noise, ξ_j, of zero mean and delta-correlated.

We add electrons to the system of Morse particles assuming that the electrons are in reality in 3d space though the lattice is 1d. Thus, for the electron-lattice dynamics, we have

$$H_{el} = E_n\left(q_k\right) c_n^* c_n - \sum_n V_{nn-1}\left(q_k\right)\left(c_n^* c_{n-1} + c_n c_{n-1}^*\right), \quad (2)$$

where n denotes the site where one electron is "placed" on the lattice, the complex numbers c_n give the n-th component of the wave function and $p_n = |c_n|^2$ gives the probability of finding the electron residing at the site n. The bound state energy at site n may depend on the relative distances between neighbors. We will use the ansatz

$$E_n = E_n^0 + \chi_0\left(q_{n+1} - q_{n-1}\right). \quad (3)$$

This is a translation-invariant modification of the linear shift used by Holstein and Kalosakas *et al.* [33,34]. In view of its minor role, in what follows, we shall neglect the second term in Equation (3). With this approximation, we want to keep the effect of energy shifts rather small, concentrating more on the influence on the transfer elements. The quantity V_{nn-1} defines the transfer matrix element or overlapping integral responsible for the transport of the electron along the chain (considering only nearest neighbors). This matrix is the key ingredient, allowing for the coupling of the electron to the lattice displacements, and hence the lattice vibrations, phonons or solitons. Following Slater [35–37], we take

$$V_{nn-1} = V_0 \exp\left[-\alpha_0\left(q_n - q_{n-1}\right)\right], \quad (4)$$

where the parameter α_0 accounts for the strength of the coupling added to parameter V_0. If we measure all energies in units 2D, it is convenient to measure also the energy levels in these V_0 which gives

$$E_n = 2D\epsilon_n. \tag{5}$$

For the sake of universality, it is best to rescale quantities and consider a dimensionless problem. We take as unit of time, Ω_{Morse}^{-1}, where $\Omega_{Morse} = \left(2DB^2/M\right)^{1/2}$ denotes the frequency of harmonic oscillations (linear, first-order approximation to the Morse exponential). For displacements, we take B^{-1} as the unit, for momenta we take $(2MD)^{-1/2}$, hence for the interaction force we have $\alpha_0 V_0/2BD$, and α_0 is measured in $\left(B^{-1}\right)$ units. Then, expecting no confusion in the reader, denoting the new dimensionless quantities with the same symbols as the old ones, the dynamics of the Hamiltonian system Equations (1) and (2), is given by the following equations for the components of the electronic wave function electron c_n, and the lattice vibrations, q_n

$$i\frac{dc_n}{dt} = \epsilon_n c_n - \tau\left\{\exp\left[-\alpha\left(q_{n+1} - q_n\right)\right]c_{n+1} + \exp\left[-\alpha\left(q_n - q_{n-1}\right)\right]c_{n-1}\right\}, \tag{6}$$

$$\begin{aligned}\frac{d^2q_n}{dt^2} = & \left\{1 - \exp\left[-\left(q_{n+1} - q_n\right)\right]\right\}\exp\left[-\left(q_{n+1} - q_n\right)\right] - \\ & - \left\{1 - \exp\left[-\left(q_n - q_{n-1}\right)\right]\right\}\exp\left[-\left(q_n - q_{n-1}\right)\right] \\ & + \alpha V\left\{\left(c_{n+1}^* c_n + c_{n+1}c_n^*\right)\exp\left[-\alpha\left(q_{n+1} - q_n\right)\right] - \left(c_n^* c_{n-1} + c_n c_{n-1}^*\right)\exp\left[-\alpha\left(q_n - q_{n-1}\right)\right]\right\},\end{aligned} \tag{7}$$

where

$$V = V_0/2D,\ \alpha = \alpha_0/B,\ \tau = V_0/\Omega_{Morse}\hbar. \tag{8}$$

Not counting the energy levels, there are four parameters, the last one relates the time scales of the electron and lattice dynamics. Needless to say, in general, the two time scales are not the same (which in frequency terms refer to ultraviolet/electronic *versus* infrared/acoustic), for most cases with electrons and phonons. Instead of the ansatz Equation (3), other equivalent expressions may be used [12,38–40]. For purposes of illustration, we take the following parameter values:

$$\begin{aligned}&\sigma = 4.0\,Angstrom,\ B = 4.45\,Angstrom^{-1},\ \alpha = 1.75B,\ D = 0.1eV,\ V_0 = 0.05eV,\\ &\Omega_{Morse} = 3.04\times10^{12}s^{-1},\ c_{sound} = 12.1\,Angstrom/ps,\ \Omega_{electron} = V_0/\hbar = 0.608\times10^{14}s^{-1},\ \tau = 10.\end{aligned} \tag{9}$$

These numerical values are relevant, e.g., for electron transport along hydrogen bonded polypeptide chains such as α-helices [9,25–27,38–42]. Let us note what Muto *et al.* [43] give for DNA other values, which, in our model, correspond to:

$$\sigma = 3.4\,Angstrom,\ B = 2.1\,Angstrom^{-1},\ D = 0.23eV,$$

$$c_{sound} = 16.9\ Angstrom/ps,\ \Omega_{Morse} = 5.0 \times 10^{12} s^{-1}.$$

3. Electrons Surfing on Solitons—A Guided Tour of Electrons from Donor to Acceptor

We shall estimate how the path of an electron may be influenced by a lattice soliton which was generated by an external perturbation of the lattice. The added, excess electron is placed at $t = 0$ at a donor located at site $n = 100$ at time $t = 0$ (Figure 1). Due to the electron-lattice interaction ($\alpha = 1.75$), we observe soliton-assisted ET. The electron is dynamically bound to the soliton thus creating the travelling supersonic or slightly subsonic solectron excitation. Indeed when the electron-lattice interaction is operating, we see that the electron moves with the soliton with a slightly subsonic velocity $v_{el} = (130/160)v_{sound}$ and is running to the right border of the square (n, t) plot. Let us assume that an acceptor is located there. This means that the electron is guided by the soliton from donor to acceptor. Note that, in transport processes, several solitons may be involved including those moving in opposite direction. Therefore, the above given soliton velocity is an upper bound for the soliton-assisted ET. In order to study the superposition of several solitons acting on one electron, we studied, in another experiment, the evolution of an electron in the presence of two solitons. Figure 2 illustrates the splitting of the electron probability density. The existence of bound states between electrons and lattice deformations in 1d-lattices has been first studied in the continuous picture by Davydov and his school [41,44].

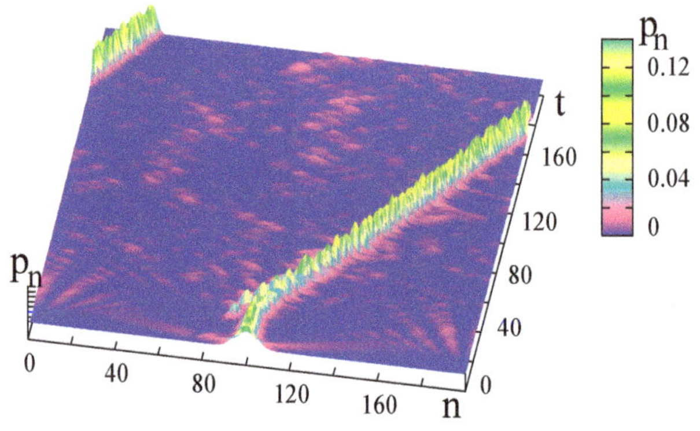

Figure 1. Evolution of one electron in probability density starting at position 100 and a soliton starting at position 70. Then, 20–40 time units after the start the electron is catched by the soliton and forms with it a bound state (called solectron) which moves with slightly subsonic velocity.

In principle, the above solectron effect may be used as a way to manipulate or otherwise to control the transport of electrons between donor and acceptor. Clearly, in our case, we may have a polaron-like effect due to the electron-phonon (or soliton) interaction coupled to a genuinely added lattice solitonic effect due to the anharmonicity of the lattice vibrations. This permits soliton rather than phonon-assisted long range hopping.

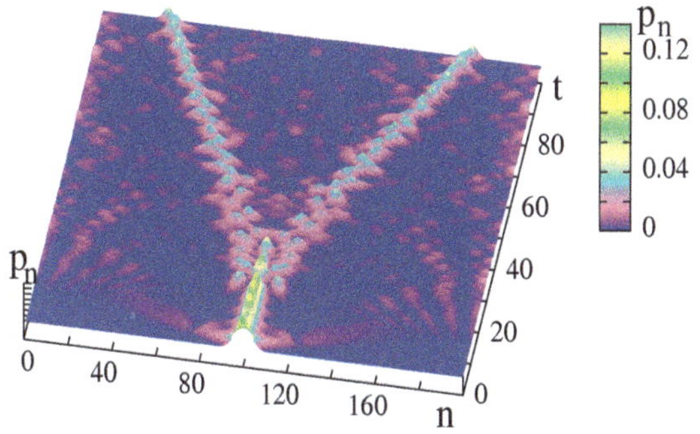

Figure 2. Evolution of an electron starting at position 100 between two solitons emitted at positions 70 and 130, respectively. The electron probability density splits between the two solitons.

4. Dynamics of Electrons Interacting with Thermal Solitons

Let us study now a lattice heated to some temperature T. Then, thermal solitons are excited due to the influence of thermal energy-rich collisions. However, we do not have just one or two solitons, but many of them are generated by the heat bath [38–40]. To be expected is that this completely changes the picture. In previous work, we have shown that we do not find any more long-living soliton excitations in the chain [24]. Instead, we see up to the range of physiological temperatures many small local soliton portions in the system which have a finite lifetime up to a few picoseconds. The electron probability density may split between all of them.

The general picture is now that the electron probability density is concentrated in the local "hot spots" created by the local soliton thermal excitations. In order to study the influence of this effect on donor-acceptor ET in more detail, we performed a series of computer simulations. In this series of experiments, we release an electron into an already heated lattice by means of the friction and noise sources. These sources are switched off at an instant that we now denote $t = 0$. In Figure 3, we represent the electron probability density developing in a lattice heated up to $T = 0.1$. We see that the electron density is for $t > 0$ confined more or less in a cone. A similar picture is obtained for $T = 0.5$, with a mere narrowing of the evolutionary cone. Clearly, the electron probability density splits into many small spots which are localized at thermally excited solitons. These "hot spots" comprise up to 10 lattice (take up to 50 Å) sites and have a short life time which is in the range of a few picoseconds. The little maximum of the electron density "dies" with the soliton or without destruction of the latter and moves eventually to the next nearby soliton. The whole process is time dependent as the "hot spots" are created and annihilated in the thermal process. Note that the spots denote only probabilities; in reality, the electron is localized at one of the spots.

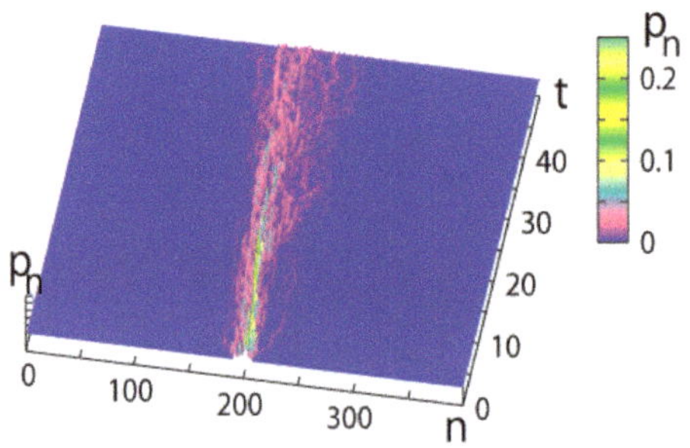

Figure 3. Evolution of the electron probability density of an electron released into a heated lattice ($T = 0.1$, $\tau = 10$, $\alpha = 1.75$) gets concentrated at places of local soliton excitations (with a size up to 10 lattice units) and survives there for a finite time (may be a few picoseconds), and then it moves to another solitonic "hot spot".

Modeling the netto-transfer of the electrons theoretically is rather difficult due to the complexity of the life of an electron after injection. At first glance at the temperatures which we study here, it looks like a diffusion-like process. Let us discuss whether the ET is a proper diffusion process. As is well known, according to a linear Schrödinger equation, the electron probability density spreads in some respect similar to a diffusion process, meaning that the mean square displacement growths linearly in time. On the other hand, as shown by Brizhik and Davydov [44], the nonlinear Schrödinger equation corresponds to processes which are much more complicated, e.g., the spreading of the density may be fractal-like and may depend on the initial conditions. This way, we cannot expect that the ET is at $T = 0$ or near to zero temperature a diffusion-like in all respects. Being aware of this complication, we take here a diffusion approach for temperatures $T > 0$, or more precisely $T > 0.1$ (in units 2D), since we expect that the thermal motion smoothes the expansion processes. Possible complications at low T including tunneling have been discussed elsewhere [13,14]. The quantum-mechanical processes at low temperatures, in particular the tunneling at $T = 0$, is certainly not diffusion-controlled. In order to decide the question of whether the ET-processes at finite temperatures are diffusion-like, we have to study in more detail the mean square displacement of the spreading of a free electron on the heated lattice.

5. Mean Square Displacement of Electrons

For simplicity, we place the injected electron initially at $t = 0$ at $n = n_0 = 0$ (for simplicity). The mean square displacement is given by

$$\left\langle n^2(t) \right\rangle = \sum n^2 c_n(t)\, c_n^*(t)\,, \tag{10}$$

where the quantities $c_n(t)$ have to be calculated according to the discrete Schrödinger Equation (6). This procedure gives, however, only the quantum-mechanical mean. Still, we need here a second average with respect to the stochastic trajectories of the lattice according to Equation (10). This way, we define a diffusion-function of time

$$d(t) = \left\langle < (n(t) - n_0)^2 > \right\rangle = \left\langle \sum n^2 c_n(t) c_n^*(t) \right\rangle. \tag{11}$$

The outer bracket means that we have to take the average over many realizations (time evolutions) of the lattice dynamics. The result has to be drawn as a function of time. We expect that this function is linear in time at least in certain temperature range, that is, $d(t) = 2D_{eff}t$. This is confirmed by computer simulations (Figure 4). For the case of a linear lattice model, the linear dependence of the mean square displacement with time was obtained by Lakhno and collaborators [45,46].

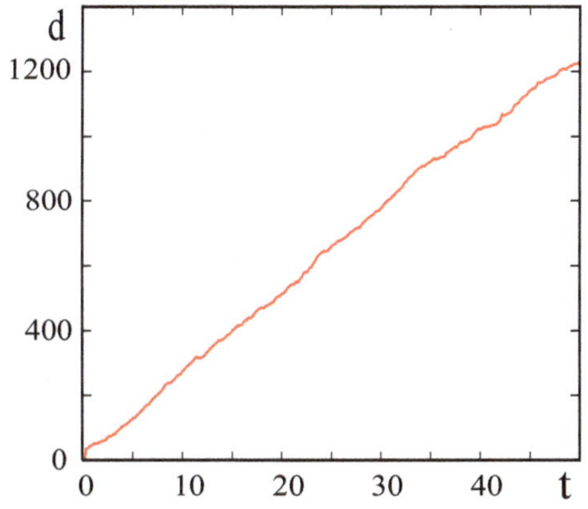

Figure 4. The diffusion function $d(t)$ over the time t for the temperature $T = 0.5$. It shows this ET process as diffusion-like.

The results of a stochastic simulation based on the master equation is shown by the thin line in Figure 5 [38–40] together with D_{eff} derived using the Schrödinger equation (fat points). We see that both approaches disagree at low temperatures. As far as we see, the (theoretical) diffusion coefficient of the hopping process at $T = 0$ (no lattice dynamics) should be τ (or may be proportional to τ). Therefore, we believe that the calculation of d Equation (11) at low $T < 0.1$ with the Schrödinger equation is incorrect due to the problems of fractality and other difficulties with the Schrödinger equation [8]. In order to avoid all these difficulties, as already said, we concentrate in this work only on $T > 0.1$, where the two approaches go together. In the TBA, we model a heat bath corresponding to a temperature T by introducing a Langevin source for the temperature and wait for complete thermalization. Then, we switch-off the heat bath, *i.e.*, the stochastic source and friction, and start with the initial conditions for velocities and positions the simulations of the coupled TBA-Hamiltonian system. This procedure guarantees for a short simulation time thermal conditions. In the case of using a master equation the method corresponds to a standard Monte-Carlo thermalization. For the diffusional regime, the diffusion theory is telling us that an electron initially concentrated at the position n_0 has an evolving probability density described by

$$\rho(n,t) = C \exp\left[-\frac{(n-n_0)^2}{2D_{eff}t}\right]. \tag{12}$$

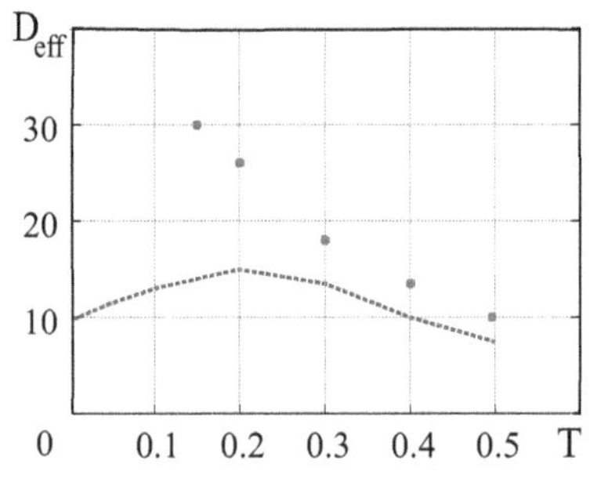

Figure 5. The effective diffusion coefficient, D_{eff}, based on thermal electron hopping between solitons leads to two different outcomes. The fat points are based on quantum-mechanical hopping modeled in TBA whereas the thin line is obtained by solving quantum-statistical master equations and include quantum-mechanical and thermal hopping effects in the stochastic description.

If we identify n_0 with the donor, this formula shows how the electron probability density evolves at some distance, in particular, also at the place where the absorber is located. For diffusion processes, the mean square displacement is given by

$$\left\langle (n(t) - n_0)^2 \right\rangle = 2D_{eff}t. \tag{13}$$

This tells us that the distance travelled by the electron goes as the square root of time t. This is still a relatively fast mechanism, though is slower than ballistic surfing on solitons. On the other hand, the density decays exponentially with the distance, this means the log of the density decays with the square of the distance from the source. As we will discuss later, many authors report a linear decay of the log of time with the distance. This remains still an open problem. Evidently, the ET donor-acceptor is not solely diffusion-controlled. Surely there are other contributions such as tunneling [4,14,47].

Let us discuss now the findings (and conjectures) on ET presented so far. Electrons injected into the lattice may form very stable bound states which move with supersonic or slightly subsonic velocity along the lattice and may transfer the electron over hundreds of lattice sites. This affects points in the direction of the observation of electron transfer over long distances with only a small loss of energy [17,18]. The problem, however, remains how stable solitons may be excited. In principle, any fast mechanical deformation or structural reconformation could be responsible for the emission of solitons running along the lattice. This, however, requires some coincidence between the electron emission and the soliton emission. It seems to be more natural to use the assistance of thermal solitons, which are always present in the nonlinear lattice.

We have seen above that the electron probability density may spread freely according to the Schrödinger equation, and then it may be trapped in solectron bound states, a process which is not diffusion-like, described by a nonlinear Schrödinger equation. The most important step is, however, the decay of the hot spot and the occupation of the electron by another hot spot. Such charge trapping in vibrational hot spots was observed also in the work of Kalosakas, Rasmussen and Bishop [34]. From the overall view, our process is diffusion-like in spite of the fact that it might be not diffusion-like in some steps. However, in the thermal average, it seems to be well approximated as "normal diffusion". As a consequence of the fact that, at least in our model, the ET is diffusion-like it is characterized by a mean-square displacement and not by a well-defined speed. From this background (of a diffusion-like character), we will develop now an absorber model.

6. A Diffusion-Type Absorber Model of the ET Donor-Acceptor

The model which we will develop now is based on the concept developed by Smoluchowski and Chandrasekhar that an acceptor reaction can be modeled as a stochastic absorber problem [48,49]. Accordingly, we study a 1d model with periodic boundary conditions but containing an absorbing boundary. In fact, we consider the (1d) chain including a barrier consisting of some sites where electron probability density is absorbed, but these sites are fully transparent for the chain excitations. The absorbing coefficient is defined by a Gaussian function

$$\rho(n) = \exp\left[-\frac{(n - N + M)^2}{2m^2}\right]. \tag{14}$$

We use $N = 100$ as the number of particles and here $M = 4$ (where $\rho = 1$ that is maximal), $m = 1$ (m defines the width of the absorbing barrier). To start, we used an absorbing gap $\rho = 1$ at $n = 100$ (that is at the right wall only), $\rho = 0$ at other sites, but then the electron is reflected mainly from such a barrier. Therefore, we used a smooth absorbing barrier, preventing reflection of the wave function of an electron from the barrier. The procedure of computer simulation is the following. We make a step of integration as usual, then multiply the component of an electron wave function by the absorbing coefficient (more precisely by $1 - \rho(n)$). It means that the absorption process in an insertion is considered to be instantaneous, without inertia.

To demonstrate the influence of thermal effects, we studied first the absorption of the electron of a solectron in a cold lattice (Figure 6) by following the electron probability density and the total probability (Figure 7)

$$P = \sum_n |c_n|^2. \tag{15}$$

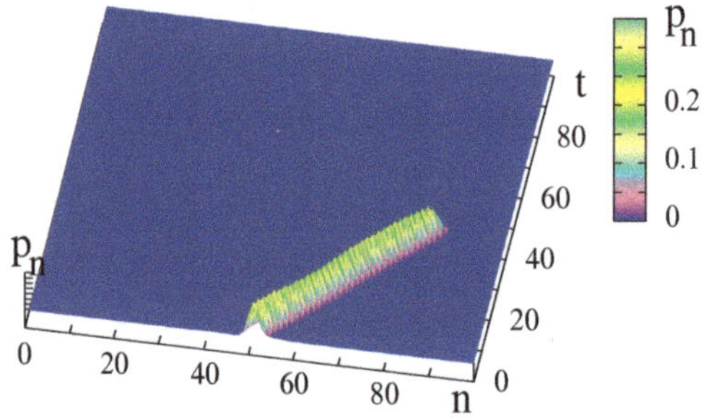

Figure 6. Evolution of the probability density, p_n, of an electron approaching an absorbing boundary.

We observe that the solectron reaches the absorbing boundary at time $t \sim 42$, then the electron is absorbed almost entirely. Note that this corresponds in our model to the "time of arrival at the acceptor" or, in other words, as the "transition time donor-acceptor". Note also that within the model a very small "part" of the electron probability density penetrates across the barrier. That the soliton keeps on moving is a specific feature of our model and may even increase its velocity when loosing its electron in the absorbing barrier. (Parameter values for the simulation are $N = 100$, $\alpha = 1.75$, $B\sigma = 1$, $V = 1$, $\tau = 10$). Next, we considered the heated lattice with the same parameter values for several temperature values. Initially, the electron is supposed to be localized (in the form of the Gaussian function) in the center of the chain.

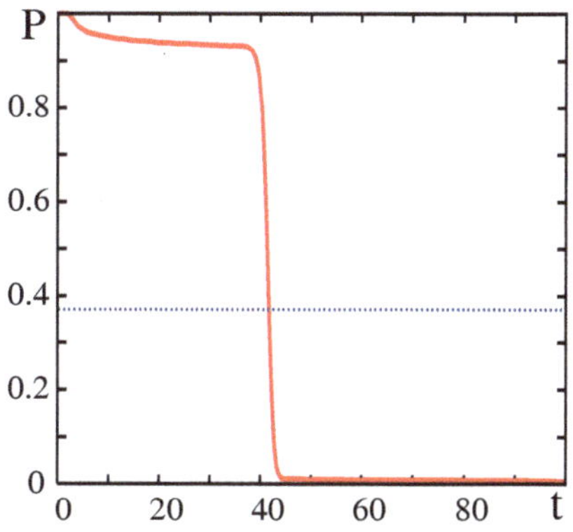

Figure 7. Total probability P(t) according to Equation (15) to find the electron in the lattice system, the probability decreases due to escape through the absorbing boundary. The blue line shows the value 1/e which is used for defining the "life-time" of residence of the electron in the lattice before being absorbed by the boundary.

The results presented in Figure 8 are based so far on one computer simulation for each temperature. Those figures show typical distributions of probability $|c_n|^2(t)$ (to find an electron at site "n", the local electron density, left) and the total density P(t) (right). However, in order to estimate the transition time in a more correct way, we have to perform a set of computer simulations with different realizations of noise influence and then calculate an average characteristic value (Figure 9).

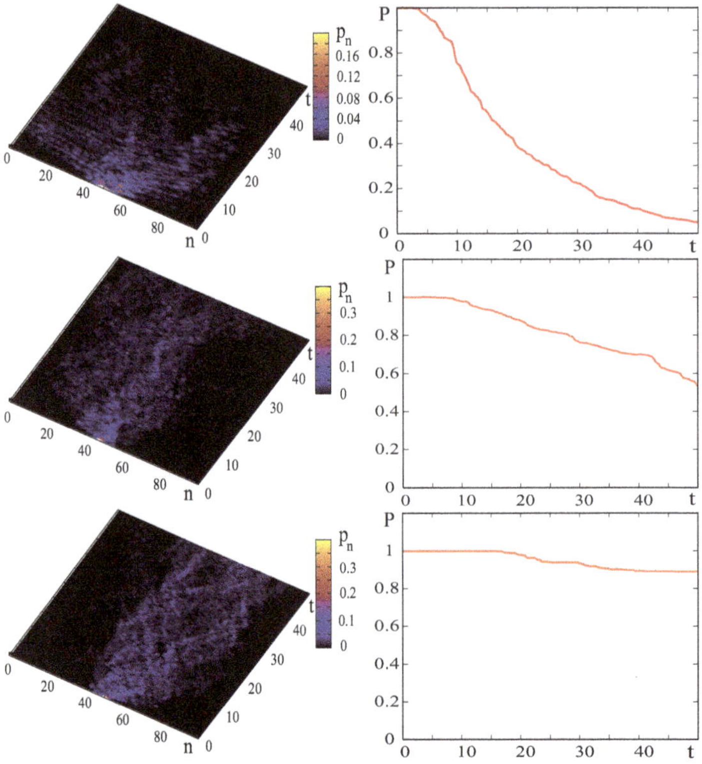

Figure 8. Heated lattice (top-down values: $T = 0.01, 0.1$ and 0.5 in 2D units). (**Left panels**): evolution of the local electron probability density (p_n). (**Right panels**): total density (P) at different temperatures in the regime of thermal solitons.

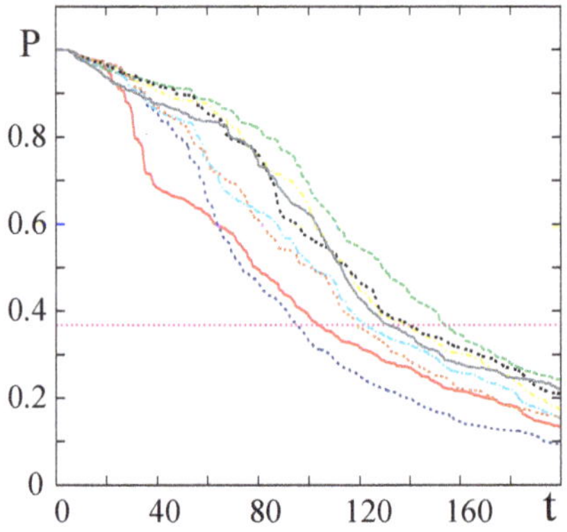

Figure 9. The same problem as in Figure 8 but for a set of noise realizations ($T = 0.15$, in 2D units).

The averaged transition time as a function of temperature is presented in Figure 10. The transition time is defined as a time when the total probability to fix an electron in a lattice becomes less than $e^{-1} \approx 0.37$ (note that sometimes authors use as criterium "1/3" instead of 1/e). One may see that there is a minimum at temperature $T \sim 0.15$, corresponding to the most effective interaction of the electron with solitons. Note that the transiton time is here a sum of two time intervals: first, the time to extract an electron from an "electron" potential well and, second, the travelling time of an electron along the lattice from the donor to the acceptor. Note also that the tunneling does not "work" at such long distances $\delta n = 40$, as the tunneling time is practically infinite [13,14]. In the case studied here, thermal transitions interfering with quantum effects dominate. The large values of the transition time t_{tr} at low temperatures observed in Figure 10 are to be explained by the fact that the first time interval grows sharply with temperature decreasing, although the electron travels along a lattice quickly. In contrast, at high temperature the electron is extracted from a well rapidly, but thermal solectrons become less stable.

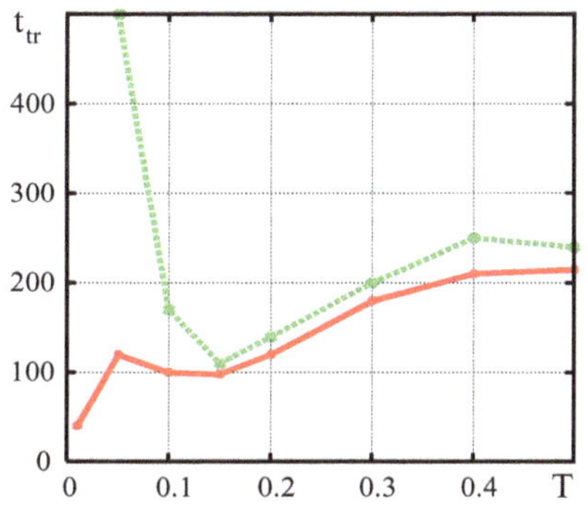

Figure 10. Transition time t_{tr} for moving from the donor to an absorbing acceptor divided by the distance $\delta n = 40$ as a function of temperature. These results were obtained using TBA. (dotted green line with the portion for $T < 0.1$ being spurious as discussed in the main text) and a master equation approach (solid red line [13]).

To exclude the influence of characteristics of an initial potential well and exclude the contribution of "extracting" time to the transition time, a set of computer simulations have been performed for the initial state of an electron in the form of a narrow localized Gaussian bell in a non-deformed lattice consisting only of an absorber at the same position (Figure 10). These results were obtained with a stochastic algorithm solver of master equations.

The results shown in Figures 9 and 10 agree for high enough temperatures $T > 0.1$ but disagree somehow for lower temperatures $T < 0.1$. This is due to the different behavior of the diffusion coefficient obtained in both formalisms (the purely quantum-mechanical and the quantum-statistical). In some sense, these two curves (Figure 10) remind the two curves presented in Figure 5; however, the details and the deeper reasons still remain unexplained. We have to take into account that the quantum mechanical diffusion (dispersion of the density) is not a physical diffusion process but just dispersion of wave functions. Our results strictly speaking are reliable only for $T > 0.1$.

7. Discussion of the ET Donor-Absorber and the Transition Times

At very low temperature $T = 0.01$, an initially localized wave function of the electron spreads along the lattice and the total probability P(t) to find the electron along the chain decreases in time approximately like an exponential function of time. Note that (1 − P(t)) is the probability of absorption. At $T = 0.1$ (weak nonlinearity), the total electron density P(t) decreases much slower and does not look like an exponential function of time. At $T = 0.2$, several steps in the dependence P(t) are observed. In this case, the electron originally localized is quickly transformed to several solectronic maxima; the steps in P(t) correspond to the absorption of these solectronic maxima. At $T = 0.5$, several solitons are excited which are moving oppositely carrying parts of the electron probability density. The probability of existence decreases rather slowly, it takes a very long time before the absorption occurs, and the electron may overcome a quite a long distance before absorption, provided the acceptor is placed far enough from the donor. In our diffusion model, the process is considered as a diffusion-controlled reaction. In the framework of the the Smoluchowski theory, the reaction rate of an electron with density n_e and one absorber with radius r_A at a fixed position is given by

$$r = n_e r_A D_e. \tag{16}$$

We see that the rates are proportional to the diffusion coefficient of the electron, D_e. The electron probability density is decreasing with the distance and develops in time according to the diffusion law Equation (12).

According to our estimate of the diffusion constant, reaction rates based on diffusion would first increase with temperature, then reach a maximum at $T \sim 0.2$ and then decay. The optimal temperature is somewhere around $T \sim 0.2$ in units of 2D. At this time, we are not aware of data which are in agreement with this prediction. Most data seem to show an exponential relation between distance and time. This point to the existence of other mechanisms for ET, such as tunneling [4,13,14,47,50,51].

Nevertheless, let us try to give some estimates based on the two mechanisms studied here: surfing on external solitons and diffusion-like transport on thermal solitons.

Different computer simulations with absorbing boundary have been shown in Figure 9. As shown by Chandrasekhar, diffusion transport is based on the mean square displacement which is given by Equation (13) [49]. He estimated the mean flow through an absorbing boundary located at x_1 and found

$$q\left(t, x_1\right) = \frac{x_1}{t\sqrt{4D_{eff}t}} \exp\left[-\frac{x_1^2}{4D_{eff}t}\right]. \tag{17}$$

Introducing here a dimensionless time τ, we get

$$q(\tau) = \frac{q\left(t, x_1\right)}{\left(4D_{eff}/x_1^2\right)} = \frac{1}{\sqrt{\tau^3}} \exp\left[-\frac{1}{\tau}\right]. \tag{18}$$

We show the flow d' as a function of time according to the Chandrasekhar model in Figure 11.

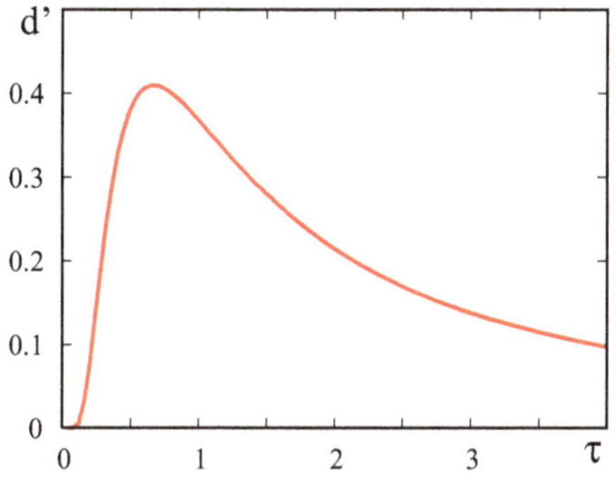

Figure 11. The (dimensionless) flow through an absorbing boundary at, say, location x_1 according to the model of Chandrasekhar as a function of (dimensionless) time.

As Figure 11 shows, the overwhelming amount of probability is absorbed within the (dimensionless) time

$$\tau = 2 = \frac{4D_{eff}t}{x_1^2},\ x_1^2 = 2D_{eff}t. \tag{19}$$

By returning to quantities with dimensions as Å and picoseconds,

$$t = 2\sigma^2 D'_{eff}\frac{t}{t_0},\ D_{eff} = \frac{\sigma^2}{t_0}D'_{eff}. \tag{20}$$

This corresponds just to the mean square displacement which is

$$x_1^2 = \left\langle \Delta x^2 \right\rangle = 2D_{eff}t = 2\sigma^2 D'_{eff}\frac{t}{t_0}. \tag{21}$$

For the dimensionless effective diffusion coefficient D'_{eff}, we estimated a value around 30 close to to the maximum with respect to the temperature. This way, we estimate in dimensional quantities a value around 80 Å^2/ps and, correspondingly, we find

$$\left\langle (x_1\,[Angstrom])^2 \right\rangle = 160\left(Angstrom^2/ps\right)t(ps) \tag{22}$$

The empirical data are usually given as a log of the reciprocal time over the traveled distance given in Å [18]. In our case, we get for the log of the reciprocal time the electron needs to travel the distance $l(t)$ the following estimate

$$\lg\left[1/t\,(s)\right] = 14.2 - 2\lg\left[l\,(t)\,/Angstrom\right] \tag{23}$$

This estimate in Figure 12 is shown by the green curve (time is measured in seconds and the distance in Å, so the 12 (ordinate) corresponds to a picosecond, the six to a microsecond).

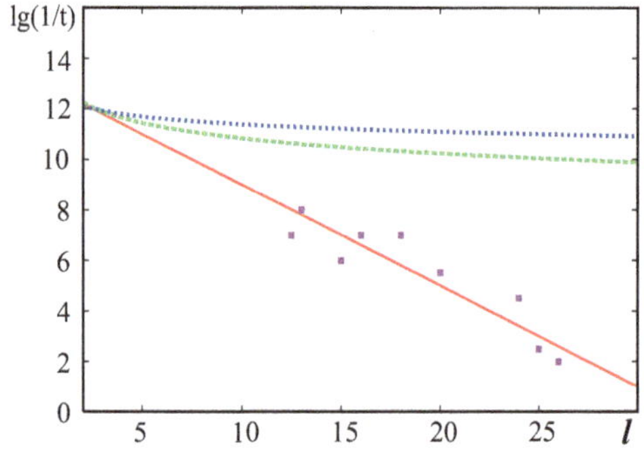

Figure 12. Estimated time t (log-scale) which an electron needs to travel a distance $l(t)$ (on the abscissa in Å units) by using several alternative mechanisms. On the y-axis, we represent (corresponding to standard plots) the log of the reciprocal time in seconds. The green curve shows the estimate obtained from our diffusion mechanism. The blue curve corresponds to the surf on an externally excited soliton which moves with sound velocity. The red curve represents an average of data points (denoted by crosses) measured for azurin and other biomolecules [18].

The estimate based on the diffusion mechanism is surely only a first rough result; in detail, things might be more complicated. Indeed, ET is more complicated than the simple diffusion-controlled reaction studied by Smoluchowski and

Chandrasekhar [48,49]. Let us consider now the other mechanism of surfing on one externally excited soliton such that the electron will travel approximately with sound velocity from the donor to the acceptors: $l(t) = ct$. We estimate the sound velocity c here as approximately 20 Å/ps. Let us mention that in some experimental work [52] a sound velocity of 16.9 Å/ps was measured for DNA. We have

$$\lg\left[1/t\,(s)\right] = 12 + \frac{20}{l\,(Angstrom)}. \tag{24}$$

This is the blue curve we have drawn in Figure 12. We see that the soliton-assisted mechanism (blue curve), which we discussed here is much faster than the estimated average of the observed data given by the lowest (red) curve where slower processes not yet understood seem to dominate in the overall balance.

8. Conclusions

We studied here several mechanisms of electron transfer in nonlinear lattices. We have shown that soliton-assisted electron transfer is one possible rather fast mechanism in a heated nonlinear lattice. We modeled the lattice dynamics with Morse interactions and the electron dynamics as well as the electron-lattice interaction in the quantum mechanical "tight binding" approximation. We have shown that, while the thermal solitons which comprise around 10 lattice sites provide the carrier, the electron "localization" on the lattice, strongly affects the ET-dynamics. The general tendency is that, due to this mechanism, the transfer is much slower and the living times of the solectrons are much higher than previously expected. A similar conclusion has been made based on a completely different method [53]. These authors conclude, that "electronic coupling is most likely determined by nonequilibrium geometries beyond a critical distance (6–7 Å in proteins and 2–3 Å in water)". In our work, some kind of "nonequilibrium geometries" are identified as local deformations due to thermal lattice soliton excitations.

In conclusion, regarding the role of solitons, we may say that electron surfing on lattice solitons generated by external sources provides an extremely fast way of ET (see the blue curve in Figure 12); however, in this case, the source of solitons is yet to be identified. A possible candidate for the source for externally excited solitons are Volkenstein's conformons [54–56]. An alternative way of ET studied here, which is more natural, is the surfing on thermal solitons. The latter provides a slower transport mechanism, due to the quick change of thermal soliton directions. However, this surf is free since thermal solitons are always present. Note that this is a significant improvement relative to the less realistic model used in [11]. A second mechanism, which is slower but still rather fast, is diffusion-based on thermal solitons (green curve in Figure 12). We estimated here the needed times for donor-acceptor transitions based on the Smoluchowski–Chandrasekhar model of absorbing boundaries. The comparison with times observed in experiments (see lowest curve and data points in Figure 12) shows that, under realistic conditions, the transition times are much longer, which means, in reality, the transitions are slowed down by other yet unknown mechanisms such as capturing by impurities. Indeed, as noted by Giese [4] and others, the question of how electrons migrate over long distances was raised about thirty years ago and still is a matter of debate. One may assume that, in reality, one has a combination of several mechanisms, between them the very slow mechanism of tunneling may play quite a significant role [4,13,14,38–40,47].

In conclusion, the computer simulations based on our simple models show that the real processes of donor-acceptor transitions may be very complex. It is to be understood with the interplay of several components such as tunneling, thermal activation, diffusion, electron surfing on lattice solitons and other effects. Clearly, the soliton-assisted transport is the fastest of them all eventually allowing the longest path (or shortest time). All this illustrates the rather difficult problem of dissipative quantum transport [47].

Acknowledgments: The authors wish to express their gratitude to L. Cruzeiro-Hanson for numerous illuminating discussions and suggestions. Further we also thank E. Brandas, L. Brizhik, D. Hennig, J. Kozak, V.D. Lakhno, G. Nicolis, G. Röpke and G. Vinogradov. This research was supported by the Russian Ministry of Education and Science under Grant No. 1008. The revision of the manuscript was completed during a research stay of the authors at the Collaborative Research Center 910 of the Technical University in Berlin. The authors gratefully acknowledge E. Schöll for hospitality and financial support.

Author Contributions: The three authors contributed equally to this work. All authors have read and approved the final manuscript.

Conflicts of Interest: The authors declare no conflict of interest.

References

1. Dekker, C.; Ratner, M. Electronic properties of DNA. *Phys. World* **2001**, *14*, 29, doi:10.1088/2058-7058/14/8/33.
2. Lakhno, V.D.; Korshunova, A.N. Electron motion in a Holstein molecular chain in an electric field. *Eur. Phys. J. B* **2011**, *79*, 147–151.
3. Slinker, J.D.; Muren, N.B.; Renfrew, S.E.; Barton, J.K. DNA charge transport over 34 nm. *Nat. Chem.* **2011**, *3*, 230–235.
4. Giese, B. Long-distance charge transport in DNA: The hopping mechanism. *Acc. Chem. Res.* **2000**, *33*, 631–636.
5. Wan, C; Fiebig, T.; Kelley, S.O.; Treadway, C.R.; Barton, J.K.; Zewail, A.H. Femtosecond dynamics of DNA-mediated electron transfer. *Proc. Natl. Acad. Sci. USA* **1999**, *96*, 6014–6019.

6. Wan, C; Fiebig, T.; Schiemann, O.; Barton, J.K.; Zewail, A.H. Femtosecond direct observation or charge transfer between bases in DNA. *Proc. Natl. Acad. Sci. USA* **2000**, *97*, 14052–14055.
7. Genereux, J.C.; Barton, J.K.; Zewail, A.H. Mechanisms for DNA charge transport. *Chem. Rev.* **2010**, *110*, 1642–1662.
8. Cruzeiro-Hansson, L.; Takeno, S. Davydov model: The quantum, mixed quantum-classical, and full classical systems. *Phys. Rev. E* **1997**, *56*, 894–906.
9. Scott, A. Davydov's Soliton. *Phys. Rep.* **1992**, *217*, 1–67, doi:10.1016/0370-1573(92)90093-F.
10. Scott, A.; Christiansen, P.L. (Eds.) *Davydov's Soliton Revisited: Self-Trapping of Vibrational Energy in Protein*; Springer: New York, NY, USA, 1991.
11. Velarde, M.G.; Chetverikov, A.P.; Ebeling, W.; Hennig, D.; Kozak, J.J. On the mathematical modeling of soliton-mediated long-range electron transfer. *Int. J. Bifurc. Chaos* **2010**, *20*, 185–194.
12. Chetverikov, A.P.; Ebeling, W.; Röpke, G.; Velarde, M.G. High electrical conductivity in nonlinear model lattice crystal mediated by thermal excitation of solectrons. *Eur. Phys. J. B* **2014**, *87*, 153, doi:10.1140/epjb/2014-50213-3.
13. Chetverikov, A.P.; Ebeling, W.; Velarde, M.G. On the temperature dependence of fast electron transport in crystal lattices. *Eur. Phys. J. B* **2015**, *88*, 202, doi:10.1140/epjb/e2015-60495-4.
14. Chetverikov, A.P.; Cruzeiro, L.; Ebeling, W.; Velarde, M.G. On Electron Transfer and Tunneling from Donor to Acceptor in Anharmonic Lattices. In *Quodons in Mica: Nonlinear Localized Travelling Excitations in Crystals*; Archilla, J.F.R., Jimenez, N., Sanchez-Morcillo, V.J., Garcia-Raffi, L.M., Eds.; Springer: Berlin, Germany, 2015; pp. 267–289.
15. Marcus, R.A.; Sutin, N. Electron transfer in chemistry and biology. *Biochim. Biophys. Acta* **1985**, *811*, 265–322.
16. Marcus, R.A. Electron Transfer Reactions in Chemistry. Theory and Experiment. Available online: https://www.uni-ulm.de/fileadmin/website_uni_ulm/nawi.inst.251/Lehre/ws0910/marcus.pdf (accessed on 9 March 2016).
17. Gray, H.B.; Winkler, J.R. Electron tunneling through proteins. *Q. Rev. Biophys.* **2003**, *36*, 341–372.
18. Gray, H.B.; Winkler, J.R. Long-range electron transfer. *Proc. Natl. Acad. Sci. USA* **2005**, *102*, 3534–3539.
19. Likhachev, V.N.; Astakhova, T.Yu.; Vinogradov, G.A. Electron ping-pong on a one-dimensional lattice: Wave packet motion up to the first reflection. *Theor. Math. Phys.* **2013**, *175*, 681–698.
20. Likhachev, V.N.; Astakhova, T.Yu.; Vinogradov, G.A. Electron ping-pong on a one-dimensional lattice: Multiple reflections of the wave packet and capture of the wave function by an acceptor. *Theor. Math. Phys.* **2013**, *176*, 1087–1099.
21. Velarde, M.G.; Ebeling, W.; Chetverikov, A.P. On the possibility of electric conduction mediated by dissipative solitons. *Int. J. Bifurc. Chaos* **2005**, *15*, 245–251.

22. Velarde, M.G.; Ebeling, W.; Hennig, D.; Neissner, C. On soliton-mediated fast electric conduction in a nonlinear lattice with Morse interactions. *Int. J. Bifurc. Chaos* **2006**, *16*, 1035–1039.
23. Velarde, M.G.; Ebeling, W.; Chetverikov, A.P.; Hennig, D. Electron trapping by solitons. Classical *versus* quantum mechanical approach. *Int. J. Bifurc. Chaos* **2008**, *18*, 521–526.
24. Velarde, M.G.; Ebeling, W.; Chetverikov, A.P. Thermal solitons and solectrons in 1D anharmonic lattices up to physiological temperatures. *Int. J. Bifurc. Chaos* **2008**, *18*, 3815–3823.
25. Hennig, D.; Neissner, C.; Ebeling, W.; Velarde, M.G. Effect of anharmonicity on charge transport in hydrogen-bonded systems. *Phys. Rev. B* **2006**, *73*, 024306.
26. Hennig, D.; Chetverikov, A.P.; Ebeling, W.; Velarde, M.G. Electron capture and transport mediated by lattice solitons. *Phys. Rev. E* **2007**, *76*, 046602.
27. Velarde, M.G. From polaron to solectron: The addition of nonlinear elasticity to quantum mechanics and its possible effect upon electric transport. *J. Comput. Appl. Math.* **2010**, *233*, 1432–1445.
28. Ashcroft, N.W.; Mermin, N.D. *Solid State Physics*; Holt, Rinehart and Winston: New York, NY, USA, 1976.
29. Chetverikov, A.P.; Ebeling, W.; Velarde, M.G. Nonlinear ionic excitations, dynamic bound states, and nonlinear currents in a one-dimensional plasma. *Contrib. Plasma Phys.* **2005**, *45*, 275–283.
30. Chetverikov, A.P.; Ebeling, W.; Velarde, M.G. Dissipative solitons and complex currents in active lattices. *Int. J. Bifurc. Chaos* **2006**, *16*, 1613–1632.
31. Chetverikov, A.P.; Ebeling, W.; Velarde, M.G. Non-linear excitations and electric transport in dissipative Morse-Toda lattices. *Eur. Phys. J. B* **2006**, *51*, 87–99.
32. Toda, M. *Theory of Nonlinear Lattices*, 2nd ed.; Springer: Berlin, Germany, 1989.
33. Kalosakas, G.; Aubry, S.; Tsironis, G.P. Polaron solutions and normal-model analysis in the semiclassical Holstein model. *Phys. Rev. B* **1998**, *58*, 3094–3104.
34. Kalosakas, G.; Rasmussen, K.O.; Bishop, A.R. Charge trapping in DNA due to intrinsic vibrational hot spots. *J. Chem. Phys.* **2003**, *118*, 3731–3735.
35. Slater, J.C. *Quantum Theory of Molecules and Solids*; McGraw-Hill: New York, NY, USA, 1974.
36. Hennig, D. Solitonic energy transfer in a coupled exciton-vibron system. *Phys. Rev. E* **2000**, *61*, 4550–4555.
37. Launay, J.P.; Verdaguer, M. *Electrons in Molecules*; Oxford University Press: Oxford, UK, 2014.
38. Chetverikov, A.P.; Ebeling, W.; Velarde, M.G. Local electron distributions and diffusion in anharmonic lattices mediated by thermally excited solitons. *Eur. Phys. J. B* **2009**, *70*, 217–227.
39. Chetverikov, A.P.; Ebeling, W.; Velarde, M.G. Electron dynamics in tight-binding approximation: The influence of thermal anharmonic lattice excitations. *Contrib. Plasma Phys.* **2009**, *49*, 529–535.

40. Chetverikov, A.P.; Ebeling, W.; Velarde, M.G. Bound states of electrons with soliton-like excitations in thermal systems—Adiabatic approximations. *Condens. Matter Phys.* **2009**, *12*, 633–645.
41. Davydov, S.A. *Solitons in Molecular Systems*, 2nd ed.; Reidel: Dordrecht, Holland, 1991.
42. Christiansen, P.L.; Scott, A.C. (Eds.) *Davydov's Soliton Revisited: Self-Trapping of Vibrational Energy in Proteins*; Plenum Press: New York, NY, USA, 1983.
43. Muto, V.; Scott, A.C.; Christiansen, P.L. Thermally generated solitons in a Toda lattice model of DNA. *Phys. Lett. A* **1989**, *136*, 33–36.
44. Brizhik, L.S.; Davydov, A.S. Soliton excitations in one-dimensional molecular systems. *Phys. Status Solidi* **1983**, *115*, 615–630.
45. Lakhno, V.D.; Fialko, N.S. Hole mobility in a homogeneous nucleotide chain. *JETP Lett.* **2003**, *78*, 336–338.
46. Lakhno, V.D.; Fialko, N.S. HSSH-model of hole transfer in DNA. *Eur. Phys. J. B* **2005**, *43*, 279–281.
47. Dittrich, T.; Hänggi, P.; Ingold, G.L.; Kramer, B.; Schön, G.; Zwerger, W. *Quantum Transport and Dissipation*; Wiley-VCH: Weinheim, Germany, 1998.
48. Smoluchowski, M. Three lectures on Diffusion, Brownian Motion, and Coagulation. *Physik Zeitung* **1916**, *17*, 557–571, 585–599. (In German)
49. Chandrasekhar, S. Stochastic problems in physics and astronomy. *Rev. Mod. Phys.* **1943**, *15*, 1–89, doi:10.1103/RevModPhys.15.1.
50. Bollinger, J.M. Electron relay in proteins. *Science* **2008**, *320*, 1730–1731.
51. Hopfield, J.J. Electron transfer between biological molecules by thermally activated tunneling. *Proc. Nat. Acad. Sci. USA* **1974**, *71*, 3640–3644.
52. Hakim, M.B.; Lindsay, S.M.; Powell, J. The speed of sound in DNA. *Biopolymers* **1984**, *23*, 1185–1192.
53. Balabin, I.A.; Beratan, D.N.; Skoutis, S.S. Persistence of structure over fluctuations in biological electron-transfer reactions. *Phys. Rev. Lett.* **2008**, *101*, 158102.
54. Volkenstein, M.V. The conformon. *J. Theor. Biol.* **1972**, *34*, 193–195.
55. Kemeny, G.; Goklany, I.M. Polarons and conformons. *J. Theor. Biol.* **1973**, *40*, 107–123.
56. Volkenstein, M.V. *Biophysics*; Mir Publishers: Moscow, Russia, 1983.

Nonlinear Phenomena of Ultracold Atomic Gases in Optical Lattices: Emergence of Novel Features in Extended States

Gentaro Watanabe, B. Prasanna Venkatesh and Raka Dasgupta

Abstract: The system of a cold atomic gas in an optical lattice is governed by two factors: nonlinearity originating from the interparticle interaction, and the periodicity of the system set by the lattice. The high level of controllability associated with such an arrangement allows for the study of the competition and interplay between these two, and gives rise to a whole range of interesting and rich nonlinear effects. This review covers the basic idea and overview of such nonlinear phenomena, especially those corresponding to extended states. This includes "swallowtail" loop structures of the energy band, Bloch states with multiple periodicity, and those in "nonlinear lattices", *i.e.*, systems with the nonlinear interaction term itself being a periodic function in space.

Reprinted from *Entropy*. Cite as: Watanabe, G.; Venkatesh, B.P.; Dasgupta, R. Nonlinear Phenomena of Ultracold Atomic Gases in Optical Lattices: Emergence of Novel Features in Extended States. *Entropy* **2016**, *18*, 118.

1. Introduction

Following a long series of developments in the experimental techniques of atomic and optical physics, the Bose–Einstein condensation (BEC) of cold alkali atomic gases was realized in 1995 (see, e.g., [1–3] and references therein). The creation of this new state of quantum matter has opened up a new research field, the physics of ultracold atomic gases. The novelty of this system lies in its high controllability: various system parameters such as the dimensionality, the configuration of the external potentials, and the strength and the sign of the inter-atomic interaction can be manipulated dynamically as well as statically. In addition, this system has high measurability: since both the spatial and temporal microscopic scales of this system are relatively large, real time observation and direct imaging are possible. With these unique features, ultracold atomic gases serve as an unprecedented playground of the quantum world.

Due to the emergence of the superfluid order parameter, BECs acquire a nonlinear character originating from the interparticle interaction. Here, nonlinearity means that the basic equation governing the state of the system depends on the state itself. A variety of phenomena caused by nonlinearity such as solitons [4–7] and matter-wave mixing [8], *etc.* have been predicted and realized in BECs. Remarkably, the strength of the nonlinearity in BECs of cold atomic gases is controllable. This is because here s-wave scattering length a_s, the parameter characterizing the interatomic interaction, can be tuned using the Feshbach resonance. Therefore, the realization of BECs of cold atomic gases has opened up a new horizon for the study of nonlinear phenomena (see, e.g., [9]).

With further development of technology and tools, superfluidity has been realized using cold fermionic atoms as well. It has been shown that, by increasing the interatomic attraction using Feshbach resonances, the state of an atomic Fermi gas can be varied in a controlled manner from a Bardeen–Cooper–Schrieffer (BCS) superfluid of delocalized Cooper pairs to a BEC of tightly-bound dimers [10]. Furthermore, these two limits are smoothly connected without a phase transition: the so-called BCS-BEC crossover [11–14]. The experimental confirmation of the BCS-BEC crossover is one of the prime achievements in the field of cold atomic gases. Using BCS-BEC crossover, we can understand both the Bose and Fermi superfluids from a unified perspective.

Another important development in the field of cold atomic gases is the realization of the external periodic potential called an "optical lattice": Pairs of counter-propagating laser fields detuned from atomic transition frequencies, act as a free-of-defect, conservative potential for atoms via the optical dipole force. The realization of optical lattices has opened up the connection between the physics of cold atomic gases and solid state/condensed matter physics (see, e.g., [14–17] for reviews), and especially enables the simulation of theoretical models of solid state physics using cold atoms.

As a consequence of the competition and interplay between the effects of the periodic potential and nonlinearity, rich phenomena are expected to emerge in cold atomic gases in optical lattices. Especially, equipped with Feshbach resonance, a knob for controlling the strength of the nonlinearity, cold atomic gases in optical lattices allow us to enter a regime in which the effect of the nonlinearity is comparable to (or even dominates over) that of the periodic potential. Such a strongly nonlinear regime beyond the tight-binding approximation has not been well-explored in conventional solid state physics. For example, a loop structure called "swallowtail" in the Bloch energy band [18,19] is a representative novel phenomenon emerging in this regime. In addition, using cold atomic gases, direct observations of the resulting nonlinear phenomena are possible, which is also a difficult task using solids.

In this short review article, we discuss nonlinear phenomena of superfluid cold atomic gases in optical lattices. Especially, we consider extended states and focus on the following phenomena: the swallowtail band structure, Bloch states with multiple periods of the applied optical lattice potential called multiple period states, and those in nonlinear lattices, *i.e.*, systems with a periodically modulated interaction strength in space. This article is complementary to the existing review article on nonlinear phenomena in lattices [20], which focuses mainly on localized states. Superfluidity is the most important macroscopic quantum phenomena and superfluid flow in a periodic potential is ubiquitous in many other systems, such as superconducting electrons in superconductors and even in astrophysical environments such as superfluid neutrons in "pasta" phases in neutron star crusts (see, e.g., [21,22] and references therein). Through the study of cold atomic gases in optical lattices, one may also expect to get deeper insights into these other systems.

This article is organized as follows. In Section 2, we explain the setup of our system and basic theoretical formalism employed in the later discussions. In Sections 3–5, we provide a comprehensive overview of the selected nonlinear phenomena in optical lattices starting with a simple physical explanation for each topic: swallowtail loops in Section 3, multiple period states in Section 4, and nonlinear lattices in Section 5. Finally, summary and prospects are given in Section 6.

2. Theoretical Framework

2.1. Setup of the System

In the present article, we discuss superfluid flows of either fermionic or bosonic atoms in the presence of the externally imposed periodic potentials. For the external periodicity, we mainly consider one of the most typical cases: one-dimensional (1D) sinusoidal potential of the form,

$$V(\mathbf{r}) = V(x) = sE_R \sin^2 q_B x \equiv V_0 \sin^2 q_B x \,, \tag{1}$$

either in quasi-1D or 3D systems. Here, $E_R = \hbar^2 q_B^2/2m$ is the recoil energy, m is the mass of atoms, $q_B = \pi/d$ is the Bragg wave number (note that q_B is different from the fundamental vector of a 1D reciprocal lattice, $2\pi/d$, by a factor of 2), and d is the lattice constant, $V_0 \equiv sE_R$ is the lattice height, and s is the dimensionless parameter characterizing the lattice intensity in units of E_R. For simplicity, we also assume that the superflow is in the same direction as the periodic potential (*i.e.*, x direction). Throughout the present article, we set the temperature $T = 0$.

The systems which we discuss in this article consist of a large number of particles (the number of particles per site is also large) at temperatures close to absolute zero. One of the most convenient ways to deal with such many-body systems is to use the mean-field approximation. In this formalism, one focuses on a particular single particle, and the interactions produced by all the other particles are replaced by an averaged interaction described by the "mean field". Thus the complicated many-body problem is effectively reduced to a far simpler one-body problem.

The mean-field theory provides a minimal framework to study the nonlinear phenomena emerging from the presence of the superfluid order parameter. The mean-field theory enables us to predict novel nonlinear phenomena and obtain qualitative understanding of them although its validity is not always guaranteed. We resort to a mean-field description throughout as it readily fits our motivation in the present review—to provide a physical explanation of some selected, novel nonlinear phenomena of superfluids in periodic systems.

In the rest of this section, we provide a brief explanation of the theoretical framework used in the discussions in the remaining part of this article. The main purpose of this section is to provide a minimal explanation and define the notation. Therefore, interested readers are encouraged to refer to other references (e.g., [1,2,13,23]) for further details.

2.2. Bosons

The mean-field theory describing Bose–Einstein Condensates (BECs) at zero temperature is given by the Gross–Pitaevskii (GP) equation [1,2,23–26]:

$$i\hbar\frac{\partial\psi(\mathbf{r},t)}{\partial t} = \left[-\frac{\hbar^2}{2m}\nabla^2 + V(\mathbf{r}) + g|\psi(\mathbf{r},t)|^2\right]\psi(\mathbf{r},t)\,. \tag{2}$$

Here $\psi(\mathbf{r},t)$ is the superfluid order parameter (or the condensate wave function) and g is the effective coupling constant between two interacting bosons given by

$$g = \frac{4\pi\hbar^2 a_s}{m}, \tag{3}$$

where a_s is the s-wave scattering length. The average number density n is

$$n = \frac{N}{\mathcal{V}} = \frac{1}{\mathcal{V}}\int |\psi(\mathbf{r})|^2\, d\mathbf{r}, \tag{4}$$

where N is the total number of particles and $\mathcal{V}$ is the volume of the system. Note that the GP equation can be viewed as the dynamical equation that results from a governing Hamiltonian known as the GP energy functional given by:

$$E[\psi] = \int d\mathbf{r}\left(\frac{\hbar^2}{2m}|\nabla\psi|^2 + V(\mathbf{r})|\psi|^2 + \frac{g}{2}|\psi|^4\right). \tag{5}$$

The stationary solution of Equation (2) is given by

$$\mu\psi(\mathbf{r}) = \left[-\frac{\hbar^2}{2m}\nabla^2 + V(\mathbf{r}) + g|\psi(\mathbf{r})|^2\right]\psi(\mathbf{r}), \tag{6}$$

where μ is the chemical potential.

Nonlinearity of the GP equation (the third term in the rhs of Equations (2) and (6)) originates from the interaction between bosonic atoms. Many previous studies have shown that GP equation describes BECs of dilute, weakly interacting bosons at zero temperature quite successfully (see, e.g., [23] and references therein).

2.3. Fermions

A useful method for treating superfluid Fermi gases is the standard BCS mean-field theory of superconductivity. Such a mean-field theory for inhomogeneous systems is given by the Bogoliubov-de Gennes (BdG) equations [13,27]:

$$\begin{pmatrix} H'(\mathbf{r}) & \Delta(\mathbf{r}) \\ \Delta^*(\mathbf{r}) & -H'(\mathbf{r}) \end{pmatrix} \begin{pmatrix} u_i(\mathbf{r}) \\ v_i(\mathbf{r}) \end{pmatrix} = \epsilon_i \begin{pmatrix} u_i(\mathbf{r}) \\ v_i(\mathbf{r}) \end{pmatrix} . \tag{7}$$

Here $H'(\mathbf{r}) = -\frac{\hbar^2}{2m}\nabla^2 + V(\mathbf{r}) - \mu$. Also, $v_i(\mathbf{r})$ and $u_i(\mathbf{r})$ are the quasiparticle amplitudes, associated with the probability of occupation and unoccupation of a paired state denoted by an index i, while ϵ_i is the corresponding eigen-energy. The quasiparticle amplitudes $v_i(\mathbf{r})$ and $u_i(\mathbf{r})$ satisfy the normalization condition $\int d\mathbf{r}\,[u_i^*(\mathbf{r})u_j(\mathbf{r}) + v_i^*(\mathbf{r})v_j(\mathbf{r})] = \delta_{i,j}$. Δ is the order parameter (or the pairing field), which reduces to the pairing gap in the single quasiparticle spectrum in the region of $\mu > 0$ for the uniform system. The pairing field $\Delta(\mathbf{r})$ and the chemical potential μ in Equation (7) are self-consistently determined from the gap equation,

$$\Delta(\mathbf{r}) = -g\sum_i u_i(\mathbf{r})v_i^*(\mathbf{r}) , \tag{8}$$

and the average number density

$$n = \frac{N}{\mathcal{V}} = \frac{1}{\mathcal{V}}\int n(\mathbf{r})\,d\mathbf{r} = \frac{2}{\mathcal{V}}\sum_i \int |v_i(\mathbf{r})|^2 d\mathbf{r} . \tag{9}$$

Since Δ depends on $\{u_i\}$ and $\{v_i\}$, the BdG Equations (7) are nonlinear for nonzero interatomic interaction parameter g.

The superfluid Fermi systems bear a direct analogy with traditional superconducting systems, and likewise g, the contact interaction, plays similar role as the weakly attractive interaction term in the BCS-model. Only, now g can be both small or large, and its value can be externally tuned using Feshbach resonances by applying a magnetic or an optical field. This controllability leads to a crossover between two ends: a weakly attractive BCS-like superfluid and a condensate of tightly bound bosonic molecules of a pair of fermionic atoms; popularly called the BCS-BEC crossover [11,12].

For contact interactions, the right-hand side of Equation (8) has an ultraviolet divergence, which has to be regularized by replacing the bare coupling constant g with the two-body T-matrix related to the s-wave scattering length [28]. A standard scheme [28] is to introduce a cutoff energy $E_c \equiv \hbar^2 k_c^2/2m$ in the sum over the BdG eigenstates and to replace g by the following relation:

$$\frac{1}{g} = \frac{m}{4\pi\hbar^2 a_s} - \sum_{k<k_c} \frac{1}{2\epsilon_k^{(0)}}, \tag{10}$$

with $\epsilon_k^{(0)} \equiv \hbar^2 k^2/2m$.

2.4. Discrete and Continuum Models

The systems of cold atomic gases can be studied by solving the GP equation (bosons) or the BdG equations (fermions) for the full continuum model. Let us, for the sake of simplicity, consider quasi-1D bosonic systems. So instead of $V(\mathbf{r})$, we think of a potential in x direction only: $V(x)$. If there is a periodicity in the form of $V(x)$, or, if g itself is a periodic function of x, one approach is to try the Bloch solutions $\psi(x) = e^{ikx}\phi(x)$, where $\hbar k \equiv P$ is the quasimomentum of the superflow and $\phi(x)$ is a periodic function with the same periodicity as the externally imposed periodicity by $V(x)$ or $g(x)$. One can expand $\phi(x)$ in terms of plane waves to give the following form for the order parameter:

$$\psi(x) = e^{ikx}\phi(x) = e^{ikx} \sum_{l=-l_{\max}}^{l_{\max}} a_l e^{i2\pi l x/d}, \tag{11}$$

to find the Bloch solutions. The normalization condition yields $\sum_l |a_l|^2 = 1$.

Instead of going for the full solution, one easier approach is to map the system to a discrete model, borrowed from the idea of tight-binding model in solid-state physics. In this approach the density of bosonic/fermionic atoms is assumed to be concentrated around the minima of the optical lattice potential. For example, in a quasi-1D periodic potential, the condensate wave function can be approximated by a superposition of wave functions $\phi_j(x)$ localized at the lattice sites, denoted by j, which are normalized as $\int |\phi_j(x)|^2 dx = 1$. Thus, $\psi(x,t) = \sum_j \psi_j(t)\phi_j(x)$. The coefficient ψ_j is dependent on the site index j.

The Hamiltonian for such a discrete model is

$$H = -K\sum_j (\psi_j^* \psi_{j+1} + \psi_{j+1}^* \psi_j) + \frac{U}{2}\sum_j |\psi_j|^4. \tag{12}$$

In the case of the periodic solution with the same periodicity as that of the lattice (lattice constant d), the normalization condition is given by

$$\int_{-d/2}^{d/2} |\psi(x)|^2 dx = \nu , \tag{13}$$

where ν is the filling factor (number of particles per site) with

$$\nu = |\psi_j|^2 . \tag{14}$$

In evaluating the normalization condition, one neglects the overlap between ϕ_j's localized at different sites.

The first term in the Hamiltonian describes hopping between the nearest-neighbor sites. The hopping parameter K is given by $K = -\int \phi_j [-\frac{\hbar^2}{2m}\nabla^2 + V(x)]\phi_{j+1} dx$. The on-site interaction parameter U characterizes the interaction energy between two atoms on the same site, and gives the nonlinear term. This U is connected to the interatomic interaction parameter g by $U = g\int |\phi_j(x)|^4 dx$. In a manner similar to the continuum model one can also derive a dynamical equation for the amplitudes ψ_j from an extermisation of the energy functional corresponding to Equation (12). Such a dynamical equation is known in literature as the discrete nonlinear Schrödinger equation (analogous to Equation (2) for the continuous case) [29,30] and has served as an important tool to study ultracold atoms in optical lattices.

2.5. *Energetic and Dynamical Stability*

Once one can solve for the system, either from the full continuum model or the discrete version, the next step is to study the energetic and dynamical stability of the stationary solutions. Energetic stability guarantees that the stationary states are a local energy minimum of the energy functional (Equation (5)) and dynamical stability means that the time evolution of the system is stable with respect to small perturbations (this issue will be discussed in detail later in Section 3.2.2). The standard approach in this context is the linear stability analysis described below [1,31–33].

Let $\delta\phi_q(x)$ be the deviation from the stationary Bloch wave solution $\phi(x)$ for a given quasimomentum $\hbar k$ of the superflow. This can be written in the following form:

$$\delta\phi_q = u(x,q)e^{iqx} + v^*(x,q)e^{-iqx} . \tag{15}$$

Here $\hbar q$ is the quasimomentum of the perturbation. The energy deviation from the stationary states can be written in the following form

$$\delta E = \int dx \begin{pmatrix} u^* & v^* \end{pmatrix} M(q) \begin{pmatrix} u \\ v \end{pmatrix} . \tag{16}$$

The matrix $M(q)$ is Hermitian and gives the curvature of the energy landscape around the stationary solution. The system is energetically stable as long as all of the eigenvalues of $M(q)$ are positive. When even one of the eigenvalues is non-positive, the solution is no longer a local minimum and the system is energetically unstable. This is often termed as "Landau instability".

For the same perturbation $\delta\phi_q$, the time-evolution of the system for each k is found to be:

$$i\frac{\partial}{\partial t}\begin{pmatrix} u \\ v \end{pmatrix} = \sigma_z M(q) \begin{pmatrix} u \\ v \end{pmatrix}, \tag{17}$$

with

$$\sigma_z = \begin{pmatrix} 1 & 0 \\ 0 & -1 \end{pmatrix} . \tag{18}$$

Unlike $M(q)$, the matrix $\sigma_z M(q)$ is not Hermitian. If the eigenvalue of $\sigma_z M(q)$ is complex, the perturbation corresponding to an eigenvalue with positive imaginary part grows exponentially in time: in this case the stationary solution is dynamically unstable. If the imaginary part is zero, the stationary state remains stable (*i.e.*, dynamically stable). Therefore, by noting the eigenvalues of $M(q)$ and $\sigma_z M(q)$, one can learn whether the system belongs to an energetically stable region to start with, and how it evolves during the course of time. This is important because for BECs and superfluid Fermi gases, this stability translates to the sustenance of superfluidity in the system.

3. Swallowtail Loops in Band Structure

In this section we consider one unique manifestation of the nonlinearity (see Equation (2)) governing the dynamics of BECs in optical lattices—the so-called swallowtail loops in band structures. In the first Section 3.1 we provide the basic physical idea and the context in which swallowtail structures arise in the energy dispersions. Moreover, we discuss the key implication of having such swallowtail dispersions—breakdown of adiabaticity even at very slow driving. The purpose of this subsection will be to communicate the central physical picture succinctly, hence we shall sacrifice chronology and use the most clear presentation (in our opinion) following [19,32,34,35]. In the second Section 3.2 we present a more detailed account of the various theoretical results on swallowtail band structures including their energetic and dynamical stability. In the third Section 3.3 we present some experimental results that consider the effects of such nonlinear structures. In the following Section 3.1, we present a brief account of more recent developments that extend the swallowtail phenomena to situations beyond the standard s-wave interacting Bosons in optical lattices including dipole-dipole interactions, superfluid Fermi gases, *etc.* We conclude the section with some brief remarks indicating future prospects.

3.1. Basic Physical Idea: The Nonlinear Landau–Zener Model and Variational Ansatz for Condensate Wavefunction in Optical Lattices

The GP Equation (2) for the order parameter describing a BEC at zero temperature differs from the Schrödinger equation for a single particle in one key aspect — the presence of the interaction term $g|\psi(\mathbf{r},t)|^2$ in addition to the externally applied potential term $V(\mathbf{r})$. When the order parameter $\psi(\mathbf{r})$ is expanded in terms of a given complete basis set of single particle wavefunctions, the nonlinear term in the GP equation essentially leads to an effective potential that is dependent on the occupation probability of the different single particle states. One may anticipate that in the limit that the nonlinearity is comparable to the external potential, *i.e.*, $gn \sim V(\mathbf{r})$, the resulting dynamics as well as the stationary solutions of the GP Equation (6) can be very different from the equivalent ones for the non-interacting $g = 0$ case which are simply governed by the eigen-energies of the Schrödinger equation.

The simplest model to see the structures that emerge in the limit of large nonlinearities is the nonlinear two-level system introduced in [34]:

$$i\frac{\partial}{\partial t}\begin{pmatrix} a \\ b \end{pmatrix} = H(\gamma)\begin{pmatrix} a \\ b \end{pmatrix}, \tag{19}$$

$$H(\gamma) = \begin{pmatrix} \frac{\gamma}{2} + \frac{c}{2}\left(|b|^2 - |a|^2\right) & v/2 \\ v/2 & -\frac{\gamma}{2} - \frac{c}{2}\left(|b|^2 - |a|^2\right) \end{pmatrix}. \tag{20}$$

Here γ gives the level separation and $v/2$ is the coupling between the levels. With $\gamma = \alpha t$, and $g = 0$, the above Hamiltonian is the well known Landau–Zener (LZ) model [36,37]. Equation (19) represents an extension of the LZ model representing the dynamics of a BEC with an order parameter whose overlap with the two relevant single particle levels has the amplitude a and b. Let us now consider the main results of such a model. In the limit of the small $c = 0.1$, as shown in Figure 1 left, the adiabatic energy levels (or the chemical potential μ in the language of our theoretical framework) is qualitatively similar to the linear case with the characteristic avoided crossing. When the nonlinearity exceeds the coupling strength v, the adiabatic energy levels are drastically changed with the development of the characteristic looped structures. Note that the shape of the loop for the chemical potential is not a swallowtail, the energy function has such a swallowtail structure.

While the modification of the energy level structure due to the strong nonlinearity is novel, what makes the looped energy structures particularly interesting from a physical point of view is the crucial implication of the looped structure for the transition probability between adiabatic energy levels. In the linear case ($c = 0$), the transition probability for the LZ model is given by $r_0 = \exp(-\pi v^2/2\alpha)$. In this case in the so-called adiabatic limit of $\alpha \to 0$, the transition probability vanishes. On the other hand in the nonlinear case, as shown in Figure 1 right, for strong enough nonlinearity such that $c > v$ there is a finite transition probability even in the adiabatic limit. A simple explanation for this behavior is discernible from the looped energy structure in the third panel of Figure 1 left—starting from the lower energy state initially, in the adiabatic limit there is very little tunneling as the system passes the point "X" and continues upwards along the loop to reach the final point "T". At this point the system has to make a non-adiabatic transition to either the upper or lower level irrespective of how slow the sweep rate α is. This breakdown of adiabaticity is the key implication as well as an indication of the loop structure arising from the nonlinearity.

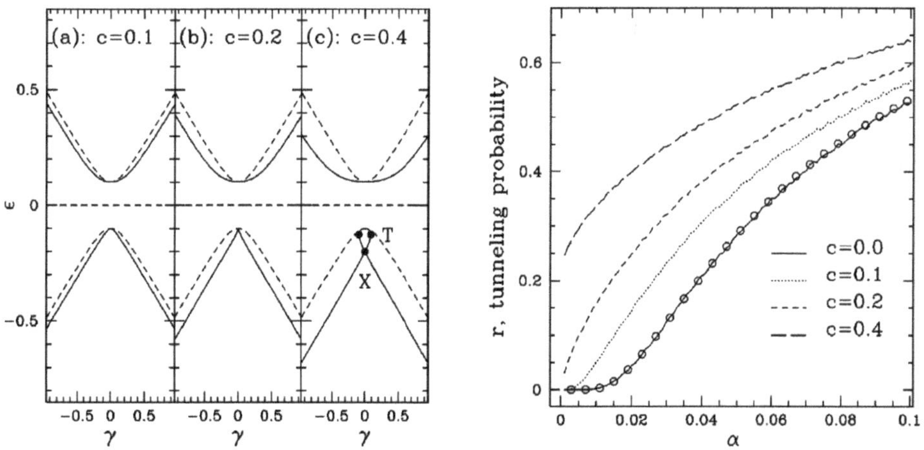

(**Left**) Energy levels. (**Right**) Transition probability.

Figure 1. (**Left**) Adiabatic energy levels (or the chemical potential μ) as a function of the nonlinearity c for the nonlinear LZ problem showing the emergence of loops for $c > v$. The dashed lines represent the adiabatic energy levels for the $c = 0$ case. (**Right**) Tunneling probability as a function of drive speed α at different values of c. For the linear case the result of the classic formula $r_0 = \exp(-\pi v^2/2\alpha)$ is displayed by the open circles. The figures are taken from [34].

Having captured the basic setting in which loops arise via the nonlinear LZ model, we move on now to the system central to this review—BEC in an optical lattice. We are interested in the stationary states of the GP equation that satisfy Equation (6). In analogy to the LZ model, when the nonlinearity gn is comparable to the applied lattice potential strength $V(x)$ we can expect interesting energy dispersions to arise. To be specific, henceforth we follow the treatment in [19,32] and consider a one-dimensional potential of the form:

$$V(x) = V_0 \cos(2\pi x/d), \tag{21}$$

and look for stationary states of the Bloch form, *i.e.*, $\psi(x) = e^{ikx} f(x)$ with $f(x+d) = f(x)$. Here k denotes the quasimomentum of the condensate and is the proper quantum number in a periodic system. A key quantity of interest is the energy per unit volume given by

$$\mathcal{E} = \frac{1}{d} \int_{-d/2}^{d/2} dx \left[\frac{\hbar^2}{2m} |\nabla \psi|^2 + V_0 \cos(2\pi x/d) |\psi|^2 + \frac{g}{2} |\psi|^4 \right]. \tag{22}$$

In the absence of interactions, *i.e.*, $g = 0$, the energy per unit volume is arranged into the usual band structure which repeats after every reciprocal lattice momentum $k = 2\pi/d$ and has gaps at the zone edges $k = r\pi/d$ with r an odd integer in the extended zone representation of the bands (or at $k = \pm\pi/d$ in the reduced zone schemes).

Equation (11) gives the full plane wave expansion for a Bloch state but a variational ansatz restricting to just three plane wave states, *i.e.*, $l_{\max} = 1$ can already capture much of the physics as shown in [32]. Since the relative phases of the plane wave amplitudes (a_0, a_1, a_{-1}) do not change the energy, they can be chosen as real and using their normalization property restricted to the form:

$$a_0 = \cos\theta, \quad a_1 = \sin\theta \cos\phi, \quad a_{-1} = \sin\theta \sin\phi. \tag{23}$$

The trial wavefunction (11) with the coefficients of the form (23) is inserted into the energy per unit volume expression (22) and extremized with respect to the parameters θ and ϕ. Also the recoil energy $E_0 = \hbar\pi^2/2md^2$ ($E_0 = E_R$ in the notation of this review article) serves as a convenient unit for different energies in the problem. Let us first consider the situation close to the Brillouin zone edge $k = \pm\pi/d$. Using intuition from the nearly free particle models for periodic potentials, the ansatz (23) may further be simplified by taking $\phi = \{0, \pi/2\}$ [19] giving:

$$\psi(x) = \sqrt{n}\, e^{ikx} \left(\cos\theta + \sin\theta\, e^{-i2\pi x/d} \right) \tag{24}$$

with n the average particle density. In fact it was shown in [34] that such an ansatz can indeed be used to map the problem to that of the nonlinear LZ discussed earlier. At the zone edge $k = \pi/d$, upon extremization of energy density functional (22) yields the solutions $\cos 2\theta = 0$ or $\sin 2\theta = V_0/(gn)$. When $gn < V_0$, the only possible solutions are $\theta = \pi/4$ or $\theta = 3\pi/4$ representing the zone edge solutions of the lowest and first excited band and are qualitatively similar to their counterparts in the linear, $g = 0$, limit. The condensate current density J, given by the derivative of energy $\mathcal{E}$ with respect to k, is zero for such solutions. When the interaction is strong enough such that $ng > V_0$, the other solution with $\theta = \sin^{-1}[V_0/(gn)]/2$ is also allowed. Moreover this solution has no linear counterpart and has non-zero current even at the zone edge with:

$$J = \pm \frac{\hbar\pi}{md}\sqrt{n^2 - \frac{V_0^2}{g^2}}\,. \tag{25}$$

For values of k away from the zone edge with $g > V_0/n$, the two solutions originally located at $\theta = \pi/4$ and $\sin^{-1}[V_0/(gn)]/2$ approach each other and finally merge giving rise to the typical loop structure depicted in Figure 2 top. Also note that the band-gap at $k = \pi/d$ in the weak lattice limit $V_0 \ll E_0$ is given by V_0 and the condition to have looped energy dispersion is that the interaction energy per particle gn be greater than this band gap.

Remarkably, the variational trial wavefunction (24), with θ values such that $\mathcal{E}$ is extremized, is identical with an exact analytical solution found in [18,38]. While this initially leads one to suspect that loops are somehow a restricted phenomenon subject to the availability of such exact solutions, further work in [32] dispelled this notion by demonstrating the possibility of loops even at the zone center, *i.e.*, with $k = 0$, without any known exact solutions (see Figure 2 bottom). Such solutions are also captured by the variational ansatz (23). In this context the key solution in the linear limit is the one corresponding to the top of the second band with $k = 0$ and energy $\mathcal{E}/n = 4E_0$ per particle in the absence of interactions $g = 0$ and $V_0 \ll E_0$. This solution with an equal admixture of a_1 and a_{-1}, and very small contribution from a_0 has $\theta = \pi/2$ and $\phi = 3\pi/4$ (the solution with $\phi = \pi/4$ corresponds to the third band now) persists even when $g \neq 0$. However, it was shown in [32] that, for sufficiently large values of g, this solution becomes unstable and two new solutions emerge signaling the lower and upper point of the loop at $k = 0$ depicted in Figure 2 bottom. The condition for emergence of loops at the zone center is given by:

$$gn > (16E_0^2 + V_0^2) - 4E_0\,. \tag{26}$$

In the weak lattice limit, the above condition simplifies to $gn > V_0^2/(8E_0)$ which is precisely analogous to the condition that interactions exceed the band gap similar to the criterion to have loops at the zone edge. Figure 2 bottom illustrates the zone-edge and zone-center loops computed using the ansatz (23). Furthermore it was found in [32] that the size of the loops, *i.e.*, their extent in k increases monotonically as the ratio gn/V_0 is increased. Most interestingly in the limit of vanishing lattice potential $V_0 \to 0$, the swallowtail loops extend over the entire Brillouin zone and the upper edge of the swallowtail becomes degenerate with the states in the bands above. As we discuss more in detail in the next subsection, this behavior stems from the fact that the states in the upper edge of the swallowtail correspond to periodic soliton solutions of the GP equation in free space whose degeneracy is lifted by the application of a periodic potential.

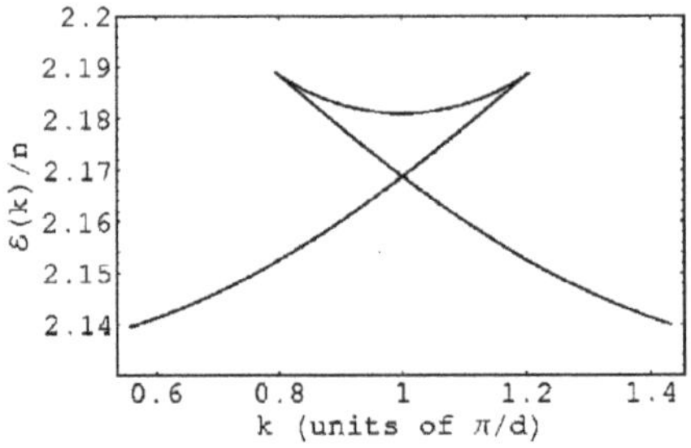

(**Top**) Zone-edge loop.

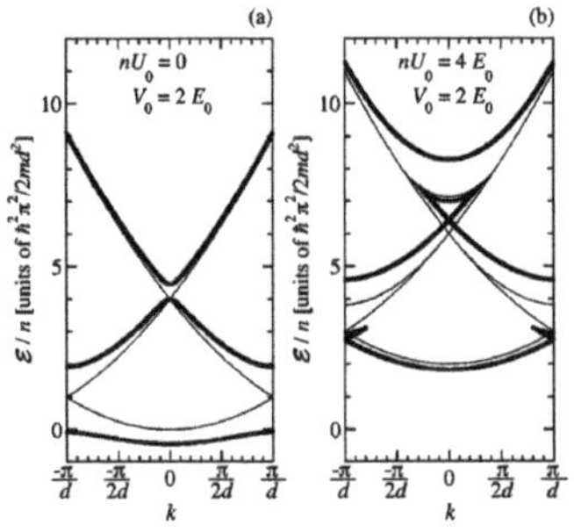

(**Bottom**) Zone-center & zone-edge loops.

Figure 2. (**Top**) Swallowtail loop structure of the energy per particle as function of k computed from the variational ansatz (24) for $V_0 = 3\hbar(4V_0/m)^{1/2}\pi/d$ and $n = 1.2V_0/U_0$. Figure was taken from [19]. (**Bottom**) Energy per particle obtained using the variational ansatz (23) demonstrating the presence of swallowtail loops at both the zone edge and zone center for sufficiently strong nonlinearity. U_0 in the figure is g in our notation. Moreover, the loops persist and extend over the entire band in the limit $V_0 \to 0$. Taken from [32].

Finally, the swallowtail loop structures discussed so far may be intuitively viewed as generic features that arise in hysteretic systems as clarified in the work of [35]. Moreover the specific case of loops in the energy band structure of a BEC in an optical lattice can also be understood as a manifestation of superfluidity and extended to other analogous systems such as superfluids in annular rings that have been realized experimentally [39]. The key insight of [35] can be explained from Figure 3. In Figure 3a, the approximately sinusoidal energy dispersion (lowest band) of a quantum particle in a periodic potential is shown. The corresponding Bloch eigenstates at different quasimomentum $\hbar k$ have periodic probability distributions commensurate with applied lattice potential. When a uniform external force is applied, the particle adiabatically follows this energy dispersion and performs periodic Bloch oscillations. In contrast for a BEC with interactions obeying the GP equation, *i.e.*, a superfluid in a periodic potential, the interaction term tends to prefer uniform density distributions. Hence for strong enough interactions, as shown in Figure 3b, the adiabatic band structure tends to be more similar to the quadratic free particle dispersion. This behavior has been referred to as the screening of the lattice potential by the superfluid in [35]. As a result, the velocity of the superfluid does not go to zero at the zone edge leading to non-zero current as expressed by Equation (25). As shown in Figure 3b, the velocity cannot increase indefinitely and the dispersion terminates (gray circle in Figure 3b) when the velocity becomes comparable to the Landau critical velocity of the superfluid giving rise to the swallowtail structure. Clearly adiabatic evolution along such a trajectory has to breakdown beyond this terminal point and forcing the system across this point from left to right and back will not restore the initial state—which is a clear sign of hysteresis. Also looking at the vicinity of the zone edge, it is clear that there are two possible minima for the energy which naturally leads to a saddle point separating the two minima given by the upper branch of the loop (the dotted lines in Figure 3b).

(a) (b)

Figure 3. (**a**) The dashed line schematically shows the energy dispersion for a free particle which is modified to the solid line showing the (lowest) band structure on the application of a periodic potential of period L. (**b**) Schematic plot of energy bands for a superfluid in an optical lattice that screens out the lattice to maintain non-zero flow velocity at the zone edge. Plots taken from [35].

The presence of multiple minima is a general characteristic of hysteretic systems. The number of minima for the superfluid in a periodic potential can be "controlled"

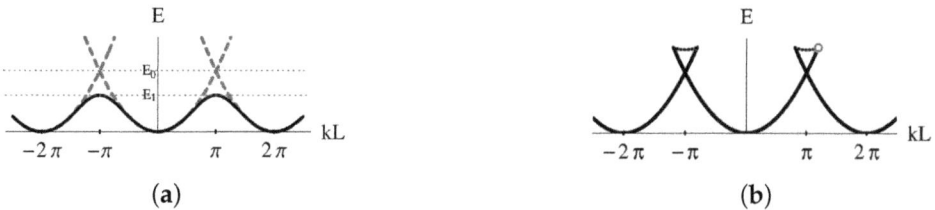

by varying the quasimomentum $\hbar k$ or the interaction strength g that sets the size of the swallowtail loop (remember that the swallowtail loops disappear if the g is below some critical value). In the purview of catastrophe theory [40], which is the study of singularities of gradient maps (any physical theory where a generating function is minimized to identify the stationary states of the system such as Fermat's principle in optics or the extremization of the GP energy functional here are examples of gradient maps), the swallowtail structures of energy bands represents a cusp catastrophe where two control parameters (g and k) may be varied to change the number of extrema of the energy by 2.

3.2. *Swallowtail Loops Structures for Bosons in Optical Lattices*

Having provided a physical picture of swallowtail loops in the previous subsection we proceed now to a comprehensive overview of the main results. We will divide this subsection into two parts. In the first part we provide an account of the different results obtained regarding the occurrence of swallowtail loop structures for BECs in optical lattices and in the second part focus on the energetic and dynamical stability of the solutions.

3.2.1. Occurrence of Loop Solutions

The phenomenon of swallowtail loops in energy band structures arising from the GP equation were first investigated by Wu and Niu in [34] for the nonlinear LZ introduced in detail in the previous subsection. In this work it was also pointed out that the nonlinear LZ model naturally arises near the zone edge in the dispersion for a BEC in a periodic optical lattice potential. The key results from this pioneering work are noted in Figure 1. Following this work, an exact solution for the GP equation in a periodic potential of the form $V(x) = -V_0 \mathrm{sn}^2(x,\kappa)$ with $\mathrm{sn}(x,\kappa)$ denoting the Jacobian elliptic sine function (with elliptic modulus $0 \le \kappa \le 1$) was discovered in [38]. These solutions for $\kappa = 0$ (Equation (2.1) of [18] or Equation (10) of [38] with elliptic modulus $k = 0$ in their notation) take the following form in our notation:

$$\psi_{\mathrm{exact}}(x) = \frac{\sqrt{c-v}+\sqrt{c+v}}{2\sqrt{c}} e^{i\pi x/d} + \frac{\sqrt{c-v}-\sqrt{c+v}}{2\sqrt{c}} e^{-i\pi x/d} \tag{27}$$

with $c = gn/(8E_0)$ and $v = V_0/(8E_0)$, which is of the Bloch wave form for the zone edge $k = k_L = \pi/d$. When $c > v$ (which is also the condition for loops to appear at the zone edge), Equation (27) gives the solution with finite current (see Equation (25)).

While on the one hand the exact solution lent more credibility [18] to the looped dispersions found from numerical solutions in [34], they also led to the suspicion that such solutions may not be a general feature. This was quickly dispelled by a simple variational calculation for the loops at zone edge by Diakonov *et al.* [19], followed by a more comprehensive analysis by Machholm *et al.* [32] demonstrating loops at the band center which we described in the previous subsection. Machholm *et al.* [32] also provided a thorough analysis of the width of the swallowtail loops, defined as the extent in quasimomentum space in an extended zone scheme, as the key parameters gn and V_0 are varied for both the zone-edge loops (Figure 4a) and zone-center loops (Figure 4b).

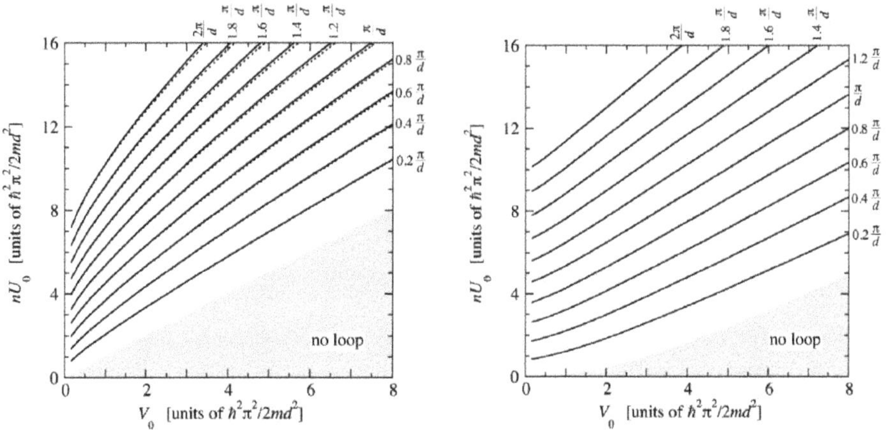

(**a**) Zone-edge loops. (**b**) Zone-center loops.

Figure 4. Contour plot of width of swallowtail loops from a numerical minimization using the variational wavefunction (11) as a function of interaction (nU_0 is ng in the notation of this review) and optical lattice depth V_0 for zone-edge loops (**a**) and zone-center loops (**b**). Dashed lines in (a) are for truncation with $l_{\text{max}} = 2$ and solid lines for $l_{\text{max}} = 3$. Taken from [32].

The results in Figure 4 were calculated by a numerical minimization of the GP energy functional (5) with the ansatz (11). An unexpected feature of the results in Figure 4 is that the width of the swallowtail loops remain non-zero even in the limit $V_0 \to 0$. In this limit the swallowtail solutions correspond to periodic soliton solutions of the GP equation. The zone edge solution with $k = \pi/d$ represents an equally spaced array of dark solitons (in one dimension, a dark soliton's wave function vanishes at some point in space whereas for a gray soliton there is a density dip of non-zero value at some point in space) with one soliton per lattice spacing d. The wave number of the solution $k = \pi/d$ can then be justified as the average phase change per unit length giving the right change in phase of π across the dark solitons in each period. When $V_0 = 0$, soliton array solutions with density dips located at $x = rd$ and $x = (r + 1/2)d$ (with r integer) are degenerate and correspond to the highest energy state in the first band and the one immediately above in the second band, respectively. The lattice potential breaks this degeneracy as the solution with soliton centers at potential maxima $x = rd$ have lower energy giving rise to the energy gap at the center of the swallowtail loop. The loops at $k = 0$ can also be understood with an analogous argument. The solutions with $k \neq 0$ correspond to arrays of gray solitons. The phase change across a gray soliton is less than π and they move with some finite velocity v_{soliton} in the absence of the potential. In order to create a stationary state from such solutions at finite V_0, one has to imagine boosting the condensate velocity by $-v_{\text{soliton}}$ giving a spatial dependent phase to the condensate wave function. The wave vector corresponding to such solutions now depends both on the density and the phase shift implied by the finite velocity boost. Moving away from $k = 0$ or π/d, the minimum density and v_{soliton} increases eventually going to zero and the sound speed $(gn/m)^{1/2}$ respectively leading to the loop branch merging with the free particle dispersion as shown in Figure 2 bottom. This manner of understanding the emergence of swallowtail loop structures from the soliton solutions provides a complementary physical picture of the phenomenon.

The work by Mueller in [35] provides a general way to understand swallowtail loops from the point of view of hysteresis and superfluid response. A corollary of such an approach is that it is possible to extend the phenomena of loops to systems beyond the standard system in this review (BEC in a periodic potential). Amongst the various examples discussed in [35], the case of a BEC in an annular trap is of particular interest in the context of the experiment [39]. The Hamiltonian in the rotating frame describing a bosonic superfluid (mass m) in a 1D ring of length L, rotating at a frequency Ω is

$$\frac{H}{\hbar^2/2mL^2} = \sum_j (2\pi j + \Phi)^2 c_j^\dagger c_j + \frac{\tilde{g}}{2} \sum_{j+k=l+m} c_j^\dagger c_k^\dagger c_l c_m + \lambda \sum_j \left(c_j^\dagger c_{j-1} + c_j^\dagger c_{j+1} \right). \tag{28}$$

In the above, c_j stands for the annihilation operator for bosons with angular momentum $j\hbar$ in a quantized picture or the amplitude of occupation of the same mode within a mean-field GP-like picture, $\Phi = 2mL^2\Omega/\hbar$ is the rotation speed in a dimensionless form, $\tilde{g} = 4\pi a_s L/d_\perp^2$ is the effective interaction for a trap perpendicular to the ring with harmonic oscillator length $d_\perp$, and λ is an impurity term that breaks the rotational symmetry coupling different angular momenta. This can arise naturally due to imperfections in the container or be generated, for instance, by applying a laser potential externally. The key point is that there is a one-to-one correspondence between the Hamiltonian (28) and the GP energy functional (5) after substituting the Bloch ansatz (11) with Φ playing the role of quasi-wave number k and λ playing the role of the periodic potential strength leading to swallowtail loops as shown in Figure 11 in [35]. We will revisit Equation (28) in Section 3.3.

Seaman *et al.* [41] and Dong and Wu [42] provided an interesting insight into the swallowtail loop structures for both repulsively ($g > 0$) and attractively ($g < 0$) interacting BECs for the special case of a Kronig–Penney periodic potential which is of the form,

$$V(x) = V_0 \sum_{j=-\infty}^{\infty} \delta(x - jd). \tag{29}$$

The Bloch states in such a potential can be solved analytically. For the repulsive interactions case, the results in [41] agreed qualitatively with the earlier numerical and approximate calculations but had unique features such as the fact that the critical interaction strength to have loops $gn > 2V_0$ for *all* bands of the energy spectrum unlike the sinusoidal band case discussed in [32]. In the case of attractive interactions with $g < 0$, they found the loop structures occurred in the upper branch at the band gaps, starting from the second band as shown in Figure 5. Moreover for the strongly attractive case $gn = -10E_0$ shown in Figure 5 the loop in the second band spans the entire Brillouin zone and splits from the original band.

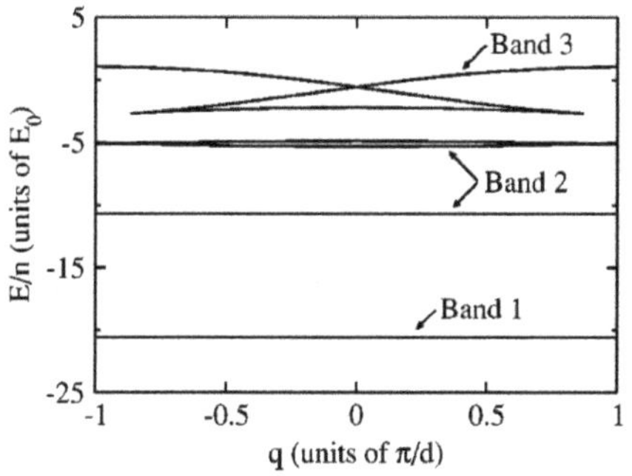

Figure 5. Energy bands with swallowtail loops in the excited bands for a Kronig–Penney potential (Equation (29)) with $V_0 = E_0$ and attractive interaction $gn = -10E_0$. Taken from [41].

3.2.2. Stability of Loop Solutions

In the discussion of swallowtail loops so far, we have completely ignored the stability properties of the solutions. In what follows, we discuss results regarding the energetic and dynamical stability of steady state solutions of the Bloch form (of which swallowtail loops are special cases) for BECs in periodic potentials. The stability of the solutions we find by extremizing the GP energy functionals (5) are crucial as they will provide us clues as to whether in an experiment the system can reach such equilibrium solutions and if they do how long can they be stable. As a part of the the theoretical framework Section 2.5, we provided the conditions for energetic stability but a physical picture of the two kinds of stability as exemplified in Figure 6a is helpful—for energetic stability the equilibrium solution has to be a local minimum of the energy functional (5) whereas for dynamical stability perturbations about the equilibrium state should not grow with time when evolved according to the time-dependent GP Equation (2).

The stability of Bloch states in the lowest band excluding loops was discussed by Wu and Niu in [31] and [43]. Figure 6b represents the results from a numerical calculation mapping out the stability of Bloch states with quasimomentum $\hbar k$ under perturbations of the Bloch wave form with quasimomentum $\hbar q$ for different values of the potential $v = V_0/(8E_0)$ and nonlinearity $c = gn/(8E_0)$. Let us denote the stability matrix, Equation (16), for a state with quasimomentum $\hbar k$ under a perturbation of quasimomentum $\hbar q$ as $M_k(q)$. For the special case of a free BEC with no lattice, *i.e.*, $v = 0$, the eigenvalues of the matrix can be computed analytically and the requirement for positivity of eigenvalues leads to the well-known Landau criterion given by

$$|k| \geq \sqrt{q^2/4 + c}. \tag{30}$$

The shaded light and dark regions in Figure 6b represent regions of energetic instability with $M_k(q)$ having negative eigenvalues. In the limit of small v the equality in the expression (30) accurately reproduces the energetic stability region shown by triangles in the plot. Another key feature to note in Figure 6b is that even at v comparable to c, as the nonlinearity c is increased, the BEC is energetically stable over an increasing area in the k-q space, which can be anticipated from expression (30).

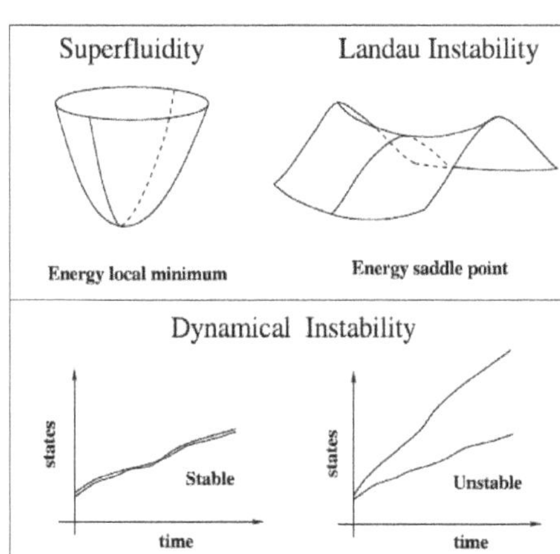

(**a**) Schematic.

Figure 6. *Cont.*

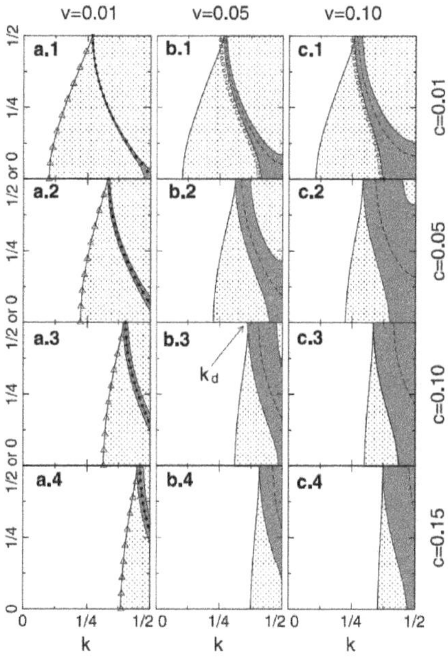

(**b**) Stability of Bloch solutions.

Figure 6. Schematic illustrations for the mid-point bounce-back scheme.

In Figure 6b, the system is dynamically unstable in the dark shaded regions. The first thing to notice is that energetic instability is a pre-requisite for dynamical instability and this can also be shown in general (see appendix of [43]). Further insight into dynamical stability can be obtained by considering the structure of the eigenvalues of the matrix $\sigma_z M_k(q)$ since the requirement for dynamical stability is to have real eigenvalues for this matrix. The states represented by the eigenvectors of $\sigma_z M_k(q)$ can also be viewed as quasiparticle excitations in the BEC, *i.e.*, phonons [43], with the positive eigenvalues giving the phonon spectrum. In general the eigenvalues of $\sigma_z M_k(q)$ can be complex but always occur in complex conjugate pairs [32,43] owing to the real nature of the matrix in momentum representation. In the case of $v = 0$, the eigenvalues of $\sigma_z M_k(q)$ are always real and given by $\epsilon_{\pm}(q) = kq \pm \sqrt{q^2 c + q^4/4}$ where k and q are measured in units of $4\pi/d$. This implies that in free space BEC, flows are always dynamically stable. However, Figure 6b shows that situations change significantly when the lattice potential is introduced as also evidenced in experiments [44–46]. In general for all parameter regimes there is a critical wave number k_d beyond which Bloch waves are dynamically unstable. At $k = k_d$, dynamical stability always sets in for $q = \pi/d$, which interestingly also corresponds to a period doubling revealing a link further explored in Section 4. At the point where dynamical instability sets in, the eigenvalues of $\sigma_z M_k(q)$ change character from real to pairs of complex conjugates, *i.e.*, with equal real parts. Hence as explained in [31,32,43], dynamical instability can be viewed as arising from a lattice induced resonance between a pair of excitations that are degenerate in the $v = 0$ limit. Thus when the instability sets in, two phonons with the sum of their momenta given by the primitive reciprocal lattice vectors $\pm G = \pm 2\pi/d$ are created from zero energy, *i.e.*, they satisfy

$$\epsilon_+(q) + \epsilon_+(2\pi/d - q) = 0\,. \tag{31}$$

This can be used to clearly justify the observation (valid at small lattice depths V_0) that instability at the critical wave number k_d always sets in with $q = G/2 = \pi/d$ and the critical vector satisfies $|k_d| = (\pi/d)(gn/2E_0 + 1/4)^{1/2}$ agreeing with the numerical results in Figure 6b.

In the stability analysis of [32], in addition to standard Bloch states, also the ones corresponding to the loop solutions (lower branch) were considered. In Figure 7 the results of this analysis is shown by plotting the largest Bloch wave vector k for which the states are energetically stable as a function of ng and V_0. States with $0 \leq k \leq \pi/d$ correspond to states with the lowest energy and $k > \pi/d$ represents lower edge-loop states. In general it was found that the range of quasimomentum values at which the system is energetically stable increases with the interaction strength gn. They also found that the wave vector of the tip of the swallowtail sets a natural limit for the wave vector at which instability sets in for parameter regimes where swallowtail loops occur. Moreover they found, in agreement with [31], with increasing k the long wavelength perturbations with $q \to 0$ become energetically unstable first. Hence a hydrodynamic description can be constructed [47–49] and it also gives analytical predictions for the instability contour as a function of V_0 and gn for the zone edge with $k = \pi/d$ which compares favorably with the numerical calculations [49]. Note that as expected the states on the upper edge of the loop are always energetically and dynamically unstable.

Finally, the discussion so far was limited to only the linear stability of equilibrium solutions but the full response of the solutions of time dependent GP Equation (2) to perturbations was also studied numerically in [41]. Here the stable lifetime of a given initial equilibrium state was defined as the time taken for the variance of the Fourier spectrum of the time-dependent order parameter from the initial Fourier spectrum (normalized to the initial spectrum) to exceed the value $1/2$, *i.e.*, the time at which the order parameter becomes very different from the initial solution and taken as an indicative time for onset of dynamical instability. They found that for weakly attractive condensates the zero quasimomentum Bloch state in the first band is long-lived under white noise perturbations but highly unstable for time periodic perturbations. For both weak and strong attractive interactions the higher band Bloch states are unstable but the first band Bloch states with non-zero quasimomentum can be stable even to harmonic perturbations owing to their negative effective mass (defined as the inverse of the Band curvature). For weak repulsive interactions, Bloch states in the lowest band are stable as long as the quasi-wave number $k < \pi/d$, as at larger quasi-wave numbers the effective mass becomes negative, for the particular choice of Kronig–Penney potential (29) used in [41]. However, in agreement with the linear stability analysis, as the interaction strength grows, larger parts of the energy band including the lower branch of the loops are stable. In a later publication the same authors [50,51] showed that there is always a small part of the loop in the repulsive case that has negative effective mass but this area's extent monotonically decreases as gn is increased. A clear discussion of the dynamical stability of attractive BECs in an optical lattice is provided in [52]. A recent review [53] also provides detailed treatment of stability of BECs in optical lattices.

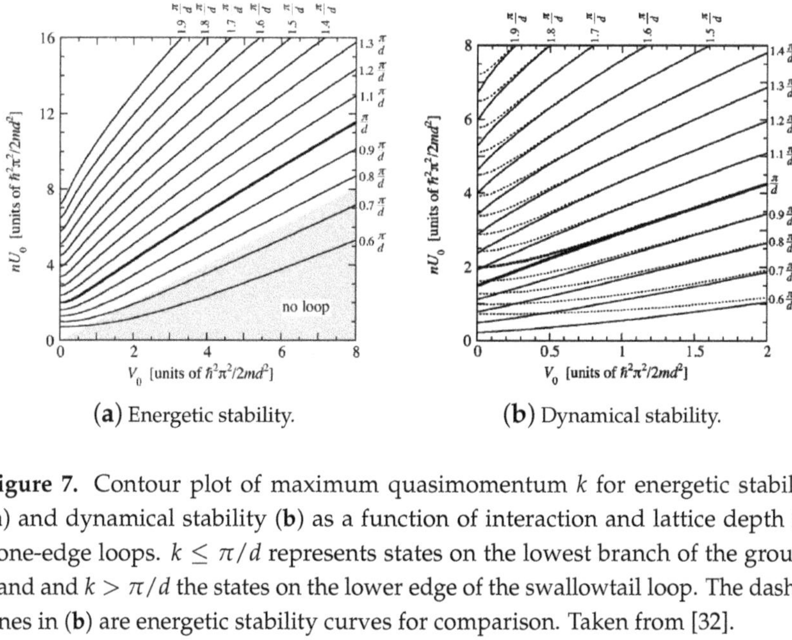

(**a**) Energetic stability. (**b**) Dynamical stability.

Figure 7. Contour plot of maximum quasimomentum k for energetic stability (**a**) and dynamical stability (**b**) as a function of interaction and lattice depth for zone-edge loops. $k \leq \pi/d$ represents states on the lowest branch of the ground band and $k > \pi/d$ the states on the lower edge of the swallowtail loop. The dashed lines in (**b**) are energetic stability curves for comparison. Taken from [32].

3.3. Experimental Realization

In the past decade and half there has been tremendous progress in the field of ultracold atoms in optical lattices [14,15] with a range of experiments tackling interesting many-body physics. In this light it must be said that specific experiments dealing with nonlinear energy dispersions in optical lattices have been few and far in between.

The earliest relevant experiment concerns the instability of superfluids in optical lattices by the group at LENS, Florence reported in Burger *et al.* [44–46]. In this experiment a cigar shaped quasi-1D BEC was localized in an harmonic trap and an optical lattice potential was also turned on. Following this, the center of the trap was suddenly shifted, corresponding to a sudden change of the quasimomentum in our language. For small shifts Δx the dynamics of the BEC was coherent but for shifts greater than a critical value $\Delta x > \Delta x_c$, the oscillations are disrupted and the dynamics became dissipative. The authors of the experiment attributed this to the energetic Landau instability of the condensate. But theoretical work from Wu and Niu [43,45,46] showed that it may be more appropriate to associate this behavior with dynamical instability especially considering that the experimental parameters fall in a regime ($c \sim 0.02$, $v \sim 0.2$) where dynamical instability is rampant (see Figure 6b) and the critical displacement Δx_c increases with decrease of lattice depth in the experiment (the energetic stability is essentially independent of lattice depth in the linearized stability treatment). A follow-up comment from the authors of the experiment [45,46] was essentially inconclusive but hinted that even the GP equation may not be a valid description for some of their experimental results and beyond mean-field effects may have to be included. Following this, in [54] a thorough theoretical treatment of the problem including 3-dimensional GP equations and effective 1-dimensional GP equations taking into account the transverse degrees of freedom was undertaken. This theoretical treatment adapted to the experiment [44] clearly showed that the onset of instability observed in the experiment was due to dynamical instability. There are two main reasons that led to some uncertainty regarding the conclusions in the experiments [44]—the inability to distinguish experimentally between dynamical and energetic stability as both processes manifest as an enhanced loss of atom number from the condensate and the inability to accurately set the initial quasimomentum of the condensate as a mean displacement from the harmonic trap center can in general lead to a mixture of quasimomenta and band eigenstates.

In the follow-up experiment from the LENS group [55,56] both these issues were succesfully resolved. In [55], a moving optical lattice was implemented by frequency detuning the two laser beams creating the lattice. The ground state in a moving lattice is simply a state with a finite quasimomentum. By varying the amount of detuning, they were able to load the BEC adiabatically into states with a given quasimomentum both in the ground and excited bands. A subsequent measurement of the loss rates as a function of quasimomentum revealed a threshold for the onset of dynamical instability in very good agreement with the theoretical expectation for the lattice depth used. In [56] they were able to also distinguish between the onset of energetic and dynamical instability by an ingenious use of a radio-frequency (RF) shield to selectively control the thermal fraction of the atomic cloud. The presence of the thermal cloud can effectively trigger energetic instability, which has generally a lower threshold in quasimomentum, by providing a dissipation channel. Hence when a large thermal fraction is present, the onset of dynamical instability is marred by energetic instability. On the other hand when the RF shield is turned on to remove the thermal fraction and experiments are performed with nearly pure condensate, the onset of dynamical instability stands out via a dramatic loss of atom number. It is important to emphasize at this point that experiments such as [44] are performed at small lattice depths v. In this regime, as evident from Figure 6b, there are significant regions of quasimomentum space that are energetically unstable but dynamically stable. On the other hand at large lattice depths (see Figure 1 of [56]), the region of quasimomentum space that exhibits only energetic instability is tiny. Thus in this so called tight-binding lattice limit, there was unambiguous agreement between experiments [57] and the theoretical [43,58] and numerical [59,60] treatments that predicted dynamical instability. It was also shown in [58] that the dynamical instability in this regime can also be viewed as a kind of modulational instability which is a general feature of nonlinear wave equations where a small perturbations of a carrier wave can exponentially grow as a result of interplay of dispersion and nonlinearity.

In [61], the behaviour of the critical quasimomentum upto which superfluidity persisits across the superfluid-Mott insulator (SF-MI) quantum phase transition was studied. Within a mean-field GP picuture, as we have discussed in Section 3.2.2, the stability of the superfluid is in general enhanced with increasing interaction and lattice depth. On the other hand, within the full quantum model, there is a critical interaction strength to tunneling ratio beyond which the system is no longer superfluid and goes into the Mott insulating phase where the critical quasimomentum is trivially equal to 0. This study maps the behaviour of the critical quasimomentum as it goes from a finite value in the SF phase to zero in the MI phase giving an accurate determination of the phase boundary. In [62] beyond dynamical instability at the zone edge is investigated experimentally and theoretically within the truncated Wigner approach which can account for beyond GP effects including the thermal depletion of the condensate. Finally, in recent experiments [63] dynamical instability of spin-orbit coupled (SOC) BECs in moving optical lattices was investigated and a manifestation of the breakdown of Galilean invariance predicted for SOC systems was evidenced by the difference in the strength of the dynamical instability (measured by atom loss rate) depending upon the direction of motion of the lattice.

In the seminal experiment of the Bloch group [64], some aspects of the nonlinear LZ tunneling phenomena originally considered in the theoretical work of Wu and Niu [34] was explored. The experimental system consisted of an array of tubes of BECs in a 2D optical lattice potential. A superlattice potential along x direction allows pair-wise coupling between tubes giving many copies of coupled double wells. In the experiment they effectively realize the nonlinear LZ energy function of the the form:

$$E[\psi_R, \psi_L] \approx \frac{\Delta}{2}\left(|\psi_R|^2 - |\psi_L|^2\right) - J(\psi_R^*\psi_L + \text{c.c.}) + \frac{U}{2}\langle\delta\hat{n}^2\rangle\left(|\psi_R|^4 + |\psi_L|^4\right), \quad (32)$$

where ψ_L and ψ_R represent the amplitude to occupy either the left or the right tube. Δ, the energy detuning between the tubes, and J, the tunnel coupling between the tubes can be controlled by varying the relative phase and lattice depth respectively of the superlattice potential along x direction.

The experimental protocol consists of preparing all the atoms initially in the left tube with the initial detuning Δ_i either chosen to be negative (ground state) or positive (excited state) and sweeping the detuning at the linear rate α and finally measuring the number of atoms in the left and right tube. As already anticipated by Wu and Niu [34], the model (Equation (32)) is exactly the same as the one introduced in Equation (19) except for the interaction term's sign is switched. As a result once the ratio of interaction to tunneling $\eta = U\langle\delta\hat{n}^2\rangle/J$ (c/v in [34]) is large, there is a loop in the upper branch as shown in lower panel of Figure 8 left. In the experiment, for sweeps along the ground state branch (gray dots in Figure 8 left) there is no adiabaticity breakdown observed. In the sweep starting with the excited state (left tube at higher energy), there is a complete breakdown of adiabaticity even for reasonably small sweep rates α (red dots in Figure 8 left). Moreover, for small sweep rates the transfer efficiency, given by number of atoms n_R in the final state, decreases with decreasing sweep rate completely opposing the expected LZ behavior. The presence of the loop in the upper branch contributes to this adiabaticity breakdown for small sweep rates, as the atoms follow the upper branch and are *self-trapped* in the middle branch of the loop, (see lower panel of Figure 8 left) which is a local maxima. They finally make a diabatic transition to the adiabatic ground state at the loop edge leading to the sharp breakdown of transfer efficiency near zero detuning seen in Figure 8 left upper panel. A confirmation of the effect of loops on the adiabaticity breakdown is provided by the non-monotonic behavior of the transfer efficiency (number of atoms in the initially empty right tube at the end) at a given sweep rate and tunnel coupling as a function of the z-lattice depth shown in top panel of Figure 8 right. At small z-lattice depths, an increase in lattice depth effectively increases the on-site interaction $U\langle\delta\hat{n}^2\rangle$, leading to larger loop sizes which leads to lower transfer efficiency but eventually beyond a certain lattice depth the suppression of on-site fluctuations $\langle\delta\hat{n}^2\rangle$ dominates leading to a decrease in the effective interaction and increase of transfer efficiency restoring standard LZ behavior. Moreover as shown in the lower panel in Figure 8 right, the experimentally determined minimum transfer efficiency agrees with a theoretical calculation for the position of the maximum loops size as a function of the z-lattice depth and tunnel coupling.

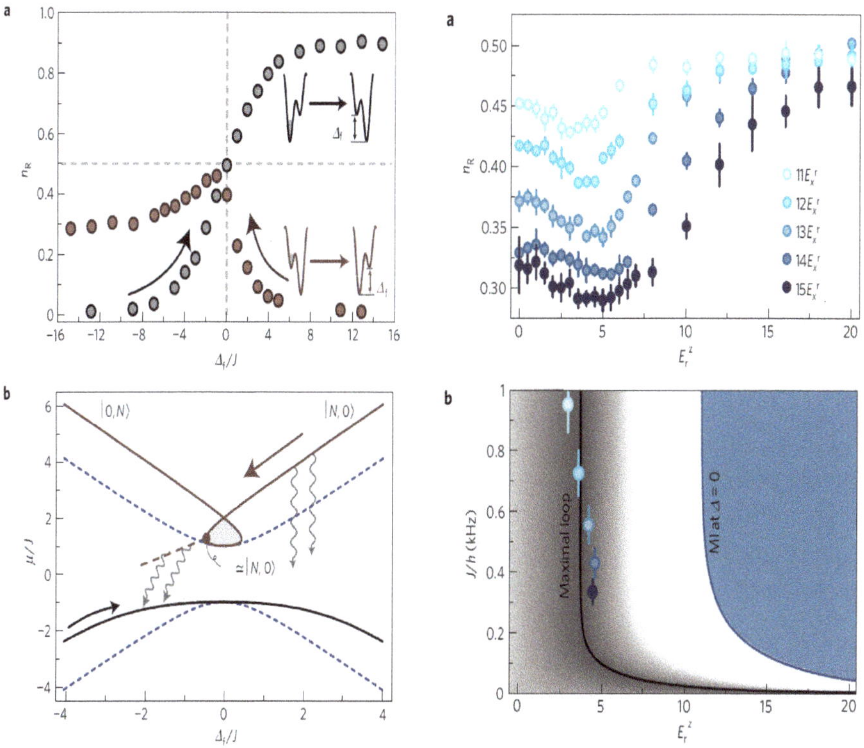

(Left) **(Right)**

Figure 8. **(Left)** (**a**) Transfer efficiency n_R from filled to unfilled well (tube) as a function of the energy detuning for ground state sweep (gray dots) and excited state sweep (red dots) for constant sweep rate of $2\pi J^2/\hbar|\alpha| = 2.1(1)$; (**b**) Adiabatic energy levels for the excited (with loops) and ground state levels including strong repulsive interaction (solid lines) and excluding interactions (dashed lines). The curved lines indicate the relaxation process enabling non-adiabatic transitions. $|n_R, n_L\rangle$ is a shorthand denoting the left- and right-well (tube) population respectively. **(Right)** (**a**) Transfer efficiency as a function of z-lattice depth at constant sweep rate 0.53(3); (**b**) Phase diagram of the metastable excited condensate branch. In the gray shaded area there is a loop in the adiabatic energy level. Data points represent minima in the measured transfer efficiency and agree well with the solid line depicted for the calculated maximum loop size. Taken from [64].

In the experiment by Eckel *et al.* [39], a physical situation approximately corresponding to the Hamiltonian in Equation (28) was realized for a BEC of ^{23}Na atoms confined in a ring shaped trap. The goal of this experiment was to observe hysteresis between quantized states of circulation of the superfluid BEC caused by the presence of swallowtail loops in the energy landscape of such a system [35,65].

In the experiment [39], they were concerned with the quantized circulation states with winding numbers $n = 0$ and $n = 1$ with frequency n times the rotational quanta $\Omega_0 = \hbar/mR^2$ and drive transitions between these states by tuning the relative angular velocity between the trap and the superfluid Ω which can be controlled by applying a repulsive rotating laser potential.

In the absence of coupling between the different circulation states, the energy landscape of the interacting superfluid forms a swallowtail loop shape as a function of the relative angular velocity Ω. At a fixed value of Ω in the swallowtail region, $n = 0$ (red line in Figure 9 left) and $n = 1$ (blue line in Figure 9 left) states form the minima of a double-well energy landscape with a barrier state (green dashed line) separating them. If the system begins in the $n = 0$ state and its angular velocity is increased, the flow will be stable as long as $\Omega < \Omega_{c+}$ when it reaches the edge of the swallowtail and after this it will make a transition to the lower energy $n = 1$ state. Beginning with the $n = 1$ state, a similar stable flow can exist as long as $\Omega > \Omega_{c-}$, leading to the hysteresis loop shown in lower panel of Figure 9 left. The rate at which the repulsive potential created using a blue detuned laser is rotated, controls the flow velocity Ω and the strength of the potential U controls and drives transitions (via phase slips) [66] between different circulation states. Comparing to the Hamiltonian (28) the repulsive potential actuates two of the terms namely the rotation frequency $\Phi = \Omega/\Omega_0$ and the "impurity" term λ whose strength is controlled by U.

In order to observe this hysteresis loop in the experiment, the BEC is prepared initially in either the $n = 0$ or $n = 1$ state in the trap and then this repulsive trap potential with a chosen strength U_2 and different rotation velocity Ω_2 is applied for a fixed time of 2 s, followed by a time of flight image to determine the final rotational state n. Due to the swallowtail loop structure, at a given strength U_2, as shown in upper panel of Figure 9 right, the transition from 0 to 1 and *vice-versa* clearly happen at different angular momenta $\Omega_{c\pm}$. Moreover as the strength of the applied potential U_2 is increased, $\Omega_{c\pm} \to \Omega_0/2$ and the swallowtail loop size decreases and eventually vanishes. The size of the loop as a function of U_2 was determined from the experiment and is plotted in the bottom panel of Figure 9 right. The discrepancy from theoretical calculations that include relaxation effects required to accomplish the non-adiabatic transitions indicate that further work may be required to understand some quantitative aspects of the experiment. A detailed treatment of modeling the relevant excitations leading to the dissipation by vortex-antivortex pairs was already provided by the authors in [39].

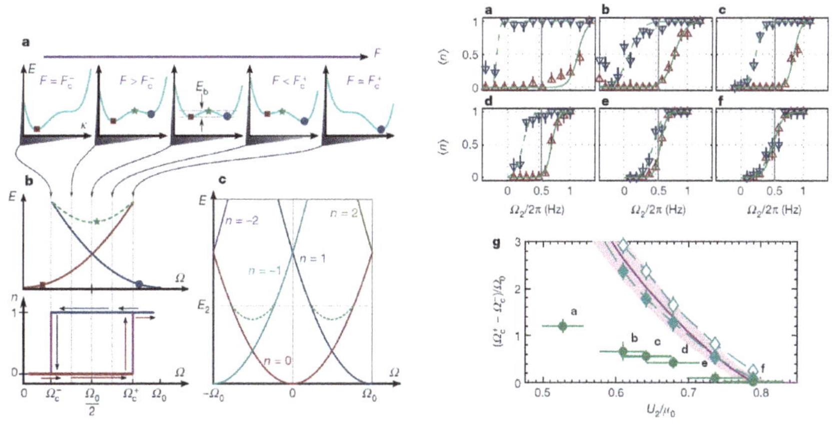

(Left) **(Right)**

Figure 9. (**Left**) (**a**) Schematic plot of the energy landscape of a hysteretic system as a function of the applied field F (for example Ω the rotation rate of the superfluid) showing stable states of different energies separated by a barrier; (**b**) Energy diagram for a superfluid as a function of Ω showing a swallowtail loop and the related hysteresis loop; (**c**) The swallowtail structure is periodic in the rotation quanta Ω_0. (**Right**) (**a–f**) Measured hysteresis loop with sigmoid fits at different values of potential strength U_2 shown in (**g**). The red up-triangles (blue down-triangles) are results from 20 shot averages while starting from $n = 0$ ($n = 1$). (**g**) Experimentally determined size of the hysteresis loop (green dots) as a function of U_2 and open and filled cyan diamonds are results of numerical GP calculation of dynamics with differing amounts of phenomenological dissipation. Taken from [39].

3.4. Other Extensions

Owing to the general nature of swallowtail shaped dispersions resulting from the interplay of atom-atom interactions and periodicity, they have been predicted to occur in a variety of systems different from the setting mainly considered in this review—namely that of BECs in an optical lattice with effectively 1D dynamics. In this subsection we catalogue these developments without going into the details owing to the restricted scope of the review.

Lin *et al.* [67] study BECs with magnetic dipole-dipole interactions in optical lattices. In this case the effective atom-atom interaction can be controlled by changing the alignment of the atomic dipoles to the optical lattice axis by applying magnetic fields. At strong enough interaction, they observe swallowtail loops whose sizes and stability may be controlled by modifying the magnetic dipole orientations. In [68], Venkatesh *et al.* study band-structures of atoms confined in optical lattices formed inside optical cavities that are continuously driven by an external laser. In the limit of very dilute gases, s-wave contact interactions do not play a role but the cavity-induced atomic interactions in the strong coupling regime of cavity quantum electrodynamics (QED) can lead to swallowtail loop structures in the atomic bands. Moreover this also corresponds to bistable solutions for the steady state photon number in the cavity. In the work of Watanabe and colleagues [69], swallowtail band structures of superfluid Fermi gases in optical lattices in the BEC-BCS crossover are investigated. They find that typically the width of the swallowtail is largest at unitarity. In addition, they find that the microscopic mechanism of the emergence of the swallowtail in the BCS side of the crossover is very different in nature from that of the BEC case: a narrow band in the quasiparticle energy spectrum close to the chemical potential plays a crucial role for the appearance of the swallowtail in the BCS side. It is also pointed out that, as a consequence, the incompressibility experiences a profound dip. Chen and Wu extend the study of interplay between interactions and band structures of superfluid systems in optical lattices to two dimensions in [70] considering BECs in honeycomb shaped optical lattices. Such optical lattices serve as analogues to the structure of graphene and support Dirac points in their band close to which the energy dispersion is linear in 2D having characteristic conical shape. In [70] the authors show that even for arbitrarily small interaction strength, the Dirac point is extended into a closed curve and a tube like structure, a 2D version of the 1D swallowtail loop, arises around the original Dirac point. Moreover in work that followed closely thereafter Hui *et al.* showed that even in the case of 2D optical lattice with double-well superlattice like geometry along one direction, swallowtail loop structures emerge for any interaction strength [71]. Thus, the possibility of having swallowtail loops structures or analogues thereof for arbitrary small interactions seems to be a more ubiquitous feature in 2D as opposed to 1D where a the nonlinearity given by interactions has to be comparable to the lattice potential. In [72], a BEC trapped in a double-well potential with an additional degree of freedom given by a single bosonic impurity atom that interacts with the condensate is considered. In this setup, as the impurity-BEC interaction strength is tuned above tunneling energy of the bosons, swallowtail loops appear in the adiabatic energy dispersions as a function of the tilt of one of the wells relative to the other. The relation between swallowtail loops and self-trapping of the condensate in one well or the other as well as relation to the Dicke model are explored.

3.5. Future Prospects

Presently, experiments showing direct evidence for looped band structures for BECs in optical lattices have not been performed. One of the impediments preventing the experimental observation of loops in optical lattices is the requirement of large atom-atom interactions so that a large part of the loop solution is energetically stable [41,50,51]. The simplest way to ascertain breakdown of adiabaticity caused by the loop is to study Bloch oscillations in optical lattices. In this regard it is quite important to control and characterize other sources of loss of adiabaticity such as LZ tunneling to higher bands and distinguish them from the effect of the loops. Clearly for this a control of atom-atom interaction from very small to large enough to obtain loops is required. In this context, some recent experiments in the group of Nägerl with the ability to tune interactions using Feshbach resonances is promising [73,74]. Also, the extended theoretical schemes for 2D optical lattices [70,71] may be easier to implement in an experiment as they do not require large atom-atom interactions to have looped energy dispersions. On the theoretical side, a clear understanding of the quantum mechanical underpinnings of the mean-field loops is already available for the case of repulsive and attractive interacting BECs in double-well potentials [64,75]. An extension of such a study to optical lattices in two dimensions or for fermionic atoms can be interesting. Another interesting theoretical consideration would be examine the idea of shortcuts to adiabaticity [76] that has received a lot of attention of late to systems where the underlying evolution equation is not linear and understand if one may conceive of protocols where the loss of adiabaticity predicted due to the loops could be avoided.

4. Multiple Period States in Cold Atomic Gases in Optical Lattices

Density structures and patterns caused by the interplay of competing effects are ubiquitous in nature. In the case of superfluids flowing in a periodic potential, non-trivial density patterns can emerge due to the interplay of spatial periodicity imposed by the external potential and the nonlinearity due to the superfluid order parameter.

According to the conventional wisdom of the Bloch theorem, in the linear system described by the Schrödinger equation, the density pattern of the stationary solution in a lattice is periodic with periodicity coinciding with that of the lattice potential. However, nonlinearity can cause non-trivial density patterns with different periodicity. For BECs in a periodic potential, it has been found that nonlinearity of the interaction term can cause the appearance of stationary states whose period does not coincide with that of the lattice; instead, a multiple of it [1,33]. Such states are called multiple (or n-tuple) period states.

In this section, we discuss multiple period states of superfluid atomic gases in optical lattices. In the Section 4.1, we provide the basic physical idea of the emergence of the multiple period states due to nonlinearity. To provide a physical picture concisely, we take the BEC case as an simple example. In the Section 4.2, we present an account of existing results of the multiple period states in BECs. In the Section 4.3, we present some theoretical results of the multiple period states in superfluid Fermi gases along the BCS-BEC crossover focusing on their unique features in contrast to the multiple period states in BECs.

4.1. Basic Physical Idea: A Simple Explanation of the Emergence of Multiple Period States by a Discrete Model

The emergence of the multiple period states in BECs can be explained by the discrete model (Equation (12)) in a simple manner [33]. In the following exposition, we follow the discussion given in the above cited paper. For clarity, here we focus on the period-2 states: states whose period is equal to twice of the lattice constant d.

The stationary states with a fixed total number of particles N can be obtained by the variation of $H' \equiv H - \mu N$ with respect to the amplitude ψ_j^* at site j, where μ is the chemical potential and $N = \sum_j |\psi_j|^2$,

$$\frac{\delta H'}{\delta \psi_j^*} = -K(\psi_{j+1} + \psi_{j-1}) + U|\psi_j|^2\psi_j - \mu\psi_j = 0 \,. \quad (33)$$

We then separate from ψ_j a plane wave part at site j, e^{ikjd}, as $\psi_j = e^{ikjd} g_j$ with $\hbar k$ being quasimomentum of the bulk superflow flowing in the same direction of the periodic potential and g_j being the complex amplitude at site j. Equation (33) becomes

$$-K(g_{j+1}e^{ikd} + g_{j-1}e^{-ikd}) + U|g_j|^2 g_j - \mu g_j = 0 \,. \quad (34)$$

Due to the boundary conditions of the period-2 states, we have $g_0 = g_2$ and $g_1 = g_3$. We solve combined two Equations (34) for $j = 1$ and 2 with these boundary conditions. Subtracting these two equations, we obtain

$$-2K \cos kd \left(\frac{|g_2|}{|g_1|} e^{i(\phi_2 - \phi_1)} - \frac{|g_1|}{|g_2|} e^{i(\phi_1 - \phi_2)} \right) + U(|g_1|^2 - |g_2|^2) = 0 \,, \quad (35)$$

with $g_j \equiv |g_j| e^{\phi_j}$.

For the linear case ($U = 0$), we can readily see that $|g_1| \neq |g_2|$ cannot satisfy Equation (35) except at $kd = \pi/2$, which corresponds to a trivial solution of $g_1 = g_2 = 0$. On the other hand, solutions with $|g_1| = |g_2|$ exist provided $\phi_2 - \phi_1 = 0$ (modulus of 2π): thus these solutions are normal period-1 states.

For the nonlinear case ($U \neq 0$), nonzero contribution from the kinetic energy part (the first term in the left-hand side of Equation (35)) can be compensated by that from the interaction energy part (the second term in the left-hand side of Equation (35)) so that this equation can be satisfied. Therefore, the emergence of the period-2 states is a purely nonlinear phenomenon. Since the second term in the left-hand side is real, the phase difference $\phi_2 - \phi_1$ should be 0 or π, namely:

$$\pm 2K \cos kd \left(\frac{|g_2|}{|g_1|} - \frac{|g_1|}{|g_2|} \right) = U(|g_1|^2 - |g_2|^2) \,. \quad (36)$$

Thus we obtain

$$\pm \frac{\cos kd}{\frac{U\nu}{2K}} = \frac{|g_1||g_2|}{\nu}, \tag{37}$$

where the filling factor $\nu \equiv (|g_1|^2 + |g_2|^2)/2$ is the average number of particles per cell (in the present case of the period-2 states, the cell consists of two lattice sites). Since the right-hand side of Equation (37) takes $0 \le |g_1||g_2|/\nu \le 1$, solutions with period $2d$ exist when $|\cos kd| \le U\nu/2K$ [33].

In Figure 10, we show the energy bands of the period-1 and period-2 states for $U\nu/2K = 1/2, 1$, and 2. Note that, in the case of the Figure 10a for $U\nu/2K = 1/2$, the period-2 states exist in the limited region of $1/6 \le kd/2\pi \le 1/3$. At infinitesimally small $U\nu/2K$, the period-2 states exist only at $kd/2\pi = 1/4$. As $U\nu/2K$ increases, the region in which the period-2 states exist increases and it finally extends over the whole Brillouin zone for $U\nu/2K \ge 1$ (see Figure 10b,c).

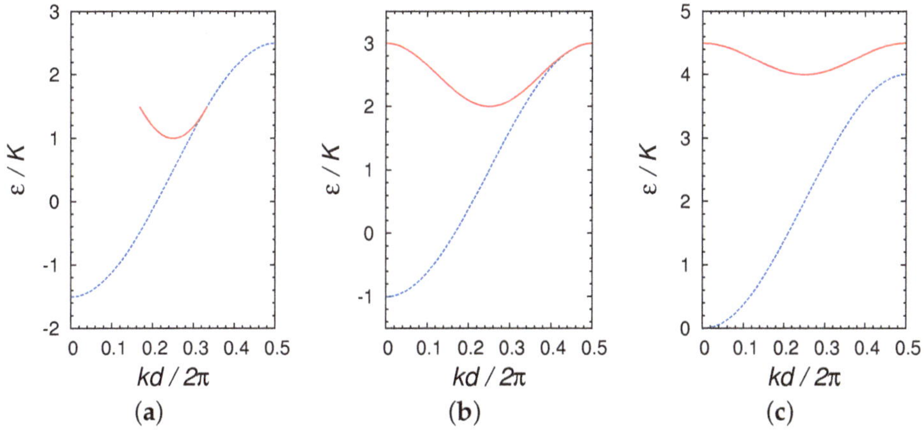

Figure 10. Energy per particle ϵ as a function of k for period-1 (blue dashed lines) and period-2 (red solid lines) states of BECs in a periodic potential obtained from the discrete model, for (**a**) $U\nu/2K = 1/2$; (**b**) $U\nu/2K = 1$; (**c**) $U\nu/2K = 2$. In the case of the (**a**), period-2 states exist in the limited region of $1/6 \le kd/2\pi \le 1/3$.

Note that there is another class of period-2 states called the phase states [33]. From Equation (35), we see that, at $|k|d = \pi/2$, $|g_1| = |g_2|$ is a solution for arbitrary phase difference $\phi_2 - \phi_1$. The periodicity of the density distribution and the energy of the phase states are the same as the normal Bloch state at $|k|d = \pi/2$, but only the phase profile has the period $2d$ [32].

4.2. Multiple Period States in BECs

Multiple period states in BECs in optical lattices were first predicted by Machholm *et al.* [33]. Using both (1) the simple discrete model within the tight-binding approximation to the mean-field GP equation and (2) the more general continuum GP equation, they studied BECs flowing along the 1-dimensional external periodic potential of the form given by Equation (1) (multiple period states of BECs in a Kronig–Penney potential (a periodic delta-function potential) were studied in [77]). They have shown the existence of the multiple period states as stationary states and have clarified that they emerge due to nonlinearity originating from superfluidity.

Figure 11 shows the lowest energy bands obtained by solving the GP equation for the continuum model. A striking difference from the energy bands obtained from the discrete model discussed in Section 4.1 is that the phase states form a band in the continuum model (see the lower thick solid lines in Figure 11) and their density profiles have period $2d$.

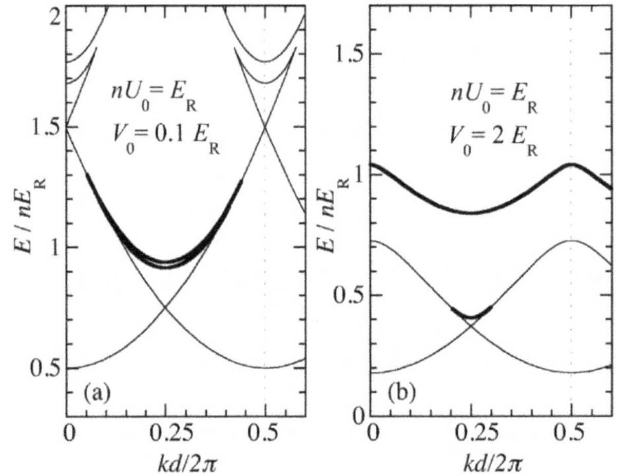

Figure 11. Energy per particle E/n of BECs in a periodic potential as a function of k for the lowest bands obtained from the GP equation for the continuum model. The bands of the period-2 states are shown by the thick solid lines. In the notations of the present article, $U_0 = g$ and $V_0 = sE_R/2$. This figure is taken from [33].

Figure 12 shows the density profiles of the lower-energy and higher-energy period-2 states. By comparing with the external potential $V(x)$ shown in the lower panel, we can see that the periodicity of these states is indeed $2d$. Here, the quasi-wave number k of the superflow is at $k = \pi/2d$ corresponding to the first Brillouin zone edge of the system with period $2d$, and thus the condensate wave function ψ has a node in each period. The density profile shown by the solid (dashed) line, which has nodes at the potential maxima (minima), is that of the period-2 state in the lower (higher) energy branch (see the lower (upper) thick solid line in Figure 11). According to the energy bands shown in Figure 11, we can also see that the lowest band of the period-2 states appears as an upper edge of the swallowtail for the period

2*d* system. Connection between the period-2 states and the swallowtail was studied in depth in [78].

It was pointed out that the lowest band of the period-2 states is closely connected with the dynamical instability [33]. The linear stability analysis has shown that, with increasing k, the dynamical instability of the normal Bloch state sets on at the quasimomentum where the band of the normal Bloch states merges with the lowest band of the period-2 states. There, the dynamical instability is caused by the growing perturbation mode with wavelength 2*d* [31,32,43,54]. The growth of the mode with wavelength 2*d* accompanying the dynamical instability has been observed experimentally as well [79]. Since the lowest band of the period-2 states appears as the saddle of the swallowtail for the period 2*d* system and it forms the upper edge of the swallowtail, these period-2 states are dynamically unstable [32] while the upper branch of the period-2 states can be dynamically stable in some region of k [33]. The lowest multiple period states can be dynamically stable by introducing long-range interactions. For example, it was demonstrated that, in dipolar BECs, multiple period states with period 2*d* and 3*d* can be dynamically stable even at $k = 0$ provided the dipole-dipole interactions are repulsive and sufficiently strong [80,81].

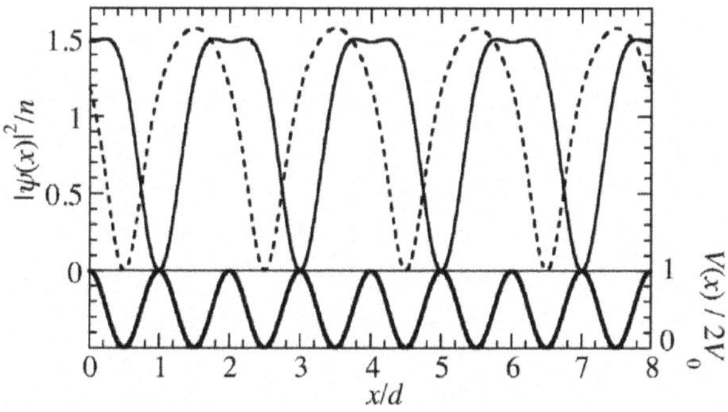

Figure 12. Density profiles $|\psi(x)|^2$ of period-2 states (upper panel) of BECs and the periodic external potential $V(x)$ (lower panel). The solid (dashed) line in the upper panel shows $|\psi(x)|^2$ for the lower-energy (higher-energy) period-2 state of Figure 11a at $kd/2\pi = 1/4$, $gn = E_R$, and $s = 0.2$. This figure is taken from [33].

4.3. Multiple Period States in Superfluid Fermi Gases

The emergence of the multiple period states in BECs in optical lattices is one of the novel nonlinear phenomena caused by the presence of the superfluid order parameter. However, in the normal repulsively interacting BECs without long-range interaction, the lowest multiple period states are higher in energy than the normal Bloch states and are dynamically unstable. Yoon *et al.* [82] have shown that the situation is very different for the multiple period states of superfluid Fermi gases in the BCS regime.

Figure 13 shows the profiles of the magnitude of the pairing field $|\Delta(x)|$ and the density $n(x)$ of the lowest period-2 states of superfluid Fermi gases in the BCS-BEC crossover obtained by solving the BdG equations for the continuum model (7). The striking difference from the BEC case shown in Figure 12 is that the feature of the period doubling shows up in the pairing field rather than the density. The difference between the regions of $-1 < x/d \leq 0$ and $0 < x/d \leq 1$ can be clearly seen in $|\Delta(x)|$ at any value of $1/k_F a_s$. On the other hand, the difference in $n(x)$ between these two regions is small in the deeper BCS side ($1/k_F a_s = -1$) (see the red line in Figure 13b) and finally disappears in the BCS limit.

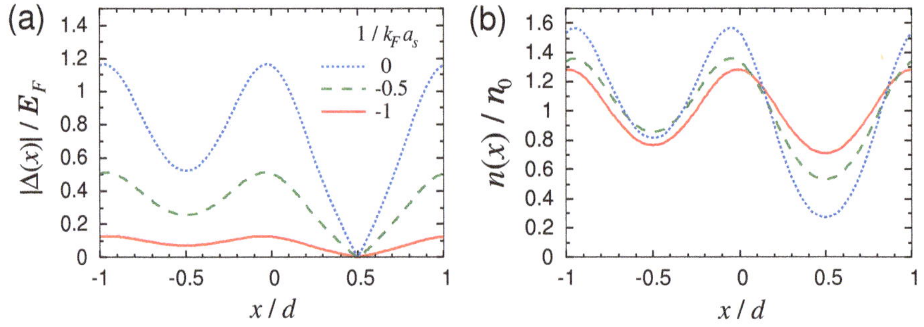

Figure 13. Profiles of (**a**) the magnitude of the pairing field $|\Delta(x)|$ and (**b**) the density $n(x)$ of the lowest period-2 states of superfluid Fermi gases in the BCS-BEC crossover: $1/k_F a_s = -1$ (red solid line), -0.5 (green dashed line), and 0 (blue dotted line). The quasimomentum P of the superflow is set at the Brillouin zone edge $P = P_{\rm edge}/2 = \hbar q_B/4$ of the period-2 states, and other parameters are set at $s = 1$ and $E_F/E_R = 0.25$. This figure is adapted from [82].

Figure 14 shows the energy bands of the lowest period-2 states in the BCS regime ($1/k_F a_s = -1$) together with that of the normal Bloch states. In the region of small P, the line of the period-2 states coincides with that of the normal Bloch states, as they are equivalent in this region, the states with period 1 being just a subset of any multiple period states with integer periods. Unlike the period-2 states in BECs, which form the concave upper edge of the swallowtail (see Figure 11), here the band of the period-2 states is convex upward. Remarkably, the lowest period-2 states are energetically more stable compared to the normal Bloch states around the Brillouin zone edge of the period-2 states ($P/P_{\rm edge} \sim 0.5$).

The manner in which the energy of the lowest period-2 states relative to that of the normal Bloch states changes along the BCS-BEC crossover can be seen in Figure 15. As we have already seen, the period-2 states are energetically more stable (*i.e.*, $\Delta E < 0$) in the deep BCS regime, where the band of the period-2 states is convex upward. With $1/k_F a_s$ increasing from the deep BCS regime, ΔE increases from a negative value and finally period-doubled states become higher in energy than the normal Bloch states (*i.e.*, $\Delta E > 0$) in the BEC side, where the band of the period-2 states forms the concave upper edge of the swallowtail.

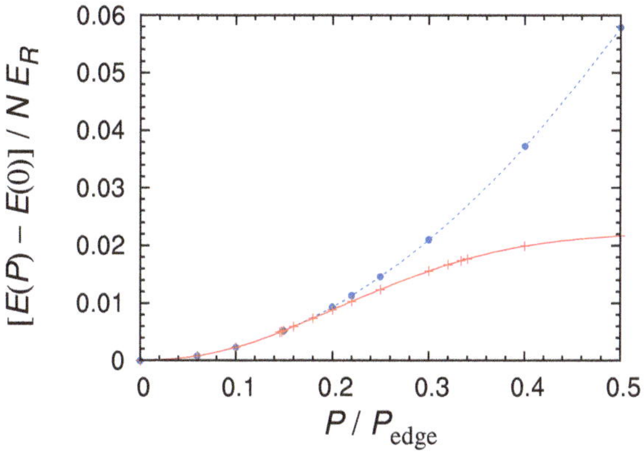

Figure 14. Energy E per particle of superfluid Fermi gases in a periodic potential as a function of the quasimomentum P. Parameter values are $s = 1$, $E_F/E_R = 0.25$, and $1/k_F a_s = -1$. The normal Bloch states with period d are shown by the blue dotted line with • symbols, and the period-2 states are shown by the red solid line with $+$. Note that the period-2 states are energetically more stable than the normal Bloch states in the region of $0.2 \lesssim P/P_{\rm edge} \leq 0.5$.

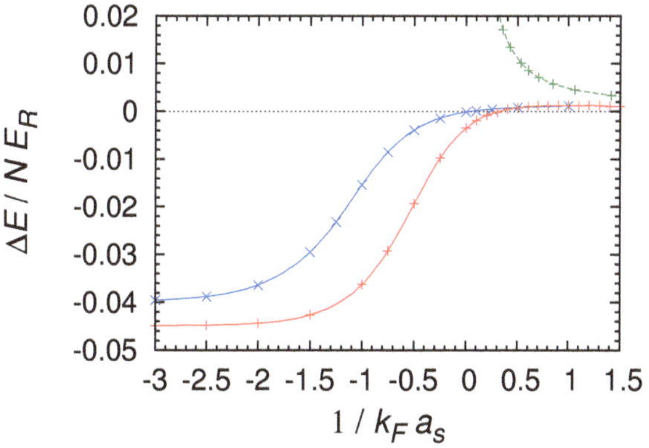

Figure 15. Difference $\Delta E \equiv E_2 - E_1$ of the total energy per particle between the period-2 states (E_2) and the normal Bloch states (E_1) at $P = P_{\rm edge}/2$ along the BCS-BEC crossover. The red solid line with $+$ is for $s = 1$ and the blue solid line with $\times$ is for $s = 2$; $E_F/E_R = 0.25$. The green dashed line shows the results by the GP equation for parameters corresponding to $s = 1$ and $E_F/E_R = 0.25$. This figure is taken from [82].

The energetic stability of the period-2 states in the BCS regime can be physically understood as follows. Let us consider the different behavior of $\Delta(x)$ and $n(x)$ for a period-2 state and a normal Bloch state at $P = P_{\text{edge}}/2$. In the case of the normal Bloch state, since $|\Delta(x)|$ is exponentially small in the BCS regime, we can distort the order parameter $\Delta(x)$ to produce a node, like the one in the period-2 state, with a small energy cost (per particle) up to the condensation energy $|E_{\text{cond}}|/N \ll E_F$, where $E_{\text{cond}} \equiv g^{-1} \int d^3r\, |\Delta(\mathbf{r})|^2$. However, making a node in $\Delta(x)$ kills the supercurrent $j = V^{-1}\partial_P E$, which yields a large gain of kinetic energy (per particle) of the superfluid flow of order $\sim P_{\text{edge}}^2/m \sim E_R$. Even if $\Delta(x)$ is distorted substantially to have a node, the original density distribution of the normal Bloch state is almost intact so that the increase of the kinetic energy and the potential energy due to the density variation is small. Therefore, the period-2 state is energetically more stable than the normal Bloch state in the BCS regime. In the above discussion, the key point is that $\Delta(x)$ and $n(x)$ can behave in a different way in the BCS regime. On the other hand, in the BEC limit, the density is directly connected to the order parameter as $n(x) \propto |\Delta(x)|^2$, and distorting the order parameter accompanies an increase of the kinetic and potential energies due to a large density variation.

In the deep BCS regime, the period-2 states are not only energetically stable, but also they can be long-lived although dynamically unstable. The black solid line in Figure 16 shows the growth rate γ of the fastest exponentially growing mode $|\eta(t)| = |\eta(0)|\, e^{\gamma t}$ of the deviation $|\Delta(x,t)| - |\Delta_0(x)|$ from the true stationary state $\Delta_0(x)$ for the period-2 states. We see that γ is suppressed with decreasing $1/k_F a_s$, which makes the period-2 states long-lived in the BCS regime. The growth rate γ corresponds to the imaginary part of the complex eigenvalue for the fastest growing mode obtained by the linear stability analysis [1,83], which is an intrinsic property of the initial stationary state independent of the magnitude of the perturbation.

On the other hand, the actual survival time τ_{surv}, the timescale for which the initial state is destroyed by the large-amplitude oscillations, depends on the accuracy of their initial preparation. The survival time can be estimated by $\tilde{\eta}(0)e^{\gamma t} \sim 1$, where $\tilde{\eta}(0)$ is the relative amplitude of the initial perturbation with respect to $|\Delta_0|$ for the fastest growing mode. In Figure 16, we show τ_{surv} for various values of $\tilde{\eta}(0)$. This result suggests that if the initial stationary state is prepared within an accuracy of 10% or smaller, this state safely survives for time scales of the order of $100\hbar/E_R$ or more in the BCS side, corresponding to τ_{surv} of more than the order of a few milliseconds for typical experimental parameters [84]: for $E_{R,b} = 2\pi \times 7.3\text{kHz} \times \hbar$ used in the experiment of [84], $1\hbar/E_R = 0.0109$ ms. In the deep BCS regime ($1/k_F a_s \ll -1$), τ_{surv} increases further and may become larger than the time scale of the experiments, so that the period-doubled states can be regarded as long-lived states and, in addition, they have lower energy than the normal Bloch states in a finite range of quasimomenta. Therefore, by (quasi-)adiabatically increasing the quasimomentum

P of the superflow starting from the ground state at $P = 0$, multiple-period states such as the period-doubled states could be realized experimentally in the deep BCS regime.

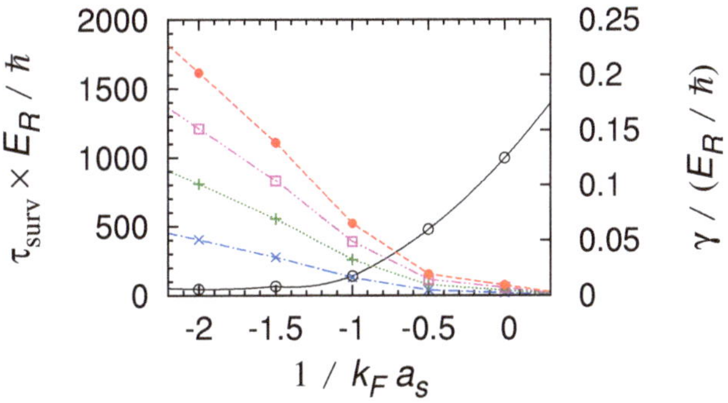

Figure 16. Growth rate γ of the fastest growing mode (black solid line) and survival time τ_{surv} of the period-2 state at $P = P_{edge}/2$ ($s = 1$ and $E_F/E_R = 0.25$). Blue dashed-dotted, green dotted, magenta dashed double-dotted, and red dashed lines show τ_{surv} for relative amplitude $\tilde{\eta}(0)$ of the initial perturbation of 10%, 1%, 0.1%, and 0.01%, respectively. This figure is taken from [82].

5. Nonlinear Lattices

This section deals with a special kind of optical lattices, called "nonlinear lattices". Here the coupling constant of the nonlinear term itself (*i.e.*, the interatomic interaction strength or the scattering length) has a space-periodic dependence. This is quite different from the systems which we have discussed so far: unlike cold atomic gases in a linear external periodic potential, those in a nonlinear lattice with periodically modulated interaction in space can be designed to have no linear periodic counterpart at all. While, in the former, properties are determined by the competition between the linear periodic potential and a nonlinear term, in the latter, periodicity and nonlinearity are generated by a single term. This leads to unique stability properties of the superfluid, and imposes additional conditions on its survival [85,86].

In the first subsection, we provide a basic sketch of how the sustenance of superfluidity depends on the geometry (homogeneous/in a linear lattice/in a nonlinear lattice) of the BEC system. In the second subsection, we explain the stability properties of BECs in a nonlinear lattice in terms of a simple discrete model. The third subsection presents the results of studies on ultracold bosons and fermions in nonlinear lattices for various parameter regimes. Finally, in the last subsection, we present a short outline of the experimental setup that constructs such a space-periodic dependence of the interatomic interaction strength.

5.1. *Dynamical Stability of the Superfluid: Special Properties of Nonlinear Lattices*

For uniform and homogeneous BECs, the dynamical stability of the superfluid is determined by the nature of the interatomic interaction. If it is repulsive (*i.e.*, the scattering length is positive), the superfluid remains dynamically stable for any value of the momentum $\hbar k$ of the superflow. On the other hand, if the interaction is attractive (*i.e.*, the scattering length is negative), the long-wavelength modes with

$$q^2 < \frac{4mng}{\hbar^2} \tag{38}$$

grow or decay in time exponentially for any value of $\hbar k$, thus invoking an instability in the system (e.g., [1]). However, shorter-wavelength modes are stable because for them, the kinetic energy dominates over the interaction energy.

The dynamical stability properties of the superfluid changes in the presence of an external periodic potential. The periodic nature of the system may lead to Bloch solutions of the form $\psi(x) = e^{ikx}\phi(x)$, where $\phi(x)$ is a periodic function with the same periodicity as that of the lattice. The quasimomentum of the superflow is given by $\hbar k$, k being the corresponding Bloch wave number. Here, for simplicity, we have assumed that the superfluid flows in the same x direction as the periodic potential. Unlike the homogeneous system, the system has a nonzero critical value of k above which the Bloch states are dynamically unstable [31] as has been seen in Section 3.2.2: in other words, the $k = 0$ state is always dynamically stable.

For a nonlinear lattice, however, this picture changes still further. The coupling constant g in the GP Equation (2) for bosons and the BdG Equations (7) and (8) for fermions now depends on the space coordinate x. It can be thought of having a form as

$$g(x) = V_1 + V_2 \cos 2k_0 x\,, \tag{39}$$

i.e., $g(x)$ consists of one constant part and one sinusoidal component. k_0 is related to the period d of the modulation by $k_0 = \pi/d$. If the nonlinear lattice is realized by an optical Feshbach resonance (details are given in the last part of this section), k_0 is equivalent to the wave number of the laser beam. This special type of periodic nonlinearity gives rise to a dynamical instability for the $k = 0$ state [85], which is in contrast with the linear lattice case (*i.e.*, the case of an external periodic potential). We shall explain this in the following subsection.

5.2. *Basic Physical Idea: The Dynamical Stability of Nonlinear Lattices*

Zhang *et al.* [85] and Dasgupta *et al.* [86] studied extended states of BECs in quasi-one-dimension with a periodically modulated interaction in space, *i.e.*, a nonlinear lattice with no periodic linear potential. It was observed that when the coefficient of the nonlinear term is purely sinusoidal (*i.e.*, $V_1 = 0$ and $V_2 \neq 0$), Bloch states at $k = 0$ are dynamically (and energetically) unstable [85]. In addition, even though $k = 0$ state is dynamically unstable, states for nonzero k could be dynamically stable in some region in $0.25 \leq k/k_0 \leq 0.5$ (this point will be seen in more detail later) [85].

Why is the $k = 0$ state unstable in the nonlinear lattice, in contrast to the stable $k = 0$ state in the linear periodic potential? Why are the states with higher values of k possibly dynamically stable even though the $k = 0$ state is unstable? To explain these, we take resort to a simple discrete model [86]. We map the 1D nonlinear lattice with $V_1 = 0$ and $V_2 \neq 0$ to a discrete model with the on-site interaction alternating between U and $-U$ (with $U > 0$):

$$H = -K\sum_j(\psi_j^*\psi_{j+1} + \psi_{j+1}^*\psi_j) + \frac{U}{2}\left[\sum_{j=\text{even}}|\psi_j|^4 - \sum_{j=\text{odd}}|\psi_j|^4\right]. \tag{40}$$

So, if we denote the distance between two adjacent sites in the discrete model as $\tilde{d}$, the actual lattice constant d of the unit cell in the original system is $2\tilde{d}$ (*i.e.*, the unit cell in the original system corresponds to a "supercell" with two sites in the discrete model).

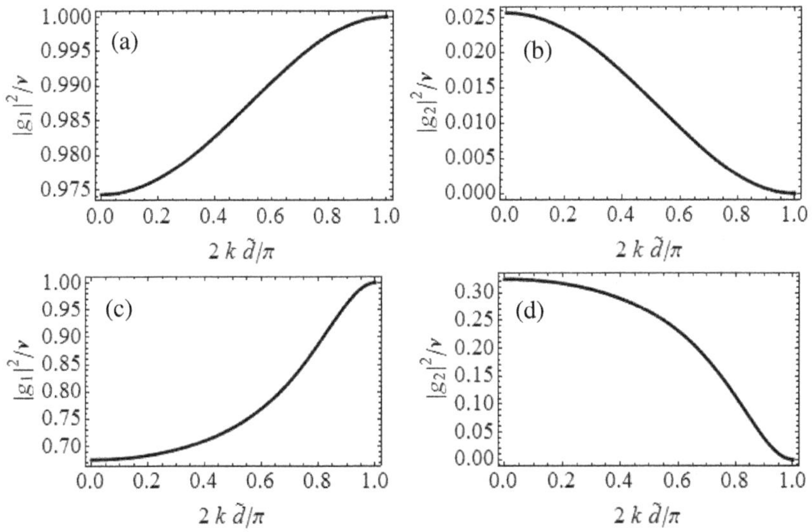

Figure 17. Density distributions in the lowest band of the normal Bloch states (*i.e.*, period-1 states whose period is one supercell) as functions of k for different values of $Uv/2K$. Panels (**a**,**b**): Populations of $|g_1|^2$ (attractive site) and $|g_2|^2$ (repulsive site) for $Uv/2K = 6$, respectively. Panels (**c**,**d**): Populations of $|g_1|^2$ and $|g_2|^2$ for $Uv/2K = 0.75$, respectively. This figure is taken from [86].

Assuming the state is in the Bloch form, we write the amplitude ψ_j at site j as $\psi_j = g_j e^{ikj\tilde{d}}$, where $\hbar k$ is quasimomentum of the superflow and g_j is the complex amplitude at site j with the periodic boundary conditions, $g_j = g_{j+2}$ (Note that the unit cell contains 2 sites.). In addition, the amplitudes g_j's are subject to the normalization condition $|g_1|^2 + |g_2|^2 = \nu$, where ν is the total number of particles in a unit cell with two sites. One can obtain the stationary solutions of g_1 and g_2 by solving the combined equations of $\delta H/\delta\psi_1^* = 0$ and $\delta H/\delta\psi_2^* = 0$ in almost the same manner as in Section 4.1. Resulting populations $|g_1|^2$ and $|g_2|^2$ in the attractive and the repulsive sites, respectively, for the lowest Bloch band are

$$\frac{|g_1|^2}{\nu} = n_+ \quad \text{and} \quad \frac{|g_2|^2}{\nu} = n_- \tag{41}$$

with

$$n_\pm = \frac{1}{2}\left\{1 \pm \left[\left(\frac{\cos k\tilde{d}}{U\nu/2K}\right)^2 + 1\right]^{-1/2}\right\}. \tag{42}$$

The populations $|g_1|^2$ and $|g_2|^2$ for two different values of $U\nu/2K$ are shown in Figure 17 as functions of k within the first Brillouin zone.

As seen from Figure 17, the population difference between the adjacent sites is the smallest at the zone center $k = 0$ while it increases as going toward the zone edge at which all the particles are accumulated in the attractive sites. To understand why $k = 0$ state is dynamically unstable, it is instructive to see the interaction energy averaged over one unit cell (with 2 sites) [85]. From Equations (41) and (42), we obtain the average interaction energy per particle for the lowest Bloch band as

$$\frac{E_{\text{int}}/K}{N} = -\frac{U}{2K\nu}(|g_1|^4 + |g_2|^4) = -\frac{U\nu}{2K}\left[\left(\frac{\cos k\tilde{d}}{U\nu/2K}\right)^2 + 1\right]^{-1/2} < 0. \tag{43}$$

Note that, in the lowest Bloch band, the average interaction energy per particle is negative for any value of k. So, roughly speaking, this situation resembles a BEC with attractive interparticle interaction, which is dynamically unstable as we have mentioned in the last subsection, and the dynamical instability of BECs in nonlinear lattices at $k = 0$ could be understood as a consequence of the net attractive interaction energy [85].

Figure 18 shows the dynamical stability diagram of the stationary states in the lowest Bloch band in the k–q plane, where q is the quasi-wave number of the perturbation on the stationary states. It is noted that there is a region at larger values of k in which the lowest Bloch states are dynamically stable (e.g., the gray-shaded region of $0.5 \lesssim 2k\tilde{d}/\pi \leq 1$ in Figure 18a) even though they are dynamically unstable at smaller values of k including the zone center at $k = 0$. To understand this somewhat counterintuitive fact, we shall take a closer look at the populations $|g_1|^2$ and $|g_2|^2$ shown in Figure 17. As we have briefly mentioned before, we note that almost all the particles are accumulated in the attractive sites and thus the repulsive sites are almost empty near the zone edge at $2k\tilde{d}/\pi = 1$: at the zone edge, only alternate sites are occupied and fragments of the BEC in these alternate sites are isolated. Since the transition amplitude between the states with populations $\{|g_1|^2, |g_2|^2\}$ and $\{|g_1|^2 \pm 1, |g_2|^2 \mp 1\}$ is $\sim \sqrt{|g_1||g_2|}K$, tunneling of particles between neighboring sites is suppressed (*i.e.*, the inter-site dynamics is frozen), and thus the dynamical instability is suppressed near the zone edge. Therefore, in nonlinear lattices, the lowest Bloch states can be dynamically stable at higher values of k near the zone edge [86]. On the other hand, at the zone center $k = 0$, the difference between the respective populations in adjacent sites is the smallest: No sites are empty and inter-site tunneling is non-negligible. Thus the suppression of the dynamical instability does not work around the zone center, rendering the system dynamically unstable due to the net attractive interaction mentioned before. Since the isolation of fragments of the BEC is a result of the attractive interaction in alternate sites and this new mechanism of dynamical stability is more effective for larger $Uv/2K$ (see wider gray-shaded area in Figure 18a than that in Figure 18b) resulting in larger net attractive interaction energy, this mechanism can be called "attraction-induced dynamical stability" [86].

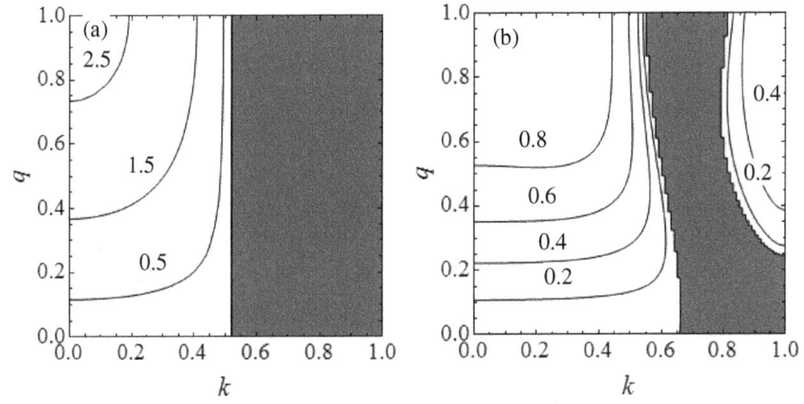

Figure 18. Dynamical stability diagrams for the normal Bloch states (*i.e.*, period-1 states) for $Uv/2K = 6$ (**a**) and 0.75 (**b**). Quasi-wave numbers k and q are in units of $\pi/2\tilde{d}$. The white regions are the dynamically unstable regions and the gray-shaded regions are the dynamically stable regions. The contours show the growth rate of the fastest growing mode, *i.e.*, the largest maximum absolute value of the imaginary part of the eigenvalues of matrix $\sigma_z M(q)$ in Equation (17) in units of K. This figure is taken from [86].

We note that, in addition to the net attractive interaction energy and the suppression of the tunneling near the zone edge, there would be other factors to determine the dynamical stability of the nonlinear lattice system. When $Uv/2K$ is sufficiently small, we observe that a dynamically unstable region appears near the zone edge (see, e.g., Figure 18b) and the dynamically stable region is located at the intermediate values of k ($0.65 \lesssim 2k\tilde{d}/\pi \lesssim 0.8$ in the case of Figure 18b). This non-trivial reentrant behavior suggests that there are several other factors that affect the stability, which are collectively responsible for the complicated stability diagram like Figure 18b.

As a final comment in this subsection, we mention that the discussion here is based on the discrete model, but the attraction-induced dynamical stability has been confirmed in the continuum model as well [86]. The main difference is that, in the continuum model, if the value of V_2 is increased beyond a certain point, the attractive interaction between intra-site particles becomes dominant and eventually leads to the collapse of fragments of the BEC. This intra-site dynamics cannot be accounted for by the discrete model, which does not include the intra-site degrees of freedom.

5.3. *Superfluid Cold Atomic Gases in Nonlinear Lattices*

Extended states of BECs in nonlinear lattices were first studied by Zhang *et al.* in [85] (Localized states such as solitons in nonlinear lattices were studied earlier in, e.g., [87,88] (see also [20] and references therein).). In this work, they considered quasi-1D BECs in nonlinear lattices described by the 1D version of the GP Equation (2) with the periodically modulated interaction strength in space given by Equation (39): $g(x) = V_1 + V_2 \cos 2k_0 x$ with V_1 and $V_2 \geq 0$. They studied stationary Bloch states in nonlinear lattices and their energetic and dynamical stability summarized in Figure 19. As we have discussed in the previous subsection, the key result is that, when $V_1 = 0$, $k = 0$ state is dynamically (and energetically) unstable for any value of $V_2 \neq 0$ (see black regions in the lower panels of Figure 19 for $c_1 = 0$). This is in contrast to BECs in a linear external periodic potential whose dynamically unstable region is restricted in the domain of $1/4 < k/2k_0 \leq 1/2$ (*i.e.*, the right half of each panel in Figure 6b). In addition, states at higher values of k near the zone edge can be dynamically stable even though $k = 0$ state is dynamically unstable. It was pointed out that the dynamical instability of the $k = 0$ state can be partially explained by the net attractive average interaction energy as we have discussed in the previous subsection. They also discussed the stability of the superfluidity due to the competition between the spatially modulated part (V_2 term) and the uniform, repulsive component (V_1 term); the latter tends superfluids to be stable. With increasing V_1 from zero for a fixed nonzero V_2, state at $k = 0$ becomes dynamically

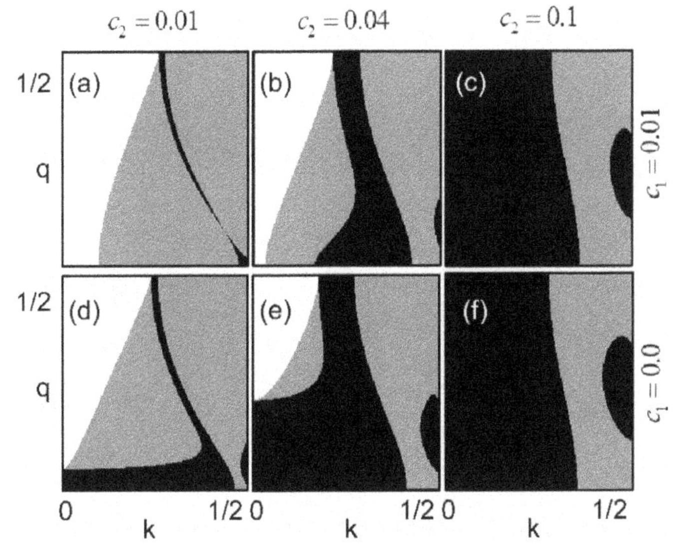

Figure 19. Stability diagrams in nonlinear lattices for various values of $c_1 \equiv mnV_1/(4\hbar^2k_0^2)$ and $c_2 \equiv mnV_2/(4\hbar^2k_0^2)$, where n is the average number density. The quasi-wave numbers k and q are in units of $2k_0 = 2\pi/d$ with d being the lattice constant of the unit cell in the notation of this review. The Bloch states are stable in the white area. In the gray area the Bloch states are energetically unstable but dynamically stable while, in the black area, they are unstable both energetically and dynamically. This figure is taken from [85].

Dasgupta *et al.* [86] studied multiple-period states of BECs in nonlinear lattices. They discussed stationary states with larger integer periodicity and their energetic and dynamical stability using the GP equation for both the discrete and continuum models. The main result of this work is that they found a new mechanism of dynamical stability called "attraction-induced dynamical stability" and provided the understanding of the dynamical stability around the Brillouin zone edge due to the isolation of fragments of the BEC in each attractive domain in the nonlinear lattice as discussed in the previous subsection. This attraction-induced dynamical stability is even better manifested for the period-2 case because the majority of the particles are stored in every second attractive domain (*i.e.*, every fourth site in the discrete model) making the fragments of the BEC more firmly separated, while for period-1 case they are stored in every attractive domain (*i.e.*, every second site in the discrete model). Figure 20 shows the dynamical stability diagrams for period-2 Bloch states calculated for the discrete model at the same values of $Uv/2K$ as Figure 18. It is noted that the growth rate of the fastest growing mode in Figure 20a is much smaller than the period-1 counterpart shown in Figure 18a; the dynamically stable region in Figure 20b is larger than the period-1 counterpart shown in Figure 18b.

Superfluid Fermi gases in nonlinear lattices were studied by Yu *et al.* [89]. They considered quasi-1D 2-component superfluid Fermi gases based on the 1D version of the BdG Equation (7) with spatially modulated interaction strength of the form $g(x) = V_1 + V_2 \cos 2k_0 x$ with $V_1 < 0$ and $V_2 \geq 0$. Note that it has been assumed that the uniform component is attractive ($V_1 < 0$) so that the system is in the superfluid phase. The properties of the Bloch states in this system for various parameter values of V_1 and V_2 are summarized in Figure 21.

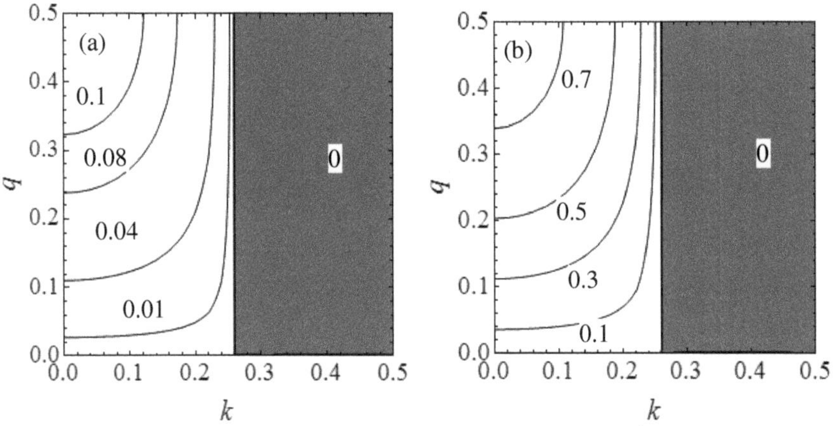

Figure 20. The same as Figure 18, but for period-2 Bloch states. Taken from [86].

The key point is the competition between the effects of the nonlinear uniform interaction by V_1 and the periodicity of the system induced by V_2. It was found that the former dominates over the latter for lager $|V_1|/V_2$ (the region denoted by "SW" in Figure 21), the Bloch band has a swallowtail loop around the Brillouin zone edge. This situation is similar to superfluid Fermi gases in a periodic potential, where the swallowtail appears due to the effect of the nonlinear interaction dominating over the periodicity of the system induced by the external potential [69]. On the other hand, for smaller $|V_1|/V_2$ (the region denoted by "FFLO-like" in Figure 21), it was predicted that the state at the Brillouin zone edge with nonzero quasimomentum of the superflow (quasimomentum per atom $P = \hbar k_0/2$ in the notation of this review article) becomes the ground state of the system. They call this state as "FFLO-like state" because of the nonzero value of the quasimomentum of the superflow (Note that the current of this state, however, is zero.). The stability of the Bloch states of superfluid Fermi gases in nonlinear lattices has yet to be studied.

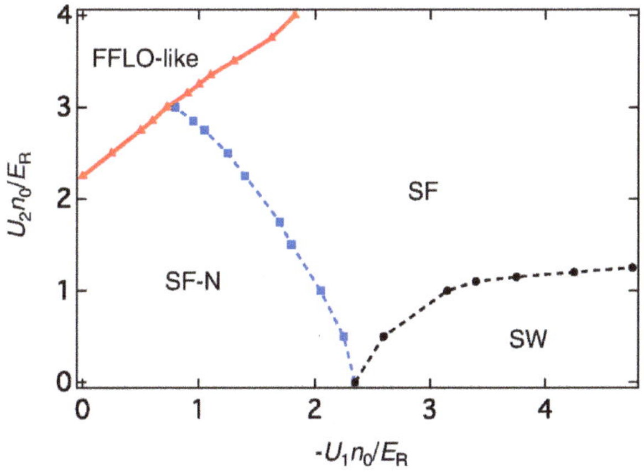

Figure 21. Phase diagram of the Bloch states of superfluid Fermi gases in nonlinear lattices in the $|U_1|$–U_2 plane, where $U_1 = V_1$ and $U_2 = V_2$ in the notation of this review. In the parameter region denoted by "SF-N", the system has a critical quasimomentum of the superflow above which the normal state has lower energy compared to the superfluid state while, in the regions denoted by "SF" and "SW", there is no such critical value and the superfluid state is always lower in energy than the normal state in the whole Brillouin zone. Particularly, in the region of "SW" at larger $|U_1|/U_2$, the energy band has a swallowtail loop around the zone edge. In the region denoted by "FFLO-like" (Fulde–Ferrell–Larkin–Ovchinnikov) at smaller $|U_1|/U_2$, the ground state has a nonzero quasimomentum of the superflow. Taken from [89].

5.4. Experimental Setup

Finally we give a brief sketch how nonlinear lattices can be created experimentally. With the aid of modern experimental techniques, it is now indeed possible to construct systems where the nonlinear term has an explicit spatial dependence. A very efficient way to do this is to employ optical Feshbach resonances (OFRs) [90,91].

Ultracold atoms offer an immense controllability over physical quantities like scattering length. It all started with magnetic Feshbach resonances, that use Zeeman shifts to make the scattering states resonate with a bound molecular state. It was later suggested [92] that lasers can be used to induce the resonance optically. This Feshbach resonance via optical fields offers additional advantages over magnetic ones: the intensity as well as the frequency of the laser beams can be rapidly and precisely controlled. In addition, high resolution (submicron level) spatial control of the scattering length is possible by creating specially structured laser fields.

In OFRs, the basic scheme is to use a laser beam (tuned near the photoassociation resonance) that couples an initial state of free atoms to a molecular bound state.

The scattering length is accordingly modified. If the OFR is driven by a standing wave with a certain periodicity, the interparticle interaction derived from it has the same periodic nature. Thus, a nonlinear lattice is generated. Using OFRs, Yamazaki *et al.* [93] demonstrated rapid, spatial modulation of the scattering length periodically at the submicron level. In this particular setup, a pulsed optical standing wave was applied to a BEC of Ytterbium atoms. The resonant wavelength was 556 nm, and the optical pulse could generate a scattering length periodically modulated in space with wavelength 278 nm.

6. Conclusions

Over the last two decades ultracold atoms in optical lattices [14–17] have emerged as a key paradigm to study the *ideal* realizations of many important problems in highly controllable settings, ranging from single particle quantum mechanical effects such as Bloch oscillations to strongly correlated many-body effects such as the superfluid-Mott insulator quantum phase transition. In the recent past the focus in the field has shifted more towards observing equilibrium and non-equilibrium quantum many-body effects and topological phenomena [94]. In this light the subject of this review, namely, the interplay of mean-field atomic interactions and the periodicity of the applied external potential serves to re-emphasize the fact that there are unique and interesting phenomena even in a much simpler setting. Our hope is that such a review can rekindle some interest in this area especially from the experimental side as to date there has not been a clear observation of swallowtail loop structures or period-doubled solutions for optical lattices. Moreover, in striving to control complex quantum systems such as ultracold atoms to an ever greater degree in order to realize complex many-body ground states [94] it becomes very important to be aware of and understand fundamental limitations on state preparation due to unavoidable adiabaticity breaking imposed by phenomena such as swallowtail loop dispersions.

In this review we focused on some interesting phenomena originating from the nonlinear, mean-field interactions of superfluid atomic gases in periodic potentials. We began with a summary of the basic theoretical description of Bose–Einstein condensates (BECs) and superfluid Fermi gases within the mean-field framework in Section 2. In Section 3 we provided a comprehensive overview of the phenomenon of swallowtail loops in the band-structure of superfluid atomic gases in an optical lattice, followed by the the discussion of Bloch states with multiple periods of the applied optical lattice potential in Section 4, and Bloch states in nonlinear lattices, *i.e.*, situations in which the nonlinear interaction term is itself a periodic function in space in Section 5. While we have covered a substantial portion of the various interesting phenomena that can arise due to the nonlinearity in the mean-field theory of superfluids, this is by no means complete. For instance, as mentioned in the introduction, we have made no attempt to describe localized soliton solutions to the Gross–Pitaevskii equation and suggest the excellent review [20] on this topic for the interested reader.

Although considerable amount of research work has already been accomplished in this field, it is still relatively young and has flourished only over the last two decades beginning with the experimental discovery of BECs. As a result we believe there are still a range of open problems that can be investigated. Some examples include, the stability of Bloch and swallowtail loop states of superfluid Fermi gases in nonlinear lattices, Bloch oscillation dynamics for both bosons and fermions in regimes with swallowtail loops, the study of quantum equivalents of mean-field nonlinear phenomena including solitons and loops, and nonlinear phenomena in more exotic systems such as BECs with spin-orbit interactions or spinor BECs *etc.*

Acknowledgments: We acknowledge Mauro Antezza, Franco Dalfovo, Elisabetta Furlan, Jonas Larson, Takashi Nakatsukasa, Duncan O'Dell, Giuliano Orso, Francesco Piazza, Lev P. Pitaevskii, Sandro Stringari, and Sukjin Yoon for collaborations. This work was supported by IBS through Project Code (IBS-R024-D1); by the Zhejiang University 100 Plan; by the Junior 1000 Talents Plan of China; by the Max Planck Society, MEST (Ministry of Education, Science and Technology) of Korea, Gyeongsangbuk-Do, and Pohang City for the support of the JRG at APCTP; and by the Basic Science Research Program through NRF by MEST (Grant No. 2012R1A1A2008028). Raka Dasgupta would like to acknowledge Department of Science and Technology, Government of India for Inspire Faculty Award (04/2014/002342).

Conflicts of Interest: The authors declare no conflict of interest.

References

1. Pethick, C.J.; Smith, H. *Bose–Einstein Condensation in Dilute Gases*, 2nd ed.; Cambridge University Press: Cambridge, UK, 2008.
2. Pitaevskii, L.P.; Stringari, S. *Bose–Einstein Condensation*; Oxford University Press: Oxford, UK, 2003.
3. Leggett, A.J. Bose–Einstein condensation in the alkali gases: Some fundamental concepts. *Rev. Mod. Phys.* **2001**, *73*, 307–356.
4. Burger, S.; Bongs, K.; Dettmer, S.; Ertmer, W.; Sengstock, K.; Sanpera, A.; Shlyapnikov, G.V.; Lewenstein, M. Dark solitons in Bose–Einstein condensates. *Phys. Rev. Lett.* **1999**, *83*, 5198–5201.
5. Denschlag, J.; Simsarian, J.E; Feder, D.L.; Clark, C.W.; Collins, L.A.; Cubizolles, J.; Deng, L.; Hagley, E.W.; Helmerson, K.; Reinhardt, W.P.; *et al.* Generating solitons by phase engineering of a Bose–Einstein condensate. *Science* **2000**, *287*, 97–101.
6. Strecker, K.E.; Partridge, G.B.; Truscott, A.G.; Hulet, R.G. Formation and propagation of matter-wave soliton trains. *Nature* **2002**, *417*, 150–153.
7. Khaykovich, L.; Schreck, F.; Ferrari, G.; Bourdel, T.; Cubizolles, J.; Carr, L.D.; Castin, Y.; Salomon, C. Formation of a matter-wave bright soliton. *Science* **2002**, *296*, 1290–1293.

8. Deng, L.; Hagley, E.W.; Wen, J.; Trippenbach, M.; Band, Y.; Julienne, P.S.; Simsarian, J.E.; Helmerson, K.; Rolston, S.L.; Phillips, W.D. Four-wave mixing with matter waves. *Nature* **1999**, *398*, 218–220.
9. Kevrekidis, P.G.; Frantzeskakis, D.J.; Carretero-González, R. (Eds.) *Emergent Nonlinear Phenomena in Bose–Einstein Condensates: Theory and Experiment*; Springer: Berlin/Heidelberg, Germany, 2008.
10. Zwierlein, M.W.; Abo-Shaeer, J.R.; Schirotzek, A.; Schunck, C.H.; Ketterle, W. Vortices and superfluidity in a strongly interacting Fermi gas. *Nature* **2005**, *435*, 1047–1051.
11. Eagles, D.M. Possible pairing without superconductivity at low carrier concentrations in bulk and thin-film superconducting semiconductors. *Phys. Rev.* **1969**, *186*, 456–463.
12. Leggett, A.J. Diatomic Molecules and Cooper Pairs. In *Modern Trends in the Theory of Condensed Matter*; Pekalski, A., Przystawa, J., Eds.; Springer: Berlin/Heidelberg, Germany, 1980; Volume 115, pp. 13–27.
13. Giorgini, S.; Pitaevskii, L.P.; Stringari, S. Theory of ultracold atomic Fermi gases. *Rev. Mod. Phys.* **2008**, *80*, 1215–1274.
14. Bloch, I.; Dalibard, J.; Zwerger, W. Many-body physics with ultracold gases. *Rev. Mod. Phys.* **2008**, *80*, 885–964.
15. Morsch, O.; Oberthaler, M. Dynamics of Bose–Einstein condensates in optical lattices. *Rev. Mod. Phys.* **2006**, *78*, 179–215.
16. Yukalov, V.I. Cold bosons in optical lattices. *Laser Phys.* **2009**, *19*, 1–110, doi:10.1134/S1054660X09010010.
17. Watanabe, G.; Yoon, S. Aspects of superfluid cold atomic gases in optical lattices. *J. Korean Phys. Soc.* **2013**, *63*, 839–857.
18. Wu, B.; Diener, R.B.; Niu, Q. Bloch waves and bloch bands of Bose–Einstein condensates in optical lattices. *Phys. Rev. A* **2002**, *65*, 025601.
19. Diakonov, D.; Jensen, L.M.; Pethick, C.J.; Smith, H. Loop structure of the lowest Bloch band for a Bose–Einstein condensate. *Phys. Rev. A* **2002**, *66*, 013604.
20. Kartashov, Y.V.; Malomed, B.A.; Torner, L. Solitons in nonlinear lattices. *Rev. Mod. Phys.* **2011**, *83*, 247–305.
21. Watanabe, G.; Maruyama, T. Nuclear pasta in supernovae and neutron stars. 2012, arXiv:1109.3511.
22. Maruyama, T.; Watanabe, G.; Chiba, S. Molecular dynamics for dense matter. *Prog. Theor. Exp. Phys.* **2012**, *2012*, 01A201.
23. Dalfovo, F.; Giorgini, S.; Pitaevskii, L.P.; Stringari, S. Theory of Bose–Einstein condensation in trapped gases. *Rev. Mod. Phys.* **1999**, *71*, 463–512.
24. Pitaevskii, L.P. Vortex lines in an imperfect Bose gas. *Sov. Phys. JETP* **1961**, *13*, 451–454.
25. Gross, E.P. Structure of a quantized vortex in boson systems. *Nuovo Cimento* **1961**, *20*, 454–477.
26. Gross, E.P. Hydrodynamics of a superfluid condensate. *J. Math. Phys.* **1963**, *4*, 195–207.
27. De Gennes, P.G. *Superconductivity of Metals and Alloys*; Benjamin: New York, NY, USA, 1966.

28. Randeria, M. Crossover from BCS Theory to Bose–Einstein Condensation. In *Bose Einstein Condensation*; Griffin, A., Snoke, D., Stringari, S., Eds.; Cambridge University Press: Cambridge, UK, 1995; pp. 355–392.
29. Trombettoni, A.; Smerzi, A. Discrete solitons and breathers with dilute Bose–Einstein condensates. *Phys. Rev. Lett.* **2001**, *86*, doi:10.1103/PhysRevLett.86.2353.
30. Kevrekidis, P.G. *The Discrete Nonlinear Schrödinger Equation*; Springer: Berlin/Heidelberg, Germany, 2009.
31. Wu, B.; Niu, Q. Landau and dynamical instabilities of the superflow of Bose–Einstein condensates in optical lattices. *Phys. Rev. A* **2001**, *64*, 061603(R).
32. Machholm, M.; Pethick, C.J.; Smith, H. Band structure, elementary excitations, and stability of a Bose–Einstein condensate in a periodic potential. *Phys. Rev. A* **2003**, *67*, 053613.
33. Machholm, M.; Nicolin, A.; Pethick, C.J.; Smith, H. Spatial period doubling in Bose–Einstein condensates in an optical lattice. *Phys. Rev. A* **2004**, *69*, 043604.
34. Wu, B.; Niu, Q. Nonlinear Landau–Zener tunneling. *Phys. Rev. A* **2000**, *61*, 023402.
35. Mueller, E.J. Superfluidity and mean-field energy loops: Hysteretic behavior in Bose–Einstein condensates. *Phys. Rev. A* **2002**, *66*, 063603.
36. Landau, L.D. Zur Theorie der Energieubertragung. II. *Physikalische Zeitschrift der Sowjetunion* **1932**, *2*, 46–51. (In German)
37. Zener, C. Non-Adiabatic Crossing of Energy Levels. *Proc. R. Soc. Lond. A* **1932**, *137*, 696–702.
38. Bronski, J.C.; Carr, L.D.; Deconinck, B.; Kutz, J.N. Bose–Einstein condensates in standing wave: The cubic nonlinear Schrödinger equation with a periodic potential. *Phys. Rev. Lett.* **2001**, *86*, 1402–1405.
39. Eckel, S.; Lee, J.G.; Jendrzejewski, F.; Murray, N.; Clark, C.W.; Lobb, C.J.; Phillips W.D.; Edwards, M.; Campbell, G.K. Hysteresis in a quantized superfluid 'atomtronic' circuit. *Nature* **2014**, *506*, 200–203.
40. Thom, R. *Structural Stability and Morphogenesis: An Outline of a General Theory of Models*; W.A. Benjamin: Reading, MA, USA, 1975.
41. Seaman, B.T.; Carr, L.D.; Holland, M.J. Nonlinear band structure in Bose–Einstein condensates: Nonlinear Schrödinger equation with a Kronig–Penney potential. *Phys. Rev. A* **2005**, *71*, 033622.
42. Dong, X.; Wu, B. Instabilities and sound speed of a Bose–Einstein condensate in the Kronig–Penney potential. *Laser Phys.* **2007**, *17*, 190–197.
43. Wu, B.; Niu, Q. Superfluidity of Bose–Einstein condensate in an optical lattice: Landau–Zener tunnelling and dynamical instability. *New J. Phys.* **2003**, *5*, 104, doi:10.1088/1367-2630/5/1/104
44. Burger, S.; Cataliotti, F.S.; Fort, C.; Minardi, F.; Inguscio, M.; Chiofalo, M.L.; Tosi, M.P. Superfluid and dissipative dynamics of a Bose-Einstein condensate in a periodic optical potential. *Phys. Rev. Lett.* **2001**, *86*, 4447–4450.
45. Wu, B.; Niu, Q. Dynamical or Landau instability? *Phys. Rev. Lett.* **2002**, *89*, 088901.

46. Burger, S; Cataliotti, F.S.; Fort, C.; Minardi, F.; Inguscio, M.; Chiofalo, M.L.; Tosi, M.P. Burger *et al.* Reply. *Phys. Rev. Lett.* **2002**, *89*, 088902.
47. Mamaladze, Y.G.; Cheǐshvili, O.D. Flow of a superfluid liquid in porous media. *Sov. Phys. JETP* **1966**, *23*, 112–117.
48. Hakim, V. Nonlinear Schrödinger flow past an obstacle in one dimension. *Phys. Rev. E* **1997**, *55*, 2835–2845.
49. Watanabe, G.; Dalfovo, F.; Piazza, F.; Pitaevskii, L.P.; Stringari, S. Critical velocity of superfluid flow through single-barrier and periodic potentials. *Phys. Rev. A* **2009**, *80*, 053602.
50. Danshita, I.; Tsuchiya, S. Comment on "Nonlinear band structure in Bose–Einstein condensates: Nonlinear Schrödinger equation with a Kronig–Penney potential". *Phys. Rev. A* **2007**, *75*, 033612.
51. Seaman, B.T.; Carr, L.D.; Holland, M.J. Reply to "Comment on 'Nonlinear band structure in Bose–Einstein condensates: Nonlinear Schrödinger equation with a Kronig–Penney potential'". *Phys. Rev. A* **2007**, *76*, 017602.
52. Barontini, G.; Modugno, M. Dynamical instability and dispersion management of an attractive condensate in an optical lattice. *Phys. Rev. A* **2007**, *76*, 041601(R).
53. Zhu, Q.; Wu, B. Superfluidity of BEC in ultracold atomic gases. *Chin. Phys. B* **2015**, *24*, 050507.
54. Modugno, M.; Tozzo, C.; Dalfovo, F. Role of transverse excitations in the instability of Bose–Einstein condensates moving in optical lattices. *Phys. Rev. A* **2004**, *70*, 043625.
55. Fallani, L.; de Sarlo, L.; Lye, J.E.; Modugno, M.; Saers, R.; Fort, C.; Inguscio, M. Observation of Dynamical Instability for a Bose–Einstein Condensate in a Moving 1D Optical Lattice. *Phys. Rev. Lett.* **2004**, *93*, 140406.
56. De Sarlo, L.; Fallani, L.; Lye, J.E.; Modugno, M.; Saers, R.; Fort, C.; Inguscio, M. Unstable regimes for a Bose–Einstein condensate in an optical lattice. *Phys. Rev. A* **2005**, *72*, 013603.
57. Cataliotti, F.S.; Fallani, L.; Ferlaino, F.; Fort, C.; Maddaloni, P. Superfluid current disruption in a chain of weakly coupled Bose–Einstein condensates. *New J. Phys.* **2003**, *5*, 71, doi:10.1088/1367-2630/5/1/371.
58. Smerzi, A.; Trombettoni, A.; Kevrekidis, P.G.; and Bishop, A.R. Dynamical Superfluid-Insulator Transition in a Chain of Weakly Coupled Bose–Einstein Condensates. *Phys. Rev. Lett.* **2002**, *89*, 170402.
59. Adhikari, S.K. Expansion of a Bose–Einstein condensate formed on a joint harmonic and one-dimensional optical-lattice potential. *J. Phys. B* **2003**, *36*, 3951–3959.
60. Nesi, F.; Modugno, M. Loss and revival of phase coherence in a Bose–Einstein condensate moving through an optical lattice. *J. Phys. B* **2004**, *37*, S101–S113.
61. Mun, J.; Medley, P.; Campbell, G.K.; Marcassa, L.G.; Pritchard, D.E.; Ketterle, W. Phase Diagram for a Bose–Einstein Condensate Moving in an Optical Lattice. *Phys. Rev. Lett.* **2007**, *99*, 150604.
62. Ferris, A.J.; Davis, M.J.; Geursen, R.W.; Blakie, P.B.; Wilson, A.C. Dynamical instabilities of Bose–Einstein condensates at the band edge in one-dimensional optical lattices. *Phys. Rev. A* **2008**, *77*, 012712.

63. Hamner, C.; Zhang, Y.; Khamehchi, M.A.; Davis, M.J.; Engels, P. Spin-Orbit-Coupled Bose–Einstein Condensates in a One-Dimensional Optical Lattice. *Phys. Rev. Lett* **2015**, *114*, 070401.
64. Chen, Y.-A.; Huber, S.D.; Trotzky, S.; Bloch, I.; Altman, E. Many-body Landau–Zener dynamics in coupled one-dimensional Bose liquids. *Nat. Phys.* **2011**, *7*, 61–67.
65. Baharian, S.; Baym, G. Bose–Einstein condensates in toroidal traps: Instabilities, swallowtail loops, and self-trapping. *Phys. Rev. A* **2013**, *87*, 013619.
66. Wright, K.C.; Blakestead, R.B.; Lobb, C.J.; Phillips, W.D.; Campbell, G.K. Driving phase slips in a superfluid atom circuit with a rotating weak link. *Phys. Rev. Lett.* **2013**, *110*, 025302.
67. Lin, Y.Y.; Lee, R.-K.; Kao, Y.-M.; Jiang, T.-F. Band structures of a dipolar Bose–Einstein condensate in one-dimensional lattices. *Phys. Rev. A* **2008**, *78*, 023629.
68. Venkatesh, B.P.; Larson, J.; O'Dell, D.H.J. Band-structure loops and multistability in cavity QED. *Phys. Rev. A* **2011**, *83*, 063606.
69. Watanabe, G.; Yoon, S.; Dalfovo, F. Swallowtail Band Structure of the Superfluid Fermi Gas in an Optical Lattice. *Phys. Rev. Lett.* **2011**, *107*, 270404.
70. Chen, Z.; Wu, B. Bose–Einstein Condensate in a Honeycomb Optical Lattice: Fingerprint of Superfluidity at the Dirac Point. *Phys. Rev. Lett.* **2011**, *107*, 065301.
71. Hui, H.-Y.; Barnett, R.; Porto, J.V.; Sarma, S.D. Loop-structure stability of a double-well-lattice Bose–Einstein condensate. *Phys. Rev. A* **2012**, *86*, 063636.
72. Mumford, J.; Larson, J.; O'Dell, D.H.J. Impurity in a bosonic Josephson junction: Swallowtail loops, chaos, self-trapping, and Dicke model. *Phys. Rev. A* **2014**, *89*, 023620.
73. Haller, E.; Hart, R.; Mark, M.J.; Danzl, J.G.; Reichsöllner, L.; Nägerl, H.-C. Inducing Transport in a Dissipation-Free Lattice with Super Bloch Oscillations. *Phys. Rev. Lett.* **2010**, *104*, 200403.
74. Meinert, F.; Mark, M.J.; Kirilov, E.; Lauber, K.; Weinmann, P.; Gröbner, M.; Nägerl, H.-C. Interaction-induced quantum phase revivals and evidence for the transition to the quantum chaotic regime in 1D atomic Bloch oscillations. *Phys. Rev. Lett.* **2014**, *112*, 193003.
75. Karkuszewski, Z.P.; Sacha, K.; Smerzi, A. Mean field loops versus quantum anti-crossing nets in trapped Bose–Einstein condensates. *Eur. Phys. J. D* **2002**, *21*, 251–254.
76. Torrontegui, E.; Ibáñez, S.; Martínez-Garaot, S.; Modugno, M.; del Campo, A.; Guéry-Odelin, D.; Ruschhaupt, A.; Chen, X.; Muga, J.G. Shortcuts to adiabaticity. *Adv. At. Mol. Opt. Phys.* **2013**, *62*, 117–169.
77. Li, W.; Smerzi, A. Nonlinear Kronig–Penney model. *Phys. Rev. E* **2004**, *70*, 016605.
78. Seaman, B.T.; Carr, L.D.; Holland, M.J. Period doubling, two-color lattices, and the growth of swallowtails in Bose–Einstein condensates. *Phys. Rev. A* **2005**, *72*, 033602.
79. Gemelke, N.; Sarajlic, E.; Bidel, Y.; Hong, S.; Chu, S. Parametric Amplification of Matter Waves in Periodically Translated Optical Lattices. *Phys. Rev. Lett.* **2005**, *95*, 170404.
80. Maluckov, A.; Gligorić, G.; Hadžievski, L. Long-lived double periodic patterns in dipolar cigar-shaped Bose–Einstein condensates in an optical lattice. *Physica Scripta* **2012**, *T149*, 014004.

81. Maluckov, A.; Gligorić, G.; Hadžievski, L.; Malomed, B.A.; Pfau, T. Stable Periodic Density Waves in Dipolar Bose–Einstein Condensates Trapped in Optical Lattices. *Phys. Rev. Lett.* **2012**, *108*, 140402.
82. Yoon, S.; Dalfovo, F.; Nakatsukasa, T.; Watanabe, G. Multiple period states of the superfluid Fermi gas in an optical lattice. *New J. Phys.* **2015**, *18*, 023011.
83. Ring, P.; Schuck, P. *The Nuclear Many-Body Problem*; Springer: Berlin/Heidelberg, Germany, 1980.
84. Miller, D.E.; Chin, J.K.; Stan, C.A.; Liu, Y.; Setiawan, W.; Sanner, C.; Ketterle, W. Critical Velocity for Superfluid Flow across the BEC-BCS Crossover. *Phys. Rev. Lett.* **2007**, *99*, 070402.
85. Zhang, S.L.; Zhou, Z.W.; Wu, B. Superfluidity and stability of a Bose–Einstein condensate with periodically modulated interatomic interaction. *Phys. Rev. A* **2013**, *87*, 013633.
86. Dasgupta, R.; Venkatesh, B.P.; Watanabe, G. Attraction-induced dynamical stability of a Bose–Einstein condensate in a nonlinear lattice. 2016, arXiv:1603.07486.
87. Sakaguchi, H.; Malomed, B.A. Matter-wave solitons in nonlinear optical lattices. *Phys. Rev. A* **2005**, *72*, 046610.
88. Abdullaev, F.; Abdumalikov, A.; Galimzyanov, R. Gap solitons in Bose–Einstein condensates in linear and nonlinear optical lattices. *Phys. Lett. A* **2007**, *367*, 149–155.
89. Yu, D.; Yi, W.; Zhang, W. Swallowtail Structure in Fermi Superfluids with Periodically Modulated Interactions. *Phys. Rev. A* **2015**, *92*, 033623.
90. Fatemi, F.K.; Jones, K.M.; Lett, P.D. Observation of optically induced Feshbach resonances in collisions of cold atoms. *Phys. Rev. Lett.* **2000**, *85*, 4462–4465.
91. Chin, C.; Grimm, R.; Julienne, P.; Tiesinga, E. Feshbach resonances in ultracold gases. *Rev. Mod. Phys.* **2010**, *82*, 1225–1286.
92. Fedichev, P.O.; Kagan, Y.; Shlyapnikov, G.V.; Walraven, J.T.M. Influence of nearly resonant light on the scattering length in low-temperature atomic gases. *Phys. Rev. Lett.* **1996**, *77*, 2913–2916.
93. Yamazaki, R.; Taie, S.; Sugawa, S.; Takahashi, Y. Submicron spatial modulation of an interatomic interaction in a Bose–Einstein condensate. *Phys. Rev. Lett.* **2010**, *105*, 050405.
94. Lewenstein, M.; Sanpera, A.; Ahufinger, V. *Ultracold Atoms in Optical Lattices: Simulating Quantum Many-Body Systems*; Oxford University Press: Oxford, UK, 2012.

MDPI AG
St. Alban-Anlage 66
4052 Basel, Switzerland
Tel. +41 61 683 77 34
Fax +41 61 302 89 18
http://www.mdpi.com

Entropy Editorial Office
E-mail: entropy@mdpi.com
http://www.mdpi.com/journal/entropy

www.ingramcontent.com/pod-product-compliance
Lightning Source LLC
Chambersburg PA
CBHW051923190326
41458CB00026B/6385
9783038423065